T0306101

REAL-TIME STABILITY ASSESSMENT IN MODERN POWER SYSTEM CONTROL CENTERS

Books in the IEEE Press Series on Power Engineering

REAL-TIME STABILITY ASSESSMENT IN MODERN POWER SYSTEM CONTROL CENTERS

Edited by

Savu C. Savulescu

IEEE PRESS SERIES ON POWER ENGINEERING

Mohamed E. El-Hawary, *Series Editor*

IEEE Press

A JOHN WILEY & SONS, INC., PUBLICATION

Copyright © 2009 by the Institute of Electrical and Electronics Engineers, Inc. All rights reserved.

Published by John Wiley & Sons, Inc., Hoboken, New Jersey
Published simultaneously in Canada.

No part of this publication may be reproduced, stored in a retrieval system or transmitted in any form or by
any means, electronic, mechanical, photocopying, recording, scanning or otherwise, except as permitted
under Section 107 or 108 of the 1976 United States Copyright Act, without either the prior written
permission of the Publisher, or authorization through payment of the appropriate per-copy fee to the
Copyright Clearance Center, Inc., 222 Rosewood Drive, Danvers, MA 01923, (978) 750-8400, fax (978)
750-4470, or on the web at www.copyright.com. Requests to the Publisher for permission should be
addressed to the Permissions Department, John Wiley & Sons, Inc., 111 River Street, Hoboken, NJ 07030,
(201) 748-6011, fax (201) 748-6008, or online at http://www.wiley.com/go/permission.

Limit of Liability/Disclaimer of Warranty: While the publisher and author have used their best efforts in
preparing this book, they make no representation or warranties with respect to the accuracy or completeness
of the contents of this book and specifically disclaim any implied warranties of merchantability or fitness
for a particular purpose. No warranty may be created or extended by sales representatives or written sales
materials. The advice and strategies contained herein may not be suitable for your situation. You should
consult with a professional where appropriate. Neither the publisher nor author shall be liable for any loss
of profit or any other commercial damages, including but not limited to special, incidental, consequential,
or other damages.

For general information on our other products and services please contact our Customer Care Department
within the United States at (800) 762-2974, outside the United States at (317) 572-3993 or fax (317) 572-
4002.

Wiley also publishes its books in a variety of electronic formats. Some content that appears in print,
however, may not be available in electronic formats. For more information about Wiley products, visit our
web site at www.wiley.com.

Library of Congress Cataloging-in-Publication Data is available.

ISBN 978-0470-23330-6

Printed in the United States of America.

10 9 8 7 6 5 4 3 2 1

To my children,
Felicia and Crivat

To the memory of my parents,
Felicia and Savu

CONTENTS

7 Online Security Assessment for the Brazilian System—A Detailed Modeling Approach   155
Jorge L. Jardim

PREFACE

Major blackouts have traditionally triggered outbursts of research that eventually led to significant technological breakthroughs. For example, today's real-time static security analysis tools were introduced in response to the Northeast blackout of 1965. The 1980 Barbier and Barret seminal paper on voltage collapse* was written in the aftermath of the major blackout of 1978 in France. And, more recently, the renewed interest in the real-time detection of the risk of instability can be traced to the wave of blackouts that affected U.S., U.K., and mainland Europe utilities in 2003. But this is not the only reason the industry is now moving at an accelerated pace toward assessing stability in real time.

The online calculation of the loadability limits is essential for the effective and efficient utilization of a power system network, particularly in an open access environment. In the past, the computation of these limits in power system control centers was extremely difficult, and off-line studies and time-consuming simulations were the basis for determining the maximum loadability limits, especially when taking into account stability considerations. However, the complexity of online stability assessment is rapidly being mastered, and simple, yet theoretically sound, applications that can quickly tell how far a given operating state is from instability, as well as sophisticated packages that perform comprehensive stability analysis, are now being used for daily operations.

Reflecting this trend, the state of the art in the field is documented in this single volume, which will serve both as a textbook and as a reference for practitioners. Issues addressed and questions answered in this book relate to:

Implementation

- Is the stability application seamlessly integrated on the SCADA/EMS servers? If not, on what hardware is the stability application installed, and how does it retrieve its input data from the SCADA/EMS host?
- Is the stability application running automatically after each execution of the state estimator? If not, can the stability application be manually triggered to execute

*Barbier, C. and Barret, J. P., "An Analysis of Phenomena of Voltage Collapse on a Transmission System," *RGE,* special edition CIGRE, July 1980, pp. 3–21

with the most recent output of the state estimator? Can it be invoked with the results of the dispatcher power-flow calculations?

- Are the results available to the operator before the next real-time network analysis cycle is executed? If not, what is the time delay for the operator to be informed about the stability conditions?

User interface and presentation of results

- How does one extract meaningful results from the ocean of computational output, and how does one present such information to the operator?
- Does the presentation of the stability calculation results require a user interface different from the one used by the SCADA/EMS? If the answer is "yes," how many workstations are needed to fully benefit from the stability application? If the answer is "no," do stability application displays coexist with the SCADA/EMS displays?
- Are the calculation results presented in an intuitive manner, for example, in graphics and charts that would allow the operator to instantly understand what is going on?
- Can the results of the stability calculations be trended on standard SCADA trending charts, thus enabling the operator to continuously monitor the distance between the current and past system operating state(s) from states that may potentially lead to blackouts?

Underlying technology

- What kind of stability can be addressed in real-time?
- Are simplifying assumptions required in order to fit a stability algorithm into the real-time paradigm? If the answer is "yes," how do such simplifications affect the accuracy and the reliability of the stability calculations?
- How are the usability and response time of the stability tool impacted by the solution technique?

The book begins with a SCADA/EMS primer (Chapter 1) that aims to familiarize the reader with the real-time and study-mode data environment in modern power system control centers. These installations have become quite sophisticated and offer superb application integration opportunities that were not available just a few years ago. It is thus important that we set the stage so that students, researchers, and consultants other than SCADA/EMS experts can fully benefit from the material presented in this book. This background is complemented in Chapter 2 by an overview of key stability concepts applied for real-time operations. The purpose is to identify which stability aspects are amenable to real-time assessment and monitoring, and to review how, and to what extent, the difficult goal of quantifying the stability limit, or limits, can be met online, or, even better, in real-time,* by the major stability technologies available today.

---

*Please refer to Section 1.3 in Chapter 1 for a discussion of "real-time" versus "online."

The book's subsequent material is clustered along the lines traditionally recognized in the industry—from steady-state stability, to transient stability, and to voltage stability. Within these clusters, each chapter describes an actual solution that has been successfully deployed, emphasizes the particular challenges that were faced, shows how the problems were solved, and shades light on the experimental results.

The first implementations addressed in the book relate to steady-state stability. They all predict, in real-time, the distance to instability by alternating steady-state stability checks with system-stressing calculations, but differ significantly in terms of approach and functionality. The solution developed by Long Island Power Authority (LIPA) is described in Chapter 3. LIPA has loosely integrated an off-the-shelf third-party application with its existing CIM compliant SCADA/EMS. Some of the drawbacks resulting from loose integration were mitigated with assistance from Siemens Energy—Power Technologies International (PTI), which modified the real-time network analysis sequence to trigger the stability calculation engine immediately after a new valid state estimate has been obtained. An important implementation step consisted of extensively benchmarking the real-time stability program. The benchmarking results are illustrated with actual pictures taken directly from the application's displays. The ISO in Bosnia and Herzegovina (Chapter 4) followed a phased approach. Initially, it loosely integrated the stability tool running on a PC with its interim SCADA system. Then, motivated by the good results obtained during an extended testing period, and also by the inconvenience of using two different user-interface environments, it asked Siemens Energy Automation (Siemens) to seamlessly integrate the stability software with the new, Unix-based, SCADA/EMS. At Transelectrica, in Romania (Chapter 5), the stability software was delivered by AREVA T&D and seamlessly integrated with the Windows-based SCADA/EMS. A unique feature of this implementation is the continuous monitoring of the stability reserves, both for the entire power system and for electrical areas spanned by stability constrained transmission paths.

The next group of chapters addresses transient stability. Chapter 6 describes the implementation of online transient security assessment at Southern Company Services by Powertech Labs and provides details about the simulation model, including the dynamic equivalent of the external system and the complex hardware architecture that supports the computations. The Brazilian National System Operator ONS (Chapter 7) developed in-house a complex application that uses time-domain simulation, Prony analysis, energy functions, and single machine equivalent, among other techniques. Known as Organon, it runs on a multicomputer architecture with input retrieved in real-time from the SCADA/EMS. Siemens' dynamic security assessment implementations in Kuwait and Morocco (Chapter 8) are predicated on using its own transient stability software, which runs on the SCADA/EMS network analysis server with real-time input provided by the state estimator. A similar approach was followed by the National Electricity Market Management Company (NEMMCO) in Australia, but with a commercial transient stability program being loosely integrated in house with the existing SCADA/EMS (Chapter 9).

The last two chapters are dedicated to voltage stability. Chapter 10 deals with online voltage security assessment in the control center of the Hellenic Transmission System Operator. Both the application and its underlying technology are extensively described and typical results available to operators are presented. In addition, the chapter describes the use of online voltage security assessment for arming a special protection scheme against voltage collapse, and includes a theoretical overview of the quasi-steady-State (QSS) technique for fast, simplified simulation of power system long-term dynamics. Chapter 11 describes the Spica online voltage security tool that was developed in-house at the Swedish National Grid (SvK) in Sweden. This application uses a load-flow approach, computes the transfer limits by determining the PV curves and their nose points for the worst contingencies, and runs at 15 minute intervals with real-time input from the state estimator.

Since the book aims primarily at the practical aspects of implementing stability assessment in real-time, the theoretical background included in each chapter is minimal. For the benefit of readers who may not be quite familiar with the underlying theoretical techniques, appendices that describe some of the key algorithms and discuss important theoretical issues directly related to the subject matter of the book are included.

Appendix A summarizes the theoretical background of the steady-state stability technique developed by Paul Dimo in Europe in the late 1950s and early 1960s. Dimo's method was viable then, when digital computers were the exception rather than the rule, and is successfully used today both in real-time and in study mode, as illustrated by some of the actual installations described in this book. The technique is presented in detail along with numerical examples developed at the Argentinean System and Market Operator (CAMMESA) that illustrate all the algorithm steps, allow benchmarking the methodology, and should help researchers and practitioners to develop practical implementations of this methodology.

Appendix B describes the hybrid transient stability single-machine equivalent (SIME) and identifies the avenues it opens to accurate and fast transient stability assessment and, most significantly, to transient stability control. Real-time preventive control is one of the derived techniques that allows meeting steady-state and transient stability constraints together with additional objectives, possibly dictated by electricity markets. The open-loop emergency control technique is designed to mitigate the control actions taken preventively, by complementing them with emergency actions determined preventively but triggered in real time after a contingency occurrence. The emergency single-machine equivalent is yet another technique that tackles real-time closed-loop emergency control by using real-time data acquired from the system generators. Alternate open-loop and closed-loop emergency control solutions are also addressed, along with the online assessment and control of transient oscillations damping aimed at improving transient oscillations damping by proper rescheduling of power system generation.

Finally, Appendix C discusses the identification and evaluation of stability constrained transmission paths that may appear in any power system in which large MW blocks are transferred between weakly interconnected areas. This is a common scenario in electricity markets and, in order to ensure that the grid does not get too close to its stability limits, the maximum transfer capability across such transmission corridors must be frequently evaluated, especially if there is prior knowledge that the paths that interconnect the areas may be stability constrained.

It was our goal to address all the key issues that appear when extending the conventional SCADA/EMS environment with stability tools. We hope that the book provides answers to many questions and that it will assist practitioners and researchers to deploy successful real-time stability implementations.

ACKNOWLEDGMENTS

This book owes its existence to IEEE Press' Steve Welch. After the panel session, Real-Time Stability Applications in Modern SCADA/EMS, that we hosted at the IEEE Power Systems Conference and Expo in Atlanta, Georgia in 2006, he sensed the importance and timeliness of this topic and asked us to assemble an edited book that would take the panel's subject to a higher plateau. Thank you, Steve, for your insight and encouragement!

We are grateful to the industry experts and university professors who joined us in this project, replied with great enthusiasm, and subsequently devoted a significant amount of effort to ensure that their contributions would fit harmoniously into the overall scheme of the book. Particularly important was the assistance received from Dr. Sudhir Virmani who, in addition to coauthoring Chapter 1, reviewed extensive portions of the text and made useful editing suggestions. Thank you Sudhir, and many thanks to all the other authors in the book!

Many thanks to Professors M. A. Pai, Arun G. Phadke, and Gerald T. Heydt, and Dr. Eugene Litvinov. They offered valuable comments and recommendations both at the beginning, when the book was in the preliminary proposal stage, and at the end, during the tedious task of reviewing the entire book material. Thank you all!

Last, but not least, we would like to acknowledge the assistance received from Ms. Jeanne Audino, Project Editor at IEEE Press, during the text editing process.

SAVU C. SAVULESCU

New York, New York
November 2008

CONTRIBUTORS

Cornel Aldea, Senior Engineer, Operational Planning, S.C. Transelectrica S.A., Bucharest, Romania

Loris Arnold, Senior Engineer, Electric Planning and Forecasting Department, LIPA/National Grid, Hicksville, New York

Dr. Guenther Beissler, Director Software, Siemens AG, Energy Sector, Erlangen, Germany

Stephen J. Boroczky, Senior Energy Management Systems Engineer, National Electricity Market Management Company (NEMMCO), Baulkham Hills, New South Wales, Australia

Dr. Horia S. Campeanu, Manager of EMS/MMS/SCADA Systems, IT Process Department, S.C. Teletrans S.A., Bucharest, Romania

Martín Cassano, Gerencia de Programación de la Producción, Compañía Administradora del Mercado Mayorista Eléctrico S. A. (CAMMESA), Rosario, Santa Fe, Argentina

George Christoforidis, Department of System Operation and Control, Hellenic Transmission System Operator, Athens, Greece

Dr. Roland Eichler, Sales Group Manager, Siemens AG, Energy Sector, Germany

Cornel Erbasu, Head of Operation Department, Operational Unit of National Power Dispatch, S.C. Transelectrica S.A., Bucharest, Romania

Dr. Mevludin Glavic, Senior Research Fellow, Electrical Engineering and Computer Science Department, University of Liege, Liege, Belgium

Janos Hajagos, Manager, Area Planning, LIPA/National Grid, Hicksville, New York

Fred Howell, Specialist Engineer, Powertech Labs Inc., British Columbia, Canada

Dr. Jorge L. Jardim, Principal, Nexant Inc., Chandler, Arizona

Susan M. Manessis, Lead Engineer, Transmission Planning, LIPA/National Grid, Hicksville, New York

Alan Martin, Consulting Engineer EMS Power Applications, Southern Company Services, Inc., Birmingham, Alabama

Roberto D. Molina Mylius, Senior Engineer, Electric Power Systems, Compañía Administradora del Mercado Mayorista Eléctrico S. A. (CAMMESA), Santa Fe, Argentina

Kip Morison, Director, Power System Technologies, Powertech Labs Inc., British Columbia, Canada

Dr. Mania Pavella, Emeritus Professor, University of Liege, Belgium

Anie Philip, Senior Engineer, Transmission Planning, LIPA/National Grid, Hicksville, New York

Professor Marius Pomarleanu, Consultant, Bucharest, Romania

Klas Roudén, Swedish National Grid, Stockholm, Sweden

Professor Daniel Ruiz-Vega, SEPI-ESIME-Zacatenco, National Polytechnic Institute, Mexico City, Mexico

Dr. Olaf Ruhle, Product Manager, Siemens AG , Energy Sector, Erlangen, Germany

Lars Sandberg, Swedish National Grid, Stockholm, Sweden

Dr. Savu C. Savulescu, CEO, Energy Consulting International, Inc., New York, New York

Dr. Thierry Van Cutsem, Research Director, Fund for Scientific Research (FNRS), Adjunct Professor, Department of Electrical Engineering and Computer Science, University of Liege, Liege, Belgium

Dusko Vickovic, Head of IT and Telecommunication Department, Independent System Operator in Bosnia and Herzegovina, Sarajevo, Bosnia, Herzegovina

James Viikinsalo, Principal Engineer, Bulk Power Operations, Southern Company Services, Inc., Birmingham, Alabama

Dr. Sudhir Virmani, President, EPIC Engineering Inc., Cupertino, California

Professor Costas Vournas, School of Electrical and Computer Engineering, National Technical University of Athens, Athens, Greece

Lei Wang, Manager, Software Technologies, Powertech Labs Inc., Surrey, British Columbia, Canada

1

THE REAL-TIME AND STUDY-MODE DATA ENVIRONMENT IN MODERN SCADA/EMS

Sudhir Virmani and Savu C. Savulescu

1.1 INTRODUCTION

1.1.1 General Background

Most large industrial control systems need to collect data at a central location, or at distributed sites, from a range of equipment and devices in the field, and to process this data in order to make a decision regarding any action required. Electric power control systems work basically in the same way but impose particularly stringent requirements on remote data acquisition and related processes because:

- Power systems may encompass large geographical areas as almost all electric utilities have strong electrical interconnections with neighboring systems, which are generally owned and operated by different entities. Examples include: the interconnected systems in North America, such as the Western Interconnection, which consists of the power systems in all the western U.S. states plus the provinces British Columbia, Alberta, and Manitoba in Canada; and the large interconnection in mainland Europe, covering all the mainland European Union member countries plus some nonmembers.
- Interconnected power networks are therefore very large, with potentially tens of thousands of nodes and branches and thousands of generating units.

Real-Time Stability Assessment in Modern Power System Control Centers. Edited by Savu C. Savulescu
Copyright © 2009 The Institute of Electrical and Electronics Engineers, Inc.

- Power systems in general must operate synchronously and this requires that all the interconnected systems must operate cooperatively in order to maintain reliability of the entire system.
- Because of these strong interconnections, any disturbance in one part of the large network can affect the rest of the network.
- Power system disturbances can propagate very rapidly (milliseconds to seconds) and this requires high-performance control systems, some of which are local, such as protective relays that operate in milliseconds, and some at central sites such as SCADA/EMS systems, which typically operate on the time frame of seconds (monitoring and control) to several days (scheduling).
- Power system operations typically entail control requirements that can be met only by implementing complex hierarchies of regional and central/national control systems.
- Power system operations in the context of large regional or subcontinental electric markets typically require exchange of information and coordination of control actions among various entities, such as independent system operators, security coordinators, and transmission system operators, thus leading to a higher degree of coordination and control systems.

1.1.2   Anatomy of a SCADA System

The data acquisition systems that are implemented in the utility industry therefore have to be able to support these needs. Furthermore, in order to make sure that the control actions being taken are correct and safe, certain control actions performed centrally require a positive confirmation, that is, they must be supervised. This is the *supervisory control* function and, therefore, the overall system is called supervisory control and data acquisition or SCADA.

The basic elements included in, and the minimum capability of, a typical SCADA system, consist of:

- Interfaces in the field (substations) to equipment and devices located within the substation.
- Ability to scan these interfaces to obtain the values of various quantities such as real and reactive power, current, voltage, and switch and circuit breaker position. The data are either reported by exception or scanned periodically. Typical scan rates are every 1–2 seconds for generation and interchange data and circuit breaker status indications; every 2–15 seconds for line flow and voltage measurements, and every 15 minutes to one hour for energy values.
- Transmission of these data items to a central location known as the SCADA (or SCADA/EMS, as shown in Section 1.4) center.
- Processing and analyzing this information at the SCADA center and displaying it to the operator.

- Determining any control action to be taken either automatically or by operator request. The control actions required can be for controlling real power, reactive power, voltage, circuit breakers, and power flows.
- Transmitting the request for control to the field equipment.
- Monitoring the completion of the control request.
- Building the real-time database and periodically saving real-time information for archival purposes.

1.1.3 Real-Time Versus Study-Mode Processes

Most of the SCADA functions are executed in real-time. By real-time we mean that the:

- Input data reflect the most recent picture of the system conditions. In the field (substation), they come directly from devices that capture analog values and status indications; at the SCADA center they are stored into, and retrieved from, the real-time database.
- Processing is performed within very short delays typically not exceeding a couple of seconds.
- Output is usable almost instantly; again, "instantly" in this context means approximately one to two seconds.

The monitoring of data generated by a real-time process is a typical example of real-time activity. But the information generated in a SCADA system can be used in many other ways that do not qualify as real-time. For example, statistics can be built to record how many times the taps of a tap changing under load (TCUL) transformer have moved during a specified period of time. The tap changes were recorded in the real-time database immediately after they occurred, and then they were exported to some archival system and became historical data. The calculations entailed in building the statistic constitute a "study" performed with "real-time" data and, perhaps, some additional information; thus we will say that this is a "study-mode" calculation.

In the computational environment of a modern power system control center, some functions are performed only in real-time, whereas some others are performed only in study mode. However, as we will see in the next section, there are functions that can be used both in real-time and in study mode.

Let us say in passing that real-time and online are not necessarily interchangeable attributes. On line implies that the calculations are available to the operator in the SCADA/EMS system itself, hence they are online as opposed to being available on some other separate system. However, there is no guarantee that the online computational process will be fast enough to produce results that can be labeled real-time. These considerations should help the reader understand the difference between the real-time stability assessment, stability monitoring, and online stability assessment concepts that are often mentioned throughout this book.

1.1.4   Next Level of Functionality: The EMS

In order to determine the control actions required, it is necessary to simulate the operation of the power system in close to real-time. The software tools needed include what are commonly referred to as energy management system (EMS) applications. A very terse and unstructured summary of these functions is given below:

- Automatic Generation Control (AGC) to determine the real power output of all the generating units in the system to maintain interchange and frequency. In the interconnected system, each member (control area) performs the AGC function by computing the area control generation load mismatch (area control error or ACE) at the nominal frequency (60 Hz or 50 Hz) and adjusts its generation to reduce the mismatch to acceptable limits.
- Economic Dispatch (ECD or EDC) to determine the optimal level of real power output for each generator to minimize the total production cost (this function works in conjunction with the AGC function).
- Reserve Monitoring (RM) to compute the real power capability available in the system to meet changes in demand.

The above functions consider the generating units only and generally tend to ignore the network (transmission lines, transformers, reactors/capacitors) and voltage. Some ECD implementations do include an approximate model for transmission system losses.

In order to obtain a more comprehensive view of the system, the following network analysis functions are required:

- The State Estimator is used for determining the complete state of the system (voltage and phase angle at each node) based on the measurements from the field. These measurements are generally "noisy" and not available for every element. The state estimator determines the best estimate of the state using a set of redundant measurements, taking into account the measurement error characteristics and missing and bad data.
- Static security analysis, or contingency evaluation, determines the effect of possible outages such as loss of branches (transmission lines, transformers), generating units, and combinations thereof.

These two functions run both in real-time, either executed periodically, with a period of a few seconds to a few minutes, or triggered by events or operator requests, and in study-mode, that is, executed if and when needed to assess postulated scenarios closed to, or derived from, the current operating conditions.

In addition, there are other functions that are executed in study-mode, including:

- The load-flow/optimal power flow, which is used to calculate all of the system variables, with the optimal power flow being used to compute these variables based on optimizing certain system quantities (production cost, losses, voltage levels, transformer tap changes). These are initialized using the results of the state estimator, and additional data needed, such as generator data, is retrieved from the database.

- Hourly load forecasting which is used to predict the hourly total load/total demand that the system will have to supply over the next few hours to up to several days. This function enables the operator to schedule facilities for maintenance, for reconnection, and for start-up and shutdown of units
- Unit Commitment/Hydro-Thermal Scheduling is used to schedule the start-up and shutdown of generation units to meet the forecasted demand. This function typically looks ahead for 24–168 hours depending on the characteristics of the generating units being scheduled. This is a nonlinear optimization problem with both integer and continuous variables. Consequently, it is a computationally intensive function for systems with a large number of generating units that have to be scheduled. It should be noted that whereas in vertically integrated utilities the unit commitment/hydro thermal scheduling is part of the EMS/SCADA, in deregulated electricity markets, the software that performs unit commitment/hydro thermal scheduling is part of the market system. However, the functionality is the same; only the responsibility is separated.

When these EMS functions are included, one generally refers to the system as a SCADA/EMS system although historically the EMS acronym implied that the system included both SCADA and EMS functions.

Finally, there are a number of support functions required in SCADA/EMS systems such as:

- Alarm processing,
- Display generation,
- Report formatting and printing,
- Storage of real-time data in a historical information system (HIS) for archival purposes,
- Special-purpose functions for data conversion and interfacing to local devices, such as mapboards, time and frequency standards, and chart recorders, and
- Communication interfaces with the SCADA/EMS systems of other electric utilities.

As one may infer from the above, the infrastructure needed in the current-day SCADA/EMS systems is very extensive and includes:

- A large-scale telecommunication network that interconnects the field equipment to the SCADA/EMS center. The telecommunication technologies include a mix of leased telephone lines, power-line carrier, microwave radio, copper and fiber optic cable, as well as VHF and UHF radio. Most electric utilities are implementing large-scale backbone networks using fiber optic cables and moving from a radial system to a more meshed system using IP.

- Powerful computer systems at a central location or, for hierarchical control systems, at one location on top of the hierarchy and several subordinated computer systems at the other locations.
- Operator interface equipment that can respond to multiple requests for new displays within one second.
- Large data storage and long-term historical data archival capabilities that are easily accessible to operators and engineering personnel.

Last, but not least, the system must have a very high availability. At the SCADA/EMS center, 99.9% availability is required and is achieved by providing redundant computer and local communication systems.

The brief enumeration of functions and capabilities presented in the previous paragraphs suggests that in order to understand the complex data interfaces and software interactions between the SCADA/EMS system, on the one hand, and a sophisticated add-on application such as stability assessment software, on the other hand, we need to step back and follow a systematic approach aimed at identifying the:

- Overall architecture in a simple format, such as a conceptual overview diagram, that depicts the major building blocks and would make it easy to visualize the information flow between them.
- Functional architecture, for the purpose of positioning the stability assessment application in the SCADA/EMS data and functional environment.
- Implementation architecture, which provides clues about the integration, tight or loose, of software that performs real-time and study-mode stability assessment.

This analysis is briefly developed in Section 1.2.

1.1.5 The Impact of Wide-Area Monitoring Systems

Phasor measurements, a technology developed in the late 1970s and early 1980s, mainly due to the visionary work of Arun Phadke [5], are being finally deployed extensively in power system networks. To some extent, this has been facilitated by the availability of relatively inexpensive GPS receivers that enable the synchronization of the phasor measurements over large geographical areas, but it is also due to the better and more powerful electric utility telecommunication networks, largely fiber optic based, and faster low-cost processors. Direct measurement of the voltage and current phasors throughout the entire network essentially eliminates the need for state estimators in an ideal case (complete and error-free measurement set) and gives a complete picture [6] of the system on a milliseconds (2–5 cycles) time frame.

However, due to measurement errors and bad or missing data, state estimation will be necessary but will be simpler since a linear model suffices. Most phasor measurement units calculate phasor values by sampling the analog signal for each phase at sampling rates of 12 times per cycle and higher, and using discrete Fourier transform analysis to compute the positive sequence values. Phasor reporting time is synchronized throughout the power network, and the phasors are estimated from sampled data, which is referenced to the phasor reporting time. Availability of phasor values on a milliseconds time scale thus has very important consequences for power system operation since it allows monitoring of power system dynamic behavior [2].

In the technical literature, these measurements are referred to as wide-area monitoring systems and have been deployed in many countries worldwide, for example, the United States, with more concentration in the western states (California, Oregon, etc.); in Switzerland, especially on the interconnection tie lines; in many generating plants and interconnection points in Southern China; and in Canada (BC Hydro). Phasor measurements are not simply ideal for steady-state analysis; they can be used to detect possible system separation and system oscillations in close to real-time and, in principle, may allow for closed-loop control for maintaining system stability. One of the principal problems with the current EMS state estimation and related functions such as contingency analysis is the need for a model for the "external system," that is, the system outside the internal area. This is still an unsolved problem, even with greater inter-control center communication. Phasor measurements can help to largely overcome this problem since the availability of synchronized phasor measurements from the external areas can make the modeling of external networks more accurate and simpler, since a reduced-order model can be used. Furthermore, since phasor measurements give information on a millisecond time frame they can assist in the rapid detection of system separation and potential blackouts.

However, it must be noted that phasor measurements are another mechanism for monitoring the system and do not have predictive capability. In other words, one still has to perform some analysis to determine where the system will be at a future time. Furthermore, at the present time, closed-loop control for transient operation has not been realized in practice and remains a subject of active research. However, installation of phasor-measurement units is becoming more common and their use can only increase, especially as the cost of these units continues to fall. In fact, many companies are providing the ability for determining phasors as part of the digital relays, which are being used in substation automation. Thus, if a substation is being upgraded, adding phasor measurement is a low-cost option requiring only a GPS receiver.

Acceptance and deployment of phasor measurements adds exciting new options for power system operation, monitoring, and control. Such measurements have already proved their value in many cases and their use is only likely to flourish in the future. However, phasor measurements at the current time are not fully integrated with the standard SCADA system; instead, a separate phasor data concentrator is used to receive and process phasors. The data is archived and can also be sent to the SCADA database (see [1] and [3] for some recent examples and [10] for an interesting method for including phasors in SCADA/EMS-based state estimation).

1.2 SCADA/EMS ARCHITECTURES

1.2.1 Conceptual Overview

Figure 1-1 depicts the conceptual overview of the SCADA/EMS and market system facilities of a typical organization that performs both system operator and market operator functions.

In this example, the SCADA/EMS and the market system are at the top level of the hierarchy and accommodate the EMS and market applications. The SCADA functions are performed both at this level and at the next level of hierarchy corresponding to the "regional" control centers. Historical and current operational information are made available to the users with appropriate credentials via HIS and, perhaps, a dedicated WebInfo server. Raw system data are scanned from RTUs and ISAS (integrated substation automation systems) by using standard communication protocols. Processed information is exchanged over the Internet and/or over via ICCP (Inter-utility Control Center Communication Protocol), now 60870-6-TASE 2.

Figure 1-1 is simply a conceptual overview that illustrates the key building blocks of the architecture and shows the major paths for information exchange; actual implementations vary widely in detail but not in any fundamental way. For example, the SCADA/EMS block above will in practice consist of a network of computers, display devices, communication processors, local device interfaces, and bulk storage compo-

Integrated System and Market Operator
Conceptual Overview

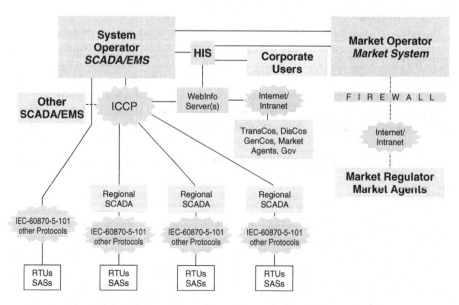

Figure 1-1. Conceptual overview illustrating the SCADA/EMS building blocks.

nents. Actual instances of implementations are given in the subsequent chapters, where specific cases are discussed (LIPA, NOS-BiH, NEMMCO, and others).

1.2.2   Functional Architecture

The basic functionality of a SCADA/EMS system has been illustrated generically in the introductory section. A more rigorous and structured approach is to consider these functions in the context of a layered architecture and, accordingly, to group them into software layers as follows: system software, support software, and application software.

The *system software* encompasses operating systems, compilers, debuggers, utilities, libraries, and so on.

The *support software* consists of communications support, database management, and user-interface software.

The *application software* typically includes the following application subsystems:

- SCADA subsystem—includes the basic data acquisition, control, and display functions that allow the operator to determine the actual status of the power system and to perform remote control actions.
- Generation control subsystem, including AGC, economic dispatch, reserve monitoring, production costing, interchange scheduling, and energy accounting.
- Network analysis subsystem—consists of a suite of programs that work in real-time and study mode, including:

  Network Topology Processor,

  State Estimator,

  Static security analysis (contingency evaluation),

  Dispatcher's Power Flow,

  Optimal Power Flow,

  Service routines such as bus load forecast and transmission losses penalty factor calculation.
- Short-term operations scheduling subsystem—composed by the short-term Load Forecasting and Unit Commitment/Hydro-Thermal Scheduling programs.

All of the above applications are used in actual operation. In addition, especially in the current deregulated environment, systems need to be prepared for auditing, accounting, and post-facto analysis. Therefore, the SCADA/EMS systems include a disturbance recording system, a historical information system (HIS), and reporting and analysis tools that enable the owner as well as outside agencies (regulators, market participants, generator companies, etc.) to view and analyze the operation that occurred.

The execution of the system, support, and application software is distributed among the SCADA/EMS system servers and workstations in order to:

- Minimize the response times of the applications
- Maximize the redundancy and availability of the information

- Allow functions and data to be accessed from any local or remote workstation included in the configuration
- Facilitate the system expansion both horizontally, by adding more processors, and vertically, by replacing existing machines and software with more powerful ones

The functional architecture described above can be considered as "standard" in the sense that it is supported by practically all the major SCADA/EMS vendors today. By contrast, in the current state of the art, real-time stability assessment is not yet a "standard" application in the sense that there is no "standard" way of performing such functionality in a SCADA/EMS context.

In order to understand the place of real-time stability assessment in this functional architecture, we need to briefly identify its input–process–output requirements:

- Input. The key input for real-time stability assessment is the current, or most recent, state of the power system, which, in turn is provided by the most recently solved state estimate; additional inputs include, but are not limited to, off-line information such as generator data and event scenarios for evaluating contingencies.
- Process. Stability applications use proprietary algorithms that are unique and cannot be used to perform other functions in the SCADA/EMS context. In some cases, as shown in Chapters 6 and 7, the stability software suite encompasses more than one application and, in order to provide an acceptable level of performance, requires a multiple computer configuration and a complex distribution of functions among processors.
- Output. Key calculation results produced by true real-time stability assessment software are amenable to real-time monitoring and must be redirected to the real-time database or, as a minimum, to historical storage, for subsequent display on standard SCADA trending charts, but most of the output of stability calculations is usually very specific and not used by other SCADA/EMS applications; also, it may entail graphics and tables not normally supported by the standard SCADA/EMS display facilities.

With this background, and without attempting, for the moment, to determine how the actual implementation is realized, the real-time stability assessment functionality can be considered as a nonstandard capability of the real-time network analysis, as shown in Figure 1-2. In this figure, the stability assessment module is labeled "real-time" but, depending upon the approach, the stability calculations may be quite involved and the time required for such calculations may be one or even several orders of magnitude higher than the completion time of the other network analysis applications. This is why the standard network analysis sequence entails the automatic execution of contingency analysis immediately after the state estimator without waiting for the stability computations to complete.

Let us also note that online or real-time stability assessment are sometimes referred to as dynamic security assessment (DSA) and typically entail evaluating contingency

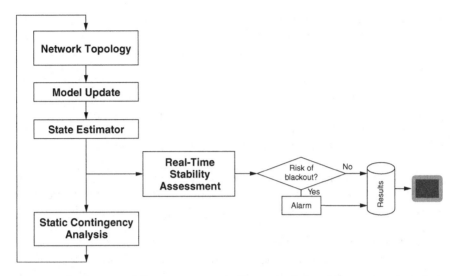

<u>Figure 1-2.</u> Real-time stability assessment in the context of real-time network analysis.

scenarios in addition to base case stability calculations; accordingly, we use the terminology "static contingency analysis" in order to differentiate it from DSA.

The stability calculations can be triggered from within the network analysis subsystem, either periodically, or by events or upon operator request, in which case we talk
about "seamless" integration. However, the functional capability depicted in Figure 1-
2 can also be implemented by "loosely" integrating the stability assessment software
with the SCADA/EMS, that is, by simply executing it periodically or on demand with
data produced by the state estimator but independently of the real-time network analysis sequence.

1.2.3 Implementation Architecture

From the early 1970s, when digital computer based SCADA/EMS systems started to
become more prevalent, until the early 1990s, SCADA/EMS systems were based on
the supplier's proprietary hardware and software. This made it virtually impossible to
add third-party software to these systems and even more unfortunately required that
the entire system be replaced when it became obsolete, even if the same vendor supplied the new system. Thus, for example, the communication protocols used to communicate with the field devices (remote terminal units or RTUs)—the user-interface
system, much of the SCADA software and all of the EMS software—were proprietary.
Often, the operating system was modified to meet the real-time control and data acquisition requirements, making it also proprietary. This forced electric utilities to rely on
the vendor to support and maintain the system.

Because of the high costs and poor performance of these systems, electric utilities
and the research institutes made a concerted effort to evolve toward more open sys-

tems, which were far less vendor dependent by developing standards for various interfaces. The growth of the internet has made the technologies of the internet de-facto standards and given additional momentum to the development of open architectures. Furthermore, electric utilities insisted that the vendors supply "integration friendly" systems that the users or their consultants could augment and modify.

Therefore, in order to meet these requirements and to be able to expand the system capabilities as the needs change and the power network evolves, almost all recent systems have a distributed Ethernet-based local area network (LAN) architecture at the central site, with a multiple-loop fiber optic backbone telecommunication system in which the individual sites are connected via a variety of communication media.

This architecture is much more flexible than the "closed" dual, quad, and multiple computer architectures used in the early digital SCADA/EMS systems. Its flexibility enables the user to not only upgrade the existing hardware but also to add third-party hardware and software without excessive effort. Incidentally, a LAN architecture is also increasingly deployed at the larger substations, with intelligent electronic devices (IEDs) replacing the traditional analog interface to field equipment.

In line with the technological developments in the computer industry, alternate architectures are being considered and in some cases deployed. These include service-oriented, remote hosting, and "service bus" architectures, which have been developed to allow access to databases by third parties using standardized models and communication protocols, as well as "cloud computing," in which one simply uses the computational resources in a "cloud" using telecommunication facilities.

The adherence to industry-wide standards such as IEC 61850 and IEC 60870 further facilitates addition of third-party hardware and software. These systems have also begun to use Microsoft operating systems, Internet type browsers and internet protocols, off-the-shelf database systems, and to provide interfaces to commercial packages such as Microsoft Word and EXCEL which contribute to the greater flexibility available for adding new hardware and software without excessive reliance on the original vendor. Figure 1-3 illustrates one of the many potential ways to implement the hypothetical system and market operator conceptual overview depicted in Figure 1-1.

The SCADA/EMS configurations described throughout this book are similar to the conceptual implementation architecture illustrated in Figure 1-3. They may differ in terms of functionality, levels of hierarchies, and so on, but they all adhere to the open architecture paradigm. The deregulation of electric utilities and the unbundling of the main activities of generation, transmission, and distribution have resulted in a clear need for such open architectures because:

1. The systems and software are constantly evolving since market needs and rules tend to change often.
2. The number and qualification of the participants in the market is large, qualifications are not uniform, and interface equipment is diverse.
3. Much more extensive audit and billing mechanisms have to be in place.
4. A single vendor cannot supply all of these capabilities nor can they support cost-effectively the almost continuous changes. This is evident in many large systems

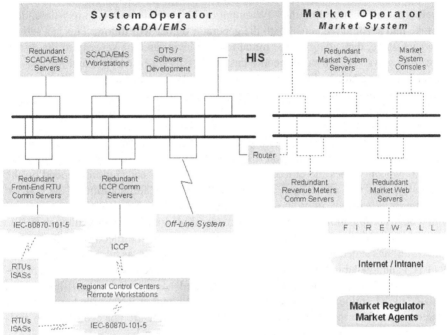

Figure 1-3. Example of SCADA/EMS conceptual implementation architecture.

installed at independent system operators, where one sees quite a mix of capabilities supplied by different vendors.

5. Participants in the generation and transmission market have to be given access to the appropriate systems.

6. A separate revenue-quality metering system of higher accuracy has to be implemented for accounting and billing purposes.

1.3 INTEGRATING STABILITY APPLICATIONS WITH THE SCADA/EMS

1.3.1 Stability Assessment in the SCADA/EMS Context

It is necessary to digress slightly in order to provide a more complete picture of power system operation and to place the rest of this book in perspective. The SCADA/EMS systems implicitly assume that the power system is essentially in a steady-state or quasi-steady-state condition. To illustrate this, when executing network analysis programs, the system is assumed to reach its postoutage operation without considering the trajectory it will follow to get from the initial state (no outage) to the final state (equipment not in service). Thus, the transient behavior is ignored. Even the AGC function, which is actually controlling the MW output of the generating units in real-time, does not take into account the actual path the generator output will follow. This is a serious drawback since, depending on the initial loading levels and the severity of the disturbance, the power system may become unstable (transient instability) and the postdisturbance steady-state condition may not be reachable. There is also a much higher frequency transient (microseconds to a few milliseconds) that occurs due to a switching operation, which is assumed to decay quickly, although it could result in cascading failures.

Unfortunately, the transient stability problem is computationally far too demanding for a true real-time implementation if one uses the conventional time-step simulation method. Not only does the outcome of transient stability calculations depend on the initial state, it is also critically dependent on the location and duration of the fault on the network. Therefore, in order to determine whether a system is stable, one is faced with a huge computational effort for even reasonably sized networks. Although NERC (North American Electric Reliability Council, now ERO—the Electric Reliability Organization) has stated that the computation of indicators such as total transfer capability (TTC) and available transfer capability (ATC) must take into account transient stability, they do not specify how this is to be computed. Most commonly, off-line studies are used to determine a "safe operating region." Although it is possible to provide the operator the ability to run transient stability studies, these are not even close to real-time; they are online only, in the sense discussed in Section 1.3.

Another type of stability is the ability of the power system to remain stable in the presence of slowly varying changes in the total demand. Specifically, one would like to know how much additional load can be carried by the transmission system starting from the current state if the load and, accordingly, generation and imports would be increased gradually. We will refer to this as steady-state stability. Unlike the transient stability problem, we are not looking at large disturbances caused by faults, loss of equipment, and so on. The steady-state stability limit thus corresponds to the maximum loadability, including both the system load and the wheeled power, if any, that the current configuration can carry without a system collapse.

For the system operator, the real-time knowledge and monitoring of the steady-state (voltage) stability limit is extremely valuable. In order to be usable as a simple indicator, this limit should be specified in terms of a "distance" to instability, that is, how far the current system is from the limit. There are a number of rather subtle issues related to the computation of this limit and these are discussed in complete detail in some of the subsequent chapters and appendices as well as the material listed in the bibliography.

In this book, we discuss the stability problem in the context of power system operation and describe all of the techniques, including implementation considerations, that are currently in use. The stability assessment implementations addressed herein are used in actual power system control centers where they are integrated with the SCADA/EMS system but with varying levels of integration. The open SCADA/EMS systems make such integration not too onerous and can be accomplished without excessive effort.

1.3.2 Data Issues

1.3.2.1 Real-Time, Study-Mode, and Planning Models

Prior to attempting to integrate a stability assessment application with a SCADA/EMS system, the software designer has to be fully aware that the power system models deployed in real-time and study mode in the SCADA/EMS, and off-line, in planning, are not identical. In fact, the discrepancies may be so significant that they may preclude the execution of stability programs with mixed models that combine both SCADA/EMS and planning data. The differences between the real-time, study-mode, and off-line models are due essentially to the:

- Bus naming and numbering conventions,
- Level of detail used to represent the power system components,
- Extent to which the internal and external areas of transmission systems are represented, and
- Making sure that the network model and state estimator results are adequate for the reactive power/voltage analysis.

A "bus" in the SCADA/EMS and a "bus" in the planning environment are different concepts. In order to avoid confusion, different names may be used, for example, in Chapter 9, where the transmission planning "buses" are called "nodes," as "nodes" in a theoretical graph.

In operations, dispatchers and system engineers work with substations in which a "bus" may either come into existence or temporarily disappear depending upon the status of breakers and switches that interconnect the various bus bars in the substation. Accordingly, the "bus names" are fixed, whereas the "bus numbers" may change each time the network topology and network model update processors triggered within the real-time network analysis sequence have identified a new system configuration. In some cases, for example, the implementation described in Chapter 5, not even the bus names are fixed; they are recreated at each run of the real-time network analysis sequence based on an algorithm that merges a fixed four-character base designating the substation with another four-character field generated dynamically. Therefore, the stability assessment program should be "smart" enough to associate correctly the off-line information, such as generator data, with the actual bus names and numbers as they appear in the current dataset.

In planning, where the substations are not represented and the one-line diagrams are fully consistent with the topological network model used by the algorithms, the bus names and numbers are fixed. This is the good news. The bad news is that both the bus naming and the bus numbering conventions deployed off-line are usually different from those used in the SCADA/EMS environment, thus making it impossible to add off-line data to cases computed in real-time and vice versa without running cross-referencing algorithms.

The level of detail to which the power system components are represented may be quite different in the SCADA/EMS and in the planning environment. The first example that comes to mind is the representation of generators. State-of-the-art stability applications used today in planning support sophisticated generator models that represent both the synchronous machine and its controls and the turbine governors. Such data are normally not available in the SCADA/EMS database, although they can be implemented specifically for the purpose of integrating the stability software. Another example, which actually goes the other way around, is offered by the so-called P-Q capability curves, which are almost always available in the SCADA/EMS database but are not normally found in the "dynamic" datasets used by off-line stability applications.

The extent to which the internal and external areas of the interconnected transmission system are represented is another modeling issue that may significantly impact the outcome of the calculations. In the planning environment, the tendency is to build large detailed power system models that include a large portion of the "external" system. Although such comprehensive models provide precise load-flow and transient-stability calculations, they place a heavy burden on the computing resources and may require special algorithms, such as the procedure described in Chapter 6, to build and update a reduced-order dynamic-equivalent model for use in online transient-stability assessment. When other types of stability are investigated, for example voltage stability via the $d\Delta Q/dV$ criterion, as shown in Chapters 3, 4, 5, and Appendix A, using a huge external model is not only unnecessary, but may adversely affect the computational results since the phenomena being investigated are local.

The SCADA/EMS databases, on the other hand, represent in detail the so-called "internal area," that is, the power system network directly supervised and controlled by the SCADA/EMS, and add a buffer zone and/or a reduced-order equivalent to model the "external areas." Expanding the internal area with buffers and equivalents for the external area is very important when simulating contingencies because line and/or generator trips within the internal area may have significant impacts on the tie lines that interconnect it with the external world. Modern state estimators are quite good at using the real-time data exported by the neighboring control centers, either directly or via ICCP, to develop state estimates that depict accurately and reliably both the internal area and the tie lines.

When performing stability assessment calculations, another subtle issue arises from the representation of tie line injections in the SCADA/EMS models. At the boundary between the internal network model and the external world, the incoming power flows (imports) are typically shown as injected generated powers. But since these are not actual generators, some mechanism must be found to represent them in the stability calculations. One way is to introduce dynamic equivalents, as described in Chapter 6. Another solution to this otherwise difficult problem is the approach described in Chapters 3, 4, and 5.

1.3.2.2   Formal and De Facto Data Format Standards
The open-system emphasis has led to increased development and deployment of software, communication, and database standards. Although much of the algorithm details are justifiably proprietary, the ability to interoperate, share data, use different hard-

ware, and communicate with devices from multiple vendors is now deemed essential in any implementation. To this end, many organizations such as the IEEE, EPRI, CI-GRE, and UCTE have been instrumental in the development of standards that have been accepted by the IEC. These standards include the 60870-5 series for RTU communication, the 60870-6 for interutility control center communication, the 61850 series for communication within substations, and others. Furthermore, increasing reliance on Internet technologies (TCP/IP, XML, browsers) and support for interfaces to Microsoft Office applications further "opens" up EMS/SCADA systems. However, here we are primarily concerned more with the interface and database standards.

In the database area, the basic data are stored in a standard off-the-shelf relational database product that supports the accepted techniques for access (e.g., SQL) and updates. Unfortunately, the "content" of the database may not be standard. For example, a transformer may be modeled differently in each implementation and may be incompatible with other implementations. Therefore, a major effort was made to develop a standard model of power system components that any user could access for his or her own application. This is referred to as the Common Information Model (CIM) and is now an IEC standard. Most systems require that CIM support be included. This does not mean that the operational data has to be in CIM format, merely that it should be possible to read and write CIM compatible data. With this facility, it is possible to share the data with third parties, something that was impossible with proprietary databases and models.

1.3.2.3   *Data Interfaces and Quality*
In spite of all the benefits of CIM, stability assessment programs, just like any advanced network analysis applications that might be seamlessly or loosely integrated with a SCADA/EMS, do not get their input directly from CIM, nor do they interface directly with the real-time database. One very important reason for this is that stability assessment software cannot handle raw data; rather, it requires on input an actual and accurate picture of the system state, that is, the *correct* values of the complex bus voltages and bus-injected complex powers as computed by the state estimator or, perhaps, by the dispatcher's power flow, which are typically assembled in an industry de facto standard solved load-flow format. The other reason is the simplicity and transparency of the integration process. Theoretically, at least, it is possible to expand the stability applications with CIM compliant front-end modules and/or SQL interfaces that could access directly the SCADA/EMS database, but such an approach would significantly increase the complexity of the integration and, furthermore, would adversely impact the portability of the stability software.

In all the integration examples described in this book, the stability assessment software gets the state estimation results in solved load-flow case format. In all these implementations, the host SCADA/EMS exports the state estimate in one of the various load-flow formats supported by the Siemens PTI/Siemens PSS/E load-flow program. Other solved load-flow formats, such as the IEEE Common Exchange Format [11], are also possible, but are seldom implemented. But the availability of the state estimate in solved load-flow case format is not the end of the road, just the beginning. This is because data must not only be available but also accurate.

The data quality of the state estimate has a direct impact on the data quality of the solved load-flow case, which, in turn, has a great impact on the outcome of the calculations performed by the stability assessment software. In this context, the "data quality of the state estimate" refers to the validity and consistency of the model and parameter data used by the state estimator; by contrast, the "quality of the state estimate" refers to how accurate (or inaccurate) was the estimation process, for example, how small (or how large) was the solution mismatch. In other words, the state estimate quality may be poor even if the data quality was good. The data quality problems that affect the state estimate can be classified as temporary versus permanent.

Temporary data-quality problems may be caused by the state estimator to compensate for analog metering errors or erroneous status-data indications. Persistent data-quality problems may be caused by incorrectly defined items in the database, such as: modeling errors, for example invalid line or transformer reactance values; or incorrect parameters, for example, upper and lower limits, specification of controlled buses, and so on. Incorrect limits may be particularly damaging, such as too high values for the upper MW and MVAr generator limits; the state estimator may be based on a solution technique that does not worry about such values, whereas the stability assessment algorithm may produce erroneous results because it just does not know whether the upper limits of the generators are too high because this is how they are in real life, or they are too high because the data are wrong.

The presence of errors in the load-flow case derived from a state estimate or a dispatcher's power flow solution can be detected by computing the MW and MVAr mismatches, but such checks may be inconclusive or even misleading. For example, extremely large mismatches can be caused by a serious bad data problem but might also be due to the fact that the SCADA/EMS exported a nonconverged load-flow case.

1.3.2.4   *User Interface and Interaction*
The layout and aesthetics of the displays, the design of messages and user interaction procedures, and the use of charts and diagrams to quickly convey the essence of large amounts of information are all matters of good software design. They apply to stability analysis programs just as they do to any other computer application. But in the case of online and real-time stability assessment, there are a few other issues that also must be addressed.

A key concern is how to convert the inherently complex calculation results to *information* that can be instantly absorbed and digested for quick and reliable decision making. The immediate answer that comes to mind is to use graphics and pictorial images. This is easier said than done, for this question transcends the user interface issue—the calculation results will be simple and easy to understand only if the underlying solution technique provides such simple and easy to use indicators that capture the distance to instability, quantify the risk of blackout, and so on.

Additional difficulties stem from the fact that, in most cases, the stability software used in the control center originates in the offline realm where the user interface hardware and software, the design of the displays, and the procedures deployed to interact with the computer are quite different from those commonly found in a SCADA/EMS.

The first barrier encountered when porting an off-line program to SCADA/EMS is that the user interface hardware and operating system can be quite different from how off-line applications are designed and implemented.

The user interface look and feel is another issue. The SCADA/EMS user typically monitors and controls different processes that run more or less concurrently and share simultaneously the user interface, as opposed to the off-line user who does only one thing at a time and can, and normally does, allow the application to take full control of the screen. Further differences arise from the style of the user computer dialog. Real-time applications are repetitive "tasks" that, once activated (automatically, triggered by events, or started manually), run transparently and require no further user intervention until the next execution cycle; facilities to set and change execution control parameters also exist but are seldom used except for system initialization. By contrast, off-line applications are designed to perform studies, where each software execution may correspond to a different scenario, with different data and different parameters that may have to be edited each time when running the program.

An easy way out of these difficulties is to install the stability software on a separate workstation, import the input data from the SCADA/EMS, and display the calculation results on the application's own user interface. But this would force the operators to navigate between two display environments—one on the SCADA/EMS console and the other one on the workstation where the stability calculation results are presented. Quite obviously, this is neither easy nor comfortable.

The best and perhaps the only acceptable approach is to control the execution of the stability application, display the key stability calculation results, and monitor the distance to instability *directly in the native SCADA/EMS user interface.* This is normally done by the SCADA/EMS vendor and can be relatively easily achieved if the stability assessment application was designed in a truly modular fashion, that is, if the modules that perform computations and the display services that create the output results are totally separated from the user dialog shell, so that they could be invoked from within the SCADA/EMS system if and when needed. Another benefit of this approach is the ability to accommodate local user requirements such as the format and language used to present calculation results and messages.

The user interface and interaction issues that were summarized in general terms in the preceding paragraphs will be addressed in more detail in the subsequent chapters of this book.

1.3.3 Performance Issues

The performance requirements for online and real-time stability assessment should represent a balance between what is truly needed, on the one hand, and what is reasonable to expect, on the other.

In order to assess "what's needed," let us note that, unlike other operating reliability issues, which can be mitigated over a period of time, stability is unique in the sense that if the risk for the system to become unstable is imminent, there is no time to react and the damages caused by the ensuing blackout could reach colossal proportions. But the risk of instability is also a matter of probabilities. Multiple contingency scenarios are by far less probable than single contingencies. So, if it could be ascertained that none of the single contingency events would cause transient instability for a relatively broad range of operating conditions, the response time requirements for online transient stability assessment do not have to be too stringent if the purpose is to assess what would happen if a very unlikely event would take place. But if the purpose of the stability calculations is to track or monitor the distance to instability in real-time regardless of how far from instability is the power system, then the solution time of the stability application has to be much shorter.

When it comes to "what is reasonable to expect" given the current state of the art, some stability computations can be performed truly in real-time, whereas some others require longer times to complete. The elapsed time expected for stability analysis depends upon the choice of methods and algorithms, size of the network model, number of contingencies to be simulated, and the computer configuration deployed to perform the calculations.

If the goal is to perform online transient stability analysis, it is not realistic to expect solution times of the order of a few seconds, and if the power system network size is large, the response time of the transient stability analysis calculations can be quite high. For example, the current performance requirements at NEMMCO, as indicated by Boroczky in the Chapter 9 of this book, are in the range of 2 to 15 minutes for online transient stability runs on computer hardware equipped with multiple 64-bit processors. For extended online simulations that encompass a suite of transient and other types of stability assessment on hardware configurations consisting of multiple computers, Morrison and coworkers (Chapter 6) quote response times of up to 30 minutes for a very large system, whereas Jardim (Chapter 7) mentions response times in the range of a couple of minutes for a medium-sized power system network.

On the other hand, it is possible today to perform real-time stability assessment and monitor the distance to instability by using special steady-state stability algorithms that are both reasonably accurate and extremely fast, with response times much shorter than the state-estimation cycle. Arnold and coworkers (Chapter 3), Vickovic and Eichler (Chapter 4), and Campeanu and coworkers (Chapter 5) describe actual real-time stability assessment and monitoring solutions that have been operational for quite some time, with which a steady-state stability case is solved in less than one second for moderately sized networks by using standard PC hardware.

The subsequent chapters of this book provide further insight into the performance requirements of stability analysis techniques and related hardware and software solutions that have successfully been implemented and are being used on a daily basis to perform online and real-time stability assessment.

What is *not* covered in this book is the possibility to perform two-stage real-time, and online stability assessment more or less like static contingency analysis, which has been implemented for a long time. The current state of the art in static security assessment consists of running a simplified form of contingency evaluation of a very large set of contingency cases as soon as a new state estimation solution has been obtained; should any contingency case result in thermal or voltage violations, it would be further investigated with a detailed load-flow calculation. Similarly for stability analysis, the idea is to execute, immediately after the state estimate becomes available, a quick stability check by using a fast algorithm such as the ones described in [7] and Appendices A and B of this book. If the system is far from the stability limit corresponding to the current operating conditions, that is, if the risk of instability is small or quasinonexistent, that would be it and no further computation would be needed until the next state estimate was computed. Otherwise, a second stage of detailed stability calculations would have to be triggered to determine whether any of the postulated contingencies could cause transient instability.

This approach has already been proposed [7] but as far as we know has not yet been tested in any actual implementation. As a further refinement, the list of contingencies could be periodically updated, say once an hour, to identify what contingencies would be truly relevant for current operating conditions. Enhancements such as these are already feasible and, we believe, it will be just a matter of time until they are implemented.

1.4 REFERENCES

There is a vast literature on SCADA/EMS, telecommunication in electric utilities, standards used in SCADA, advanced applications, and stability analysis. Here, we list a few published works that contain extensive references to the literature.

[1] Atanackovic, D., Clapauch, J. H., Dwernychuk, G., Gurney, J., and Lee, H., "First Steps to Wide Area Control," *IEEE Power and Energy Magazine,* pp. 61–68, January/February 2008.

[2] Bhargava, B., and Rodriguez, G., "Monitoring the Power Grid," *Transmission and Distribution,* December 2004.

[3] Novosel, D., Modani, V., Bhargava, B., Vu, K., and Cole, J., "Dawn of Grid Synchronization," *IEEE Power and Energy Magazine,* pp. 49–601, January/February 2008.

[4] Pavella, M., Ernst, D., and Ruiz-Vega, D. *Transient Stability of Power Systems: A Unified Approach to Assessment and Control,* Kluwer, Norwell, MA, 2000.

[5] Phadke A. G., " Synchronized Phasor Measurements in Power Systems," *IEEE Computer Applications in Power,* Vol. 6, No. 2, pp. 10–15, April 1993.

[6] Phadke, A. G., Thorp, J. S., and Karimi, K. J., " State Estimation with Phasor Measurements," *IEEE Transactions on Power Systems,* Vol. PWS1, No. 1, February 1986.

[7] Savulescu, S. C. (Editor), *Real Time Stability in Power Systems,* Springer Verlag, New York, 2006.

[8] Savulescu, S. C. (Editor), *Computerized Operation of Power Systems,* Elsevier Scientific Publishing, Amsterdam, 1976.

[9] Venikov, V. A. in *Transient Processes in Electrical Power Systems,* Edited by V. A. Stroyev, English Translation, MIR Publishers, Moscow, 1977.

[10] Zhou, M., Centeno, V. A., Thorp, J. S., and Phadke, A. G., "An Alternative for Including Phasor Measurements in State Estimators," *IEEE Transactions on Power Systems,* Volume 21, No. 4, pp. 1930–1937, Nov. 2006.

[11] IEEE Joint Working Group on Common Format for Exchange of Solved Load-Flow Data, "Format for Exchange of Solved Load-Flow Data," *IEEE Power Applied Systems,* Vol. PAS-91, No. 5, pp. 1916–1925, May, 1972.

2

OVERVIEW OF KEY STABILITY CONCEPTS APPLIED FOR REAL-TIME OPERATIONS

Savu C. Savulescu

2.1 INTRODUCTION

Modern utilities are discovering the side effects of the open transmission access the hard way. Today, they must accommodate MW transfers that can be quite different from those for which their transmission networks were originally planned. This is because energy transactions across multiarea systems of continental or subcontinental size may cause parallel flows, significant network loadings and low bus voltages. Under such deteriorated operating conditions, blackouts due to instability become possible if a major disturbance takes place, or if a large amount of MW is transferred across a stability constrained transmission corridor, or even if an otherwise insignificant topology change, such as a minor line trip, happens in a system already operating near its maximum loadability limit.

In the aftermath of the wave of blackouts that affected U.S., U.K., and mainland Europe utilities in recent years, new operating policies started to require transmission system operators, independent system operators, and reliability coordinators to compute stability limits "for the current and next-day operations processes to foresee whether the transmission loading progresses or is projected to progress beyond the operating reliability limit" [40].

This is far from being a trivial exercise primarily because, as opposed to computing thermal and voltage violations, which is straightforward and can be executed in real time, detecting *stability limits* is much more difficult. Some key questions are:

Real-Time Stability Assessment in Modern Power System Control Centers. Edited by Savu C. Savulescu
Copyright © 2009 The Institute of Electrical and Electronics Engineers, Inc.

1. How does one define and quantify the concept of "stability limit" in the first place?
2. Are stability limits computable in real time?
3. If the distance to instability can be quantified and computed in real time, should it be monitored in the SCADA/EMS environment rather than relying on off-line operating guidelines?

There are various types of stability tools that may be used for a broad range of purposes, but in the context of system operations, which is essentially a real-time process, we are primarily concerned with the risk of instability that may cause a widespread failure. The off-line assessment of the risk of system failure typically consists of executing detailed transient stability calculations on an extended collection of contingency scenarios for the purpose of determining whether *all* the postcontingency states are stable or not. When it comes to real time, detecting the risk of blackout this way, unfortunately, is easier said than done. Due to a number of intrinsic difficulties, which are briefly addressed throughout this chapter, the scope of stability assessment in system operations reflects a compromise between the:

- Depth and extent of the stability analysis,
- Level and granularity of the modeling details,
- Need and/or ability to seamlessly integrate the stability computation environment with the SCADA/EMS platform, and
- Acceptable elapsed times for performing the calculations and presenting the results.

For example, a potential approach might be as follows: for the current system state, and for a *dynamically selected set of major contingencies:*

- Compute the "distance" to the stability limit for the current operating conditions, and, respectively, for each postcontingency state by deploying an acceptable solution technique that is less computationally intensive than conventional transient stability analysis;
- Determine whether both the current and *all* the postcontingency states are within a "safe operating region," which implies that for each one of the stability limits determined previously a new "security margin" has also been calculated, as opposed to using a fixed, postulated margin of safety that would come from off-line simulations.

It is important to emphasize that the availability of a dynamically selected set of contingencies is paramount to the success of such a pragmatic approach. In other words, from the potentially huge collection of $N - 1$, $N - 2$ ... $N - X$ scenarios, one needs to be able to identify those contingency cases that may push the current system state close to or beyond its stability limits. In this context, "dynamically" means that the contingency selection procedure would be executed online, but at much finer time intervals than the time granularity of the real-time stability calculations that, presumably, would run after each successful state estimate within each real-time network analysis sequence cycle.*

In order to answer the question whether the risk of blackout should be assessed and monitored in real time, let us first note that, when it happens, instability develops almost instantly and leaves no time to react. Therefore, operating states that are vulnerable to instability must be *prevented;* waiting for them to occur while hoping that something could be done in the last minute may have catastrophic consequences. But in order to quickly devise adequate corrective actions if and when needed, one must first *predict* that there might be a risk of instability, and, since the operating conditions change continuously, the only way for the prediction to be timely and accurate is for the assessment to be performed in real time and the distance to instability to be monitored continuously. This, in turn, rests on the ability to:

- Run fast stability calculations with data that originate in the real-time SCADA/EMS environment and have been validated for completeness, accuracy, and consistency, that is, have been produced by a reliable and field-proven state estimator.
- Complete the execution of the stability application within the time span of the real-time network analysis sequence; that is, obtain the stability computation results and have them ready for display *before* the next run of the state estimator.
- Present the results in user-friendly formats that facilitate quick and reliable online decision-making.

The industry has clearly understood that operating a power system without continuously monitoring, or, at least, without periodically updating the loadability limit, is like walking on thin ice. As a result, more and more power system control centers are adopting tools that perform stability calculations in real time or, as a minimum, online.†

This book documents this trend with examples of real-life solutions that illustrate the various approaches available for performing real-time and online stability assessment in a modern power system control center. There is more than one way to tackle this problem, both because of the diversity of data and operational environments in the current SCADA/EMS implementations, and because stability per se is extremely complex and can, and perhaps should, be addressed from various angles.

In the following, we review some of the major approaches and briefly discuss their perceived strengths and limitations. In order to enhance the readability of this chapter, certain detailed algorithmic and theoretical aspects are relegated to appendices along

---

*As shown in Chapter 1, the state estimation is not an isolated process. Rather, it is part of the real-time network analysis sequence which starts with the identification of the current system model, by processing the most recent status indications and analog values; continues with the state-estimation computations; and is typically followed by contingency analysis. Network analysis sequence cycles are executed periodically but can also be triggered manually or by system events, such as the tripping of transmission and generation equipment.

†Please refer to Section 1.3 in Chapter 1 of this book for a discussion of "real-time" versus "online."

with numerical examples and benchmarking calculations aimed at further understanding the methodology.

2.2 IN SEARCH OF THE STABILITY LIMITS

2.2.1 Background

The evaluation of the operating reliability of transmission networks as required in system dispatching and operations planning is a complex undertaking. Depending upon the response time, mode of execution, and scope of analysis, the methods are referred to as static and dynamic security assessment. Typical software tools are load-flow and stability programs.

An important objective of dynamic security assessment is to determine whether the system can withstand a set of major, yet credible, contingencies. This is the field of transient stability analysis. An equally important goal is to evaluate the risk of instability if the system approaches a dangerous state slowly as a result of small topology and/or load changes accompanied by slow bus voltage changes that may trigger a voltage collapse, and/or gradual load increases that may eventually cause one or several generators to loose synchronism. In the past, this was known as steady-state stability* but today it is referred to as "voltage stability", as several authors have shown that "voltage stability" can be construed as "steady-state stability" [18] or "load stability" [20,32].

Instability in a multiarea power system may also be triggered when attempting to transfer a large MW block between weakly interconnected areas, for example, when compensating for load increases and/or generation outages in a system area by raising the generation elsewhere. In order to ensure that the grid would not get too close to its stability limits, prior to clearing such a transaction, one would first have to evaluate the maximum transfer capability across the "links," or transmission corridors, that interconnect the areas involved in such transactions. This is typically the case in longitudinal (radial) systems that span distinct system areas with significant load-generation imbalances, but may also happen in meshed transmission networks.

There are other types of instability, e.g., units going out of synchronism due to self-oscillations. Unfortunately at the present time there is no unified methodology to handle all aspects of stability. Each form of instability requires appropriate models and adequate tools tailored to the physical phenomena under evaluation.

The problem becomes even more complex when the target is a vast interconnected system because of the sheer amount of data, the large computing times, and the technical skills needed to interpret the results. Even if computational speed is achieved and

---

*In 2004, the IEEE/CIGRE Joint Task Force on Stability Terms and Definitions [41] altered the stability terms and definitions proposed in 1982 by the IEEE PES Task Force on Terms and Definitions [39] and abandoned the term "steady-state stability" as it was known in the classical stability literature. Regardless of what they are called, though, the physical phenomena that lead to instability are still the same, what is called today "voltage instability" used to be identified in the old times as a form of aperiodic steady-state instability. Please refer to Sections 2.4.1 and 2.4.2 for a cursory review of this topic and to references [1], [9], and [27] for an in-depth treatment of the subject

the stability calculations are performed in real time, or, perhaps, online (with real-time input but slower than the real-time process), the end users may have neither the time nor the background needed to assess the results.

These theoretical and practical difficulties can be overcome with approximate solution techniques that:

- Provide for quantifying the distance to instability,
- Are fast enough so that they can be used in real time,
- Are demonstrably accurate and reliable, and
- Produce the output in formats that are easy to interpret and understand.

2.2.2 Are Stability Limits Quantifiable?

The industry has taken for granted concepts such as the available transfer capability (ATC), total transfer capability (TTC), transmission reliability margin (TRM), and capacity benefit margin (CBM), but only a handful of utilities are routinely performing real-time stability computations in dispatch centers. This book provides actual examples of power system control centers in which maximum loadability calculations are automatically triggered after each and every state estimation computation, but we are not aware of market systems in which each and every postulated market transaction would be validated with fast stability checks.

According to NERC [37], the TTC is given by:

$$TTC = min(thermal\ limit,\ voltage\ limit,\ stability\ limit)$$

Thermal limits, and, to some extent, voltage limits are well known and understood. Both the thermal and the voltage limits are predictable and can even be violated for short periods of time. But "stability limits" are not clearly defined. For example, how many "stability limits" are there and how are they defined and quantified? Can they be "violated"? And, if they can, by how much and for how long?

Conceptually, the "stability limit" is a function of the system state vector: for each new system state, there is a new stability limit. But the stability limit associated with the current or postcontingency operating state is not unique; it also depends upon the trajectory followed in the computation to find such a limit. Simply stated, "stability limits" exist; are not fixed; change with the system's loading, voltages, and topology; and depend upon the procedure used to stress the system conditions until instability has been reached. It is precisely this dynamic nature of "stability limits" that makes it necessary to recompute and track them online.

However, the online evaluation of stability limits does not guarantee that a blackout can be prevented. If the power system were operating with insufficient stability margin and a disturbance would push it beyond the stability limit in effect at that particular moment, instability would be unavoidable because the phenomena develop too quickly and make it virtually impossible to react in a timely manner. Therefore, in addition to a metric that could help quantify the distance between the current conditions and a hypothetical state in which voltages may collapse and units may loose synchronism, the algorithm, or algorithms, that compute the risk of instability must be extremely fast so that the assessment can be performed truly in real time, that is, *immediately* after a new state estimate has been calculated, so that the distance to instability is monitored on a continuous basis.

The subsequent sections provide a brief overview of the three different types of solution techniques that have been implemented to date in power system control centers to address the needs for real-time stability assessment: transient stability, voltage stability, and steady-state stability.

2.3 TRANSIENT AND VOLTAGE STABILITY LIMITS

2.3.1 Transient Stability Limits

Sophisticated transient stability assessment tools are currently available to determine "whether a given condition is stable or unstable, but have not been efficient in quickly and automatically determining the stability limits, that is, how much a system, or part of a system, can be loaded before instability occurs" [15]. Since this statement was published in 1999, significant progress has been achieved in the industry and several successful online implementations of transient stability tools have been reported. The approaches that seem to have produced the most promising results are predicated on time domain simulations and single machine equivalent methods.

Time domain transient stability analysis is both accurate and flexible [19] in terms of modeling detail and can handle:

- All the known types of power system components that correspond to active injections, such as generators, loads, static VAr compensators (SVC), and FACTS devices, as well as the associated controls;
- Any type of contingency, including three-phase and single-phase faults, as well as outages of multiple transmission and active power system components;
- Any type of instability, such as first-swing or multi-swing, up-swing or back-swing, and plant or interarea mode.

The complexity of the algorithms, coupled with the extent of the modeling details, renders their online implementation difficult but not impossible. Chapters 6 and 7 of this book describe time domain transient stability analysis solutions that have been successfully implemented on dedicated multicomputer architectures loosely integrated with existing SCADA/EMS systems. Alternate approaches are addressed in Chapter 8, where the time domain transient stability analysis program was seamlessly integrated with the network analysis software by the SCADA/EMS provider, and also in Chapter 9, where the transient stability analysis package was loosely integrated by the SCADA/EMS owner. In all these cases, the stability applications use real-time data, produce results within time delays that are deemed acceptable by the users, and the overall process can be regarded as being performed "online."

On the other hand, the hybrid transient stability method called SIME (for SIngle-Machine Equivalent) opens the doors to accurate and *fast* transient stability analysis and, as shown by Pavella et al. in [19], is capable of real-time assessment and decision making. Even more importantly, this approach appears to make it possible to implement transient stability control. Chapter 7 provides a fine example of a practical implementation of SIME. The theoretical foundation and algorithm details of SIME are described in Appendix B and related references.

A number of alternate methods that, at least in theory, would make it possible to compute so-called "transient stability limits" have also been investigated and documented in [38]. To a higher or lesser degree, these methods follow similar scenarios:

- Start with a base case and a postulated major contingency.
- Derive a "severity index" for this contingency.
- Process the next major contingency.
- Repeat the process until an unstable case has been obtained. This defines the "limit."

Their limitations are similar, too:

- Computational burden, which somehow can be transcended by deploying multiple processor architectures,
- Nonconvergence of Newton–Raphson load-flow calculations near instability,
- The "stability limit" thus identified qualifies as a limit only for the particular major contingencies that were evaluated, but since many disturbances are possible, a true "system-wide transient stability limit" for a given state vector would require that a huge set of possible disturbances be examined.

In fact, a major difficulty that is intrinsic to transient stability analysis regardless of the particular computational approach stems from the fact that it tells whether the reference base case is stable *and remains stable* for each one of the contingencies evaluated, but it neither determines a "transient stability limit"* nor provides a safe margin where no contingency would cause instability.

In order to complete the search for stability limits, after both the base case and all the contingencies from the list were evaluated, and if none of them caused transient instability, the system would have to be "stressed," for example, by increasing the total MW generation, and at each step of "system stressing," the entire suite of transient stability calculations would have to be executed again. Conversely, if one or several contingencies simulated for the original base case would result in instability, the system conditions would have to be relaxed and the full suite of transient stability calculations

*The NERC document [37] mentions explicitly the "transient stability limit" but it does not say how to compute it. There is no question that a "transient stability limit" exists; the question is how to compute it. Appendix B of this book depicts the Critical Clearing Time (CCT) as a "transient stability limit," but since for different CCTs there are different MW system loadings at which transient instability might occur, the CCT should be construed as a parameter rather than a limit.

repeated until a "safe" operating state was found. Such an exhaustive search of the stability limit and safe operating margin is virtually impossible.*

On the other hand, if the current base case corresponds to a maximum expected MW demand, including the wheeled power, if any, and if none of the contingencies evaluated caused transient instability, it can be inferred that the system is safe because, presumably, the probability of an event worse than those already simulated is very small. As shown in this book, multiple computer configurations and innovative implementation schemes have made online transient stability assessment a reality.

2.3.2   Voltage Stability Limits

The realm of voltage stability, or "voltage security," assessment has been extensively addressed in the technical literature. To quote some of the best-known authors in the field, "voltage instability is much better understood today than it used to be two decades ago, in the mid 1980's," although "several of the theoretical and practical difficulties associated with the subject still remain" [26]. A detailed discussion of "voltage stability" goes beyond the scope of this book, and the interested reader is directed to references [25], [26], and [33], among other excellent references. Nevertheless, we need to briefly address this topic because, although voltage stability methods can successfully provide stability limits in the sense discussed earlier, this benefit can easily vanish if the so-called "voltage stability analysis" consists of just running load-flows until they diverge or developing P–V curves without taking into account the dynamics of the machines.

2.3.2.1   Need to Represent the Generators

In 1975, Venikov et al. [28] asserted that under "certain conditions" the singularity of the standard load-flow Jacobian may indicate steady-state instability. These "certain conditions" are neglecting the generators' internal reactances and assuming that the generators are equipped with forced-action voltage controllers capable of maintaining the voltage constant at the machine terminals [27].

This is precisely the load-flow model. In load-flow computations, the internal reactances of the generators are not represented, and the voltages are maintained constant on the machine terminals or on the high-voltage side of the step-up transformers. If the generator reactances are included in the load-flow model, the P–V buses would "move" to the internal generator nodes where the electromotive forces (emfs) are applied, and since the emfs are higher, or much higher, than 1.0 p.u., the Newton–Raphson calculations might diverge.†

Notwithstanding the above, tools based on load-flow and continuation load-flow quickly became popular and reference [28] was cited as the primary justification for studying the load-flow Jacobian matrix to determine critical load levels. However, al-

---

*Adding the CCT as a variable parameter would further increase the complexity of the search.
†Newton–Raphson calculations use the "tangent method" to develop a series of points whose convergence is not guaranteed if the solution point is far from the starting point; in other words, the calculations may diverge if the solution point is either much smaller or much higher than 1.0 p.u.

though it is true that Newton–Raphson load-flow calculations diverge near instability, the divergence may be also due to other reasons and should not be used as a stability criterion. According to Sauer and Pai [20], "for voltage collapse and voltage instability analysis, any conclusions based on the singularity of the load-flow Jacobian would apply only to the voltage behavior near maximum power transfer. Such analysis would not detect any voltage instabilities associated with synchronous machine characteristics and their controls" [20, p. 1380]. In a subsequent publication [21, pp. 10–15], Sauer and Pai have shown that the assumptions under which the standard load-flow Jacobian can be directly related to the system dynamic Jacobian are (for an m machine, n node network):

1. Stator resistance of every machine is negligible ($R_{si} = 0$, $i = 1, \ldots, m$).
2. Transient reactances of every machine are negligible ($X'_{di} = 0$, $X'_{qi} = 0$, $i = 1, \ldots, m$).
3. Field- and damper-winding time constants for every machine are infinitely large (E'_{qi} = constant, E'_{di} = constant, $i = 1, \ldots, m$).
4. There is constant mechanical torque to the shaft of each generator (T_{Mi} = constant, $i = 1, \ldots, m$).
5. Generator number one has infinite inertia. This together with (1) – (3) makes V_1 = constant, θ_1 = constant (infinite bus).
6. All loads are constant power [$P_{Li}(V_i)$ = constant, $Q_{Li}(V_i)$ = constant, $i = 1, \ldots, n$].

They further clarified the "special conditions" mentioned by Venikov and demonstrated that they actually imply the following:

1. Stator resistance is negligible ($R_{si} = 0$, $i = 1, \ldots, m$).
2. No damper windings or speed damping ($T'_{qoi} = 0$, $D_i = 0$, $i = 1, \ldots, m$).
3. Systems are high gain and fast excitation so that all generator terminal voltages are constant (V_i = constant, $i = 1, \ldots, m$).
4. Constant mechanical torque to the shaft of each generator (T_{Mi} = constant, $i = 1, \ldots, m$).
5. All loads are constant power [$P_{Li}(V_i)$ = constant, $Q_{Li}(V_i)$ = constant, $i = 1, \ldots, n$].

Also regarding voltage stability, but in a different context, C. Barbier and J. P. Barret published in 1980 a seminal paper [6] that promoted the use of the maximum power transfer theorem to identify the point of voltage collapse at any given load bus. For the elementary case of a load represented by the impedance $Z_{load}\lfloor\varphi$ fed by a constant voltage source V_{source} through a two-terminal system of impedance $Z_{line}\lfloor\beta$, Barbier and Barret showed that the current I circulating from the source toward the load and the bus voltage V_{load} at the load are given by

$$I = \frac{I_{cc}}{\sqrt{1 + \left(\dfrac{Z_{load}}{Z_{line}}\right)^2 + 2\left(\dfrac{Z_{load}}{Z_{line}}\right)\cos(\beta - \alpha)}} \tag{1}$$

$$V_{load} = \frac{Z_{load} I_{cc}}{\sqrt{1 + \left(\dfrac{Z_{load}}{Z_{line}}\right)^2 + 2\left(\dfrac{Z_{load}}{Z_{line}}\right)\cos(\beta - \alpha)}} \tag{2}$$

where $I_{cc} = (V_{source}/Z_{line})$ is the short-circuit current at the terminals of the load. When the admittance of the load increases as new loads are added to the system, the active power delivered first increases, then passes through a maximum value, and finally decreases; the voltage drop in the line impedance, created by the increase of the current I, becomes large and the increase of I is not enough to compensate the reduction of V_{load} in the product $V_{load}I$. This result is known as the maximum power transfer theorem: there is a maximum active power that can be transmitted through a two-terminal system from a constant voltage source V_{source} toward a load bus. The maximum power $P_{load_{max}}$, critical current I_{crit}, and critical voltage $V_{load_{crit}}$ correspond to $(Z_{load}/Z_{line}) = 1$ and are given by

$$P_{load_{max}} = \frac{V_{source}^2 \cos \beta}{Z_{line}\, 4 \cos^2\left(\dfrac{\beta - \alpha}{2}\right)} \tag{3}$$

$$I_{crit} = \frac{I_{cc}}{2 \cos\left(\dfrac{\beta - \alpha}{2}\right)} \tag{4}$$

$$V_{load_{crit}} = \frac{V_{source}}{2 \cos\left(\dfrac{\beta - \alpha}{2}\right)} \tag{5}$$

For the general case of a power system in which the voltages are maintained constant at the generator buses and the load at the study bus can be represented by an impedance, equations (1) and (2) become

$$I_{eq} = \frac{I_{cc}}{\sqrt{1 + Z^2 Y_{cc}^2 + 2Z' Y_{cc} \cos(\beta - \alpha)}} \tag{6}$$

$$V_A = \frac{Z' I_{cc}}{\sqrt{1 + Z^2 Y_{cc}^2 + 2Z' Y_{cc} \cos(\beta - \alpha)}} \tag{7}$$

where

A designates the study bus.

$1/Z_{line}$ in equations (1) and (2) is replaced by the Y_{cc} short-circuit admittance* of the system at the study bus.

*Please refer to Section 1.2.1 of Appendix A in this book for a detailed discussion of the short-circuit currents transformation, including the calculation of I_{cc}, I_{eq}, and Y_{cc} that appear in equation (6).

I_{eq} is an equivalent current that adds the contribution of the intermediate system loads to the load current at the study bus A.

Z' is the impedance of the equivalent load at the study bus A and results from paralleling the impedance of the load at bus A with an impedance that results from the reduction of the system by eliminating the buses where the voltages are not controlled, that is, by applying the short-circuit current transformation.

For this case, the equivalent power P_{eq} absorbed at the study bus will be maximum and the bus voltage V_A will be critical for $Z'Y_{cc} = 1$ [6, p. 690]. But can voltages be maintained constant at the generator terminals? Barbier and Barret clearly stated that the generators should be modeled via a constant emf behind an internal reactance, but this injunction probably went unnoticed, which perhaps explains why so many subsequent papers spread the idea that voltage collapse could be detected without representing the machines.

To set the record straight, this is what Barbier and Barret wrote about the representation of the generators [6, p. 681]:

> When the source substation can no longer hold its voltage constant, because it has reached its limit (rotor or stator current of a generating unit for example), there are two possibilities: either a further constant voltage point is found (such as emf behind the synchronous reactance of an alternator for operation of the latter at constant excitation . . .); or there is no constant voltage and the risk of voltage collapse is considerable. This would be the case, for example, of a system in which all the generating units are at the limit of armature current and in which the latter is maintained constant (at its maximum value) during taking over of load.

The need to represent the synchronous machines rather than considering them as pure voltage sources has been emphasized by many other authors as well, for example, Van Cutsem and Vournas [26] who noticed that "besides some voltage droop under automatic voltage regulator control, field and armature current limits must be obeyed. The former are imposed by overexcitation limiters and the latter by armature current limiters or (most often) by plant operators. These limits have a strong impact on maximum load power."

A detailed discussion of this matter along with a proposal for an approximate representation of the generators that takes into account the behavior of the automatic voltage regulators without actually representing them in detail is provided by Molina and Cassano in the Section A.1.2.3 of the Appendix A in this book.

2.3.2.2 Impact of the Load Model

Another basic assumption in equations (1) through (7), which is also frequently accepted in the voltage stability literature, is that the load can be approximated by an impedance. Ionescu and Ungureanu [12] analyzed the impact of load modeling and demonstrated that the voltage collapse process is affected by how we model the load as a function of voltage.

If the loads are modeled as *constant impedances,* successive load increases cause the generated MW to increase until the point of maximum power transfer. Then, beyond that point, the total generated power starts getting smaller and *dual power states* (same power at different voltages) are obtained, hence the "nose" shape of the well-known P–V curves. But dual states *cannot* happen in real life, and more realistic load models are needed so that the P–V graphs would stop at the point of instability.

Ionescu and Ungureanu further suggested that:

- Representing the loads as $P = $ const and $Q = $ const is a pessimistic hypothesis and produces conservative results.
- Modeling the loads as $P = $ const and $Q = BV^2$ corresponds to an average scenario and is well suited for voltage and steady-state stability assessment. A summary of the analysis developed by Ionescu and Ungureanu can be found in [13] and [23, closure to discussions].

As shown in the following, most of the aforementioned limitations and difficulties are resolved and eliminated if we revert to the classical framework of steady-state stability.

2.4   STEADY-STATE STABILITY LIMITS

2.4.1   General Considerations

The steady-state stability limit (SSSL) of a power system is "a steady-state operating condition for which the power system is steady-state stable but for which an arbitrarily small change in any of the operating quantities in an unfavorable direction causes the power system to loose stability" [39]. An earlier definition refers to this concept as the "stability of the system under conditions of gradual or relatively slow changes in load" [5]. Voltage collapse, units getting out of synchronism, and instability caused by self-amplifying small-signal oscillations are all forms of steady-state instability.

Empirically, the risk of steady-state instability is associated with low real/reactive power reserves, low voltage levels, and large bus voltage variations for small load or generated power changes. Recurring "temporary faults," whereby breakers trip without apparent reason, i.e., are disconnected by protection without being able to identify the fault, might also be indicative of steady-state instability. Breaker trips can happen when loads increase due to "balancing rotors" of generators that operate near instability trip and then get back in synchronism. In some cases, "the resynchronization happened after the rotor turned 360°, which, in turn, led to lower voltages" [9].

An interesting reading on this topic is [36]. Published in the aftermath of the August 14, 2003 blackout in the United States, EPRI's white paper begins with the statement, ". . . based on available evidence in the FirstEnergy areas, the events of August 14, 2003 did not indicate a classical voltage collapse" [36, p. 6]. Yet, subsequently, the report presents data that document:

- Unexplained line trips,
- Voltages "lower than expected,"

- "Low-voltage alarms,"
- The tripping of the 615 MW East Lake Unit 5 at 13:31:53 which ". . . dropped its reactive output from 393 MVAr to –1.8 MVAr when it exceed the maximum excitation limit,"
- Voltages continuously decaying at the bus Star 345 kV, from 0.905 p.u. (14:10 pm) to 0.899 p.u. (15:32 pm) and then to 0.878 p.u. (15:55 pm),
- Numerous line and generator trips between 16:09 and 16:29 pm, each successive line trips causing further voltage degradation.

If one places these data in the context of traditional steady-state stability, it can clearly be inferred that on that fateful day, the transmission system was slowly approaching a state where, eventually, voltages would collapse and units would loose synchronism, which actually did happen at approximately 16:29.

The phenomena encompassed by steady-state stability are extremely complex. Accordingly, specialized tools have been tailored to address natural stability versus stability that is artificially maintained or enhanced by fast voltage controllers, local stability versus global stability, aperiodic instability versus instability caused by self-amplifying small-signal oscillations, and the stability of power transfers across transmission paths between system areas, which is actually a form of aperiodic instability.

The conventional method of the small oscillations for estimating the steady-state stability [5, 9, 27] consists of examining the eigenvalues of the linearized characteristic equation associated with the system of differential equations that describe the free transient processes after a small disturbance takes place in an automatically controlled power system. The necessary and sufficient condition for steady-state stability is that all the real parts of the eigenvalues be negative [27]. The analysis encompasses the following steps:

- Describe the transient processes in the form of a system of nonlinear differential equations.
- Linearize the equations around the solution point by expanding them into a Taylor series and retaining only the linear (first-order) terms.
- Calculate the main determinant and its minors and develop the characteristic equation.
- Determine the sign of the real roots and the sign of the real part of the complex roots of the characteristic equation.

The approach is laborious* and is replaced by determining relationships between the roots and the coefficients of the characteristic equation. Venikov [27] refers to

---

*The approach described by Venikov in [27] has been rendered obsolete by straightforward eigenvalue calculations that can be performed relatively fast with the powerful computers that are available today. Nevertheless, the significance of the "practical steady-state stability criteria" (Section 4.2), which were introduced by the Russian school of stability for the purpose of easily detecting the change of the sign of the Jacobian determinant, stems from the fact that the practical steady-state stability criteria are intrinsically fast, and the theoretical conditions for their applicability are clearly defined.

these relations as "steady-state stability criteria" and classifies them as algebraic (Routh–Hurwitz) and practical.

A necessary, but not sufficient, condition for steady-state stability is derived from the Hurwitz criterion by evaluating the sign of the last term of the characteristic equation, which is the dynamic Jacobian determinant D. A change of sign from positive to negative (all Hurwitz determinants are positive) with further loading of the system indicates aperiodic, or monotonic, instability. The instability in the form of self-oscillations, however, remains unrevealed by this method.

The "algebraic steady-state stability criteria" have been known for a long time [1–4], and can form the basis for algorithms that search for the aperiodic steady-state stability limit by alternating the calculation of the dynamic Jacobian determinant with some procedure to stress the system until it becomes unstable. In fact, such an application was developed in-house at Transelectrica in Romania and is currently being used offline, as shown in Section 4.1 of Chapter 4 in this book.

For the purpose of real-time stability assessment, the so-called "practical steady-state stability criteria" (Section 2.4.2) greatly simplify the calculations and, if applied in conjunction with an adequate system-stressing procedure, allow computing the distance to instability, or "stability reserve" (Section 2.4.3.1) and evaluating the "security margin" (Section 4.3.2) quickly enough for being applicable in real-time.

2.4.2   Practical Steady-State Stability Criteria

Under certain conditions, the calculation of the dynamic Jacobian determinant can be replaced by evaluating one or several of the so-called "practical steady-state stability criteria" that were developed by the Russian school of stability as documented by Venikov in [27]. These "practical steady-state stability criteria" refer to aperiodic instability and, as shown in the previous section, cannot detect instability due to self-sustained oscillations. They are derived from the condition $D = 0$ and are valid if:

- The generators are radially connected to a nodal point; this is not generally true in actual networks, but, as we will show in the next section, it is always the case if the short-circuit currents transformation is applied to convert the power system network to a scheme of short-circuit admittances connected radially to a load bus, which becomes the "nodal point" required for the practical criteria to be valid.
- The system frequency is constant during the short period of time associated with the transient process.
- One of the following assumptions can be made:
  a. The voltage is constant at the nodal point, in which case the synchronizing power criterion $dP/d\delta$ is obtained.
  b. The power balance can be maintained at the nodal point, which leads to the reactive power steady-state stability criterion $d\Delta Q/dV$. This criterion was found to be particularly attractive in conjunction with Paul Dimo's REI methodology and has been used since early 1960s to compute the "stability reserve," which is a metric for quantifying the distance to instability.

The mathematical proof of the $d\Delta Q/dV$ criterion is provided in Annex 1-1; further insight regarding this important tool for quickly evaluating the steady-state stability conditions of a power system is provided, along with a numerical illustration, in the Section A.3.2 of Appendix A in this book.

The reactive power voltage and steady-state stability criterion was used by Paul Dimo to develop a fast and reasonably accurate technique for voltage and steady-sate stability assessment [7,8,9,22] that has been successfully deployed in real-time, as shown in Chapters 3, 4, and 5 in this book. A detailed description of this approach is provided in Appendix A.

2.4.3 Distance to Instability—Security Margin

2.4.3.1 Steady-State Stability Reserve

Approaching the search for a stability limit from the steady-state stability perspective brings promising results. To begin with, the SSSL can be defined both system-wide and for individual buses. Then, the system-wide SSSL can be quantified as the maximum total MW system grid utilization, including both internal generation and tie-line imports, right before instability. On this basis, a metric can be introduced to quantify "how far from SSSL" is a given operating state. Known as steady-state stability reserve, this metric, expressed as a percentage of SSSL, has been known and used in Europe since 1950s [7,8,9]. For example, the 1964 Special Report of the Group 32 of CIGRE states that "any network that meets the steady-state stability conditions can withstand dynamic perturbations and end in a stable operating state" [17].

A transient stability limit (TSL) can also be thought to exist but, as opposed to SSSL, it is not quantifiable through a specific formula. In order to find the TSL, transient stability simulations would have to be performed for each potential disturbance both for the base case and for a sequence of successively degraded operating states until the first unstable state has been identified. Such an approach would be too time-consuming to have practical value, but intuition suggests that a "safe" system MW grid utilization, expressed as a fixed percentage of the SSSL and referred to as a *security margin,* could be found such that, for any system state with a steady-state stability reserve higher than this value, no contingency, no matter how severe, would cause transient instability.

2.4.3.2 Security Margin

The knowledge of a "safe" amount of stability reserve, or security margin, such that transient instability would not occur, makes it possible to replace the otherwise unsolvable problem of computing the TSL with a relatively simple procedure:

- First, starting from a state estimate or solved load flow, determine the steady-state stability reserve, that is, the distance to SSSL.
- Then, for the known (and fixed) $x\%$ security margin, determine the corresponding safe system MW loading below the SSSL.

Further details regarding these concepts are available in references [22, 30, 31], and, in this book, in Section 3.3.2 of Chapter 3, Annex 4-1 of Chapter 4, and Section 5.2.1.1 of Chapter 5.

Each system has its own security margin. For example, reference [10] recommended a 20% security margin for the Romanian power system as it was in the 1970s. Reference [29] describes the procedure used by ETESA in Panama to validate the value of the security margin (15%) that is currently used in conjunction with its real-time stability assessment application.

2.5 CONCLUDING REMARKS

This chapter addressed various theoretical aspects of stability assessment in power system operations. As opposed to system planning, in which the stability studies are concerned with postulated scenarios over long periods of time, the primary concern in operations is "whether the transmission loading progresses or is projected to progress beyond the operating reliability limit" [38], both in actual dispatching, which is essentially a real-time activity, and when validating short-term scenarios, for example, ascertaining the impact of a transmission or generation outage scheduled to take place within the next couple of hours. This is consistent with the SCADA "supervisory control" function, which entails monitoring the real-time values of the system frequency, tie-line interchanges, selected bus voltages, and so on, against their prescribed operational limits.

Accordingly, the chapter focused on questions revolving around the concept of *stability limits*. What are they? Are they quantifiable? How can one compute them quickly and reliably for online decision making? We attempted to answer these questions from the perspectives of transient, voltage, and steady-state stability. The concepts of *steady-state stability reserve* and *security margin* have also been reviewed and shown to provide a solid metric for quantifying the distance to the state at which voltages may collapse and/or units may loose synchronism, and for approximating a safe operating limit at which, given the current operating conditions and a dynamically selected set of major, yet credible contingencies, there is no risk of blackout.

The successfully implemented real-time and online stability solutions described in this book rely on one or several of the techniques identified in this chapter but differ substantially in terms of:

- Seamless versus loose integration. The implementations described in Chapters 4, 5, and 8 consist of seamlessly (tightly) integrating the stability assessment engine with the SCADA/EMS, whereas the solutions discussed in Chapters 3, 6, 7, 9, 10, and 11 rely on various degrees of loose integration of a third-party or in-house-developed stability software with the existing SCADA/EMS.

- Continuous assessment versus periodic checks. Due to the intrinsic speed of the underlying algorithms, the solutions that compute the steady-state stability reserve are more efficient, can be executed automatically after each state estimation cycle, and provide for monitoring the risk of instability just like other system quantities, for example, system frequency and tie-line interchanges, are being monitored in the control center. On the other hand, the complex implementations that deploy multiprocessor configurations and/or use detailed simulation techniques typically perform periodic checks at approximately 10 to 30 minute intervals and are not amenable to closely monitoring the stability limits.

- User interaction and presentation of results. By definition, the seamlessly integrated solutions offer the significant benefit of a single-user interface environment whereby the application execution is controlled from, and the results are displayed in, the native SCADA/EMS environment. The presentation of stability calculation results also depends upon the underlying technique, in the sense that some methods naturally lead to graphical and trending displays that make it easy to quickly grasp and interpret the key information.

Additional, and very important, practical aspects related to the integration of stability applications with the SCADA/EMS are extensively addressed in Section 1.3.1, Stability Assessment in the SCADA/EMS context, and Section 1.3.2, Data Issues, in Chapter 1. Together, the first two chapters of the book set the stage for the subsequent descriptions of real-time stability assessment solutions implemented to date in modern power system control centers.

2.6 REFERENCES

[1] Arie, E., Pomarleanu, M., Bejuscu, L, and Botgros, M., "Determinarea Rezervelor de Stabilitate Statică a Sistemelor Electroenergetice Complexe," *St. Cerc. Energ. Electr.,* Vol. 23, No. 3, pp. 709–727, 1973.

[2] Constantinescu, J., "Practical Assessment of the Power System Stability Margins," *Rev. Roum. Sc. Tech., Série Elect. et Energ.,* Vol. 39, 1994, No. 2, pp. 217–222.

[3] Constantinescu, J., "Study of the Transient Processes in Large-Scale Power Systems," *Rev. Roum. Sc. Tech., Série Elect. et Energ.,* Vol. 27, 1982, No. 2, pp. 211–227.

[4] Constantinescu, J., and Homos, M., "Metoda Practică de Calcul Pentru Studiul Stabilității Statice a Sistemelor Electroenergetice Complexe," Energetica, Vol. 25, 1977, No. 9, pp. 328–336.

[5] Crary, S. B., 1945, *Power System Stability,* General Electric Series, Schenectady, New York, Copyright 1945, Third Printing October 1955.

[6] Barbier, C., and Barret, J. P., "An Analysis of Phenomena of Voltage Collapse on a Transmission System," RGE, special edition CIGRE, July 1980, pp. 3–21.

[7] Dimo, Paul, "Etude de la Stabilité Statique et du Reglage de Tension," R.G.E., Paris, 1961, Vol. 70, 11, 552–556.

[8] Dimo, Paul, "L'Analyse des Réseaux d'Energie par la Méthode Nodale des Courants de Court-Circuit. L'Image des Noeuds," R.G.E., Paris, 1962, Vol. 7, pp., 151–175.

[9] Dimo, Paul, *Nodal Analysis of Power Systems,* Abacus Press, Kent, England, 1975.

[10] Dimo, P., Manolescu, G., Iordanesscu, I., Groza, L., Ionescu, S., Albert, H., Moraite, G., and Ungureanu, B., Computation and Design of Electrical Energy Systems (Romanian edition), Editura Tehnica, 1971, Bucharest, Romania, pp. 200–201.

[11] Dobson, I., L. Liu, "Immediate Change in Stability and Voltage Collapse when Generator Reactive Power Limits are Encountered," in *Proceedings of International Seminar on Bulk Power System Voltage Phenomena II*, pp. 65–74.

[12] Ionescu, S., and Ungureanu, B., "The Dual Power States and Voltage Collapse Phenomena," *Rev. Roum. Sc. Tech., Série Elect. et Energ.*, Vol. 26, No. 4, pp. 545–562.

[13] Ionescu, S., Discussion to the paper "Fast Steady-State Stability Assessment for Real-Time and Operations Planning," *IEEE Trans. Pow. Sys.*, Vol. 8 T-PWRS, No. 4, Nov. 1993, pp. 1564–1565.

[14] Ionescu, S., "Theoretical Analysis of Algebraic Criteria for Voltage and Steady-State Stability Assessment," personal communication to Savu C. Savulescu, manuscript in Romanian language, February 9, 1994.

[15] Kundur, P., "Introduction to the Special Publication on Techniques for Power System Stability Search," in reference [38], p. 1–3.

[16] Kundur, P., and Morison, G. K., "Classes of Stability in Today's Power Systems," *IEEE Trans. Pow. Sys.*, Vol. 8 T-PWRS, No. 3, pp. 1159–1171.

[17] Magnien, M., Rapport spécial du Groupe 32 Conception et Fonctionnement des Réseaux, Conférence Internationale des Grands Réseaux Electriques à Haute Tension, CIGRE Session 1964.

[18] Navarro-Perez, R., and Prada, R. B., "Voltage Collapse or Steady-State Stability Limit," in *Proceedings of the International Seminar on Bulk Power System Voltage Phenomena II*, pp. 75–84, (Edited by L. H. Fink, 1993).

[19] Pavella, M, Ruiz-Vega, D., and Glavic, M., "SIME: A Comprehensive Approach to Transient Stability," in *Real Time Stability Assessment in Modern Power System Control Centers*, IEEE Press and Wiley, NY, 2008.

[20] Sauer, W. P., and Pai, M. A., "Power System Steady-State Stability and the Load-Flow Jacobian," *IEEE Transactions in Power Systems*, 5 T-PWRS, 4, pp. 1374–1381.

[21] Sauer, W. P., and Pai, M. A., "Relationships between Power System Dynamic Equilibrium, Load-Flow, and Operating Point Stability," in *Real Time Stability in Power Systems*, pp. 1-30, Springer Verlag, Norwell, MA, 2006.

[22] Savulescu, S. C., "Fast Assessment of the Distance to Instability. Theory and Implementation," in *Real Time Stability in Power Systems*, pp. 31–64, Springer Verlag, Norwell, MA, 2006.

[23] Savulescu, S. C., Oatts, M. L., Pruitt, J. G., Williamson, F., and Adapa, R., "Fast Steady-State Stability Assessment for Real-Time and Operations Planning," *IEEE Trans. Pow. Sys.*, Vol. 8 T-PWRS, No. 4, Nov. 1993, pp. 1557–1569.

[24] Taylor, C., *Power System Voltage Stability*, EPRI/Mc Graw-Hill, 1994.

[25] Van Cutsem, T., and Vournas, C., *Voltage Stability of Electric Power System*, Kluwer Academic Publishers, Boston, 1998.

[26] Van Cutsem, T., and Vournas, C., "Emergency Monitoring and Corrective Control of Voltage Instability," in *Real Time Stability in Power Systems*, pp. 94–118, Springer Verlag, Norwell, MA, 2006.

[27] Venikov, V. A. *Transient Processes in Electrical Power Systems*, Edited by V. A. Stroyev, English Translation, MIR Publishers, Moscow, 1977.

[28] Venikov, V. A., Stroev, V. A., Idelchick, V. I., and Tarasov, V. I., "Estimation of Electrical Power System Steady-State Stability," *IEEE Trans. on PAS*, Vol. PAS-94, No. 3, May/June 1975, pp. 1034–1041.

[29] Vergara, J. S., Thai, T. A., Cuong, N. D., Nam, N. T., Campeanu, H. S., and Savulescu, S. C., "Accuracy Testing and Real-Time Implementation of Dimo's Stability Analysis Technique," in *Real Time Stability in Power Systems,* Springer Verlag, Norwell, MA.

[30] Vickovic, D., Eichler, R., and Savulescu, S. C., "Real-Time System Stability Monitoring in the Transmission Network of Bosnia and Herzegovina," paper presented at the Power-Grid Europe Conference and Exhibition, Feria de Madrid, Madrid, Spain, 26–28 June 2007.

[31] Virmani, S., Vickovic, D., and Savulescu, S. C, "Real-Time Calculation of Power System Loadability Limit," Paper No. 576 presented at the Powertech 2007 Conference, July 1–5, 2007, Lausanne, Switzerland.

[32] Vournas, C. D., Sauer, P. W., and Pai, M. A., "Relationships between Voltage and Angle Stability of Power Systems," *Electrical Power and Energy Systems,* Vol. 18, No. 8, pp. 493–500, 1996.

[33] Vournas, C., Van Cutsem, T. "Online Voltage Security Assessment," in *Real Time Stability in Power Systems,* pp. 119–146, Springer Verlag, Norwell, MA, 2006.

[34] Wu, F. F., and Narasimhamurti, N., 1979, "Necessary Conditions for REI Reduction to be Exact," IEEE PES Winter Meeting 1979, Paper A 79 065-4.

[35] EPRI, "Power System Steady-State Stability Monitor Prototype," Final Report EPRI TR-100799, July 1992, and "Power System Steady-State Stability Monitor," Final Report EPRI TR-103169, December 1993.

[36] EPRI, "Factors Related to the Series of Outages on August 14, 2003," a White Paper, Product ID 1009317, November 20, 2003.

[37] NERC, "Available Transfer Capability Definitions and Determination," North American Electric Reliability Council, June 1996.

[38] IEEE PES, "Techniques for Power System Stability Search," A Special Publication of the Power System Dynamic Performance Committee of the IEEE PES, TP-138-0, 1999.

[39] IEEE PES Task Force on Terms and Definitions, "Proposed Terms and Definitions for Power System Stability," *IEEE Trans. on PAS,* Vol. PAS-101, No. 7, July 1982.

[40] NERC, "Policy 9—Reliability Coordinator Procedures," Version 2, Approved by Board of Trustees February 7, 2000.

[41] IEEE/CIGRE Joint Task Force on Stability Terms and Definitions, "Definition and Classification of Power System, Stability," *IEEE Trans. on PAS,* Vol. 19, No. 3, August 2004, pp. 1387–1401.

ANNEX 1-1. REACTIVE POWER STEADY-STATE STABILITY CRITERION $D\Delta Q/dV$

It will be demonstrated in the following that, under the hypothesis assumed for applying the $d\Delta Q/dV$ criterion described in Section 2.2.4.2, the dynamic Jacobian determinant associated with the system of equations that describe the steady state of a power system network and the $d\Delta Q/dV$ index vanish simultaneously. The mathematical proof is taken directly from reference [13].

Let us consider the network depicted in Figure 2-1, which meets the requirement that the generators be radially connected to a "nodal point" where the load is located. The topology of this network, in spite of its simplicity, reflects a typical radial

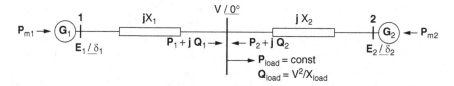

Figure 2-1. Radial network of generators connected to the nodal point.

network of short-circuit impedances (admittances) obtained after applying the short-circuit current transformation discussed in: Section 2.4.4 of this chapter, Section A.1.2.1 of Appendix A, and in [7, 8, 9, 22] and related references. The approximation entailed in neglecting the real part of the impedances that connect the generators to the nodal point is minimal and is also consistent with practical observations made by many authors regarding the nature of the short-circuit impedances (admittances), for example, references [7, 8, 9, 22, 31], as well as the analysis provided in Section A.2.3 of Appendix A. In addition, the following hypotheses are admitted during the very short period of time when the steady-state stability phenomena take place:

- The mechanical powers P_{m1} and P_{m2} at the generators' shafts are constant.
- The real part of the load P_{load} does not vary with the voltage $V\underline{|0°}$.
- The reactive part of the load* is given by $Q_{load} = V^2/X_{load}$ where X_{load} is constant.

Accordingly, the system state of this network is described as shown in the following:

- For the generators G_1 and G_2 the mechanical powers P_{m1} and P_{m2} equal the electrical MW generated powers, which are given by equations (1.a) and (1.b):

$$P_{m1} = P_1 = \frac{E_1 V}{X_1} \sin \delta_1 \tag{1.a}$$

$$P_{m2} = P_2 = \frac{E_2 V}{X_2} \sin \delta_2 \tag{1.b}$$

- The generated MW powers are entirely absorbed by the load P_{load}, as shown in equation (2):

$$P_{load} = \frac{E_1 V}{X_1} \sin \delta_1 + \frac{E_2 V}{X_2} \sin \delta_2 = const \tag{2}$$

*Further considerations about load modeling in steady-state stability analysis can be found in references [9, 12, 22, 23], Section 2.2.3.2 of this chapter, and Section A.1.2.5 in Appendix A.

- The reactive power of the load Q_{load} is supplied by:

$$Q_{load} = \frac{V^2}{X_{load}} = \frac{E_1 V}{X_1} \cos \delta_1 - \frac{V^2}{X_1} + \frac{E_2 V}{X_2} \cos \delta_2 - \frac{V^2}{X_2} \tag{3}$$

where:

X_1 and X_2 incorporate the internal reactances of the generators.

$E_1\underline{|\delta_1}$ and $E_2\underline{|\delta_2}$ are the emf (voltages behind the internal reactances) and are assumed to remain unchanged during the very short period of time when the steady-state stability criterion applies.*

- The generator G_2 is the swing bus and, therefore, there is no need for an additional equation for this node.

Equations (1), (2), and (3) contain three variables: $V\underline{|0°}$, δ_1, and δ_2. Let us now consider a perturbation ΔQ of the reactive power injected into the load bus. From (1), (2), and (3) and with $\Delta P_{m1} = 0$, $\Delta P_{m2} = 0$, $\Delta P_{load} = 0$, as well as $(\partial\, Q_{load}/\partial\delta_1) = 0$ and $(\partial\, Q_{load}/\partial\delta_2) = 0$ because Q_{load} is a function of V only, we obtain

$$\frac{\partial P_1}{\partial V}\Delta V + \frac{\partial P_1}{\partial\delta_1}\Delta\delta_1 = 0$$

$$\frac{\partial(P_1 + P_2)}{\partial V}\Delta V + \frac{\partial(P_1 + P_2)}{\partial\delta_1}\Delta\delta_1 + \frac{\partial(P_1 + P_2)}{\partial\delta_2}\Delta\delta_2 = 0 \tag{4}$$

$$\frac{\partial(Q_1 + Q_2 - Q_{load})}{\partial V}\Delta V + \frac{\partial(Q_1 + Q_2)}{\partial\delta_1}\Delta\delta_1 + \frac{\partial(Q_1 + Q_2)}{\partial\delta_2}\Delta\delta_2 = \Delta Q$$

The Jacobian matrix of the system of the system of equations (4) is given by

$$J = \begin{bmatrix} \dfrac{\partial P_1}{\partial V} & \dfrac{\partial P_1}{\partial\delta_1} & 0 \\[2mm] \dfrac{\partial(P_1 + P_2)}{\partial V} & \dfrac{\partial(P_1 + P_2)}{\partial\delta_1} & \dfrac{\partial(P_1 + P_2)}{\partial\delta_2} \\[2mm] \dfrac{\partial(Q_1 + Q_2 - Q_{load})}{\partial V} & \dfrac{\partial(Q_1 + Q_2)}{\partial\delta_1} & \dfrac{\partial(Q_1 + Q_2)}{\partial\delta_2} \end{bmatrix} \tag{5}$$

where:

$$\frac{\partial P_1}{\partial V} = \frac{E_1}{X_1}\sin\delta_1$$

---

*A detailed discussion of the simplified representation of generators in steady-state stability studies is provided in Section A.1.2.3 of Appendix A.

$$\frac{\partial P_1}{\partial \delta_1} = \frac{E_1 V}{X_1} \cos \delta_1$$

$$\frac{\partial (P_1 + P_2)}{\partial V} = \frac{E_1}{X_1} \sin \delta_1 + \frac{E_2}{X_2} \sin \delta_2$$

$$\frac{\partial (P_1 + P_2)}{\partial \delta_1} = \frac{E_1 V}{X_1} \cos \delta_1$$

$$\frac{\partial (P_1 + P_2)}{\partial \delta_2} = \frac{E_2 V}{X_2} \cos \delta_2$$

$$\frac{\partial (Q_1 + Q_2)}{\partial V} = \frac{E_1}{X_1} \cos \delta_1 + \frac{E_2}{X_2} \cos \delta_2 - 2\left(\frac{1}{X_1} + \frac{1}{X_2}\right) V$$

$$\frac{\partial Q_{\text{load}}}{\partial V} = 2 \frac{V}{X_{\text{load}}}$$

$$\frac{\partial (Q_1 + Q_2)}{\partial \delta_1} = -\frac{E_1 V}{X_1} \sin \delta_1$$

$$\frac{\partial (Q_1 + Q_2)}{\partial \delta_2} = -\frac{E_2 V}{X_2} \sin \delta_2$$

After performing the calculations and with the notations $I_1 = (E_1/X_1)$ (short-circuit current of generator G_1), $I_2 = (E_2/X_2)$ (short-circuit current of generator G_2), and $Y = (1/X_1) + (1/X_2) + (1/X_{\text{load}})$, we obtain the formula (6) for the determinant D of the Jacobian matrix:

$$D = (I_1 \cos \delta_1)(I_2 \cos \delta_2) V^2 \left(\frac{I_1}{\cos \delta_1} + \frac{I_2}{\cos \delta_2} - 2YV\right) \tag{6}$$

The system of equations (4) has a solution if, and only if, $D \neq 0$, in which case we obtain

$$\frac{\Delta Q}{\Delta V} = \frac{I_1}{\cos \delta_1} + \frac{I_2}{\cos \delta_2} - 2YV \tag{7}$$

which corresponds to the expression developed by Paul Dimo for calculating the $d\Delta Q/dV$ [7,9,22]. From (6) and (7) we infer that, at the critical (unstable) operating point, both the dynamic Jacobian determinant and $d\Delta Q/dV$ vanish simultaneously.

3

LIPA IMPLEMENTATION OF REAL-TIME STABILITY MONITORING IN A CIM COMPLIANT ENVIRONMENT

Loris Arnold, Janos Hajagos, Susan M. Manessis, and Anie Philip

3.1 INTRODUCTION

3.1.1 LIPA Power System at a Glance

The electric transmission system of the Long Island Power Authority* (LIPA) has been designed to provide adequate capacity between generation sources and load centers at reasonable cost with minimum impact on the environment. As of June 30, 2007, LIPA owns 1,292 miles of transmission and subtransmission lines that deliver power to 177 substations in its electric system. From these substations, 13,132 circuit miles distribute the power to 1.1 million customers in Nassau and Suffolk counties and the Rockaway Peninsula in Queens County, New York.

As defined by the New York Independent System Operator (NYISO), "bulk" transmission includes LIPA's 345 kV, and 138 kV systems, and subtransmission includes the 69 kV, 33 kV and 23 kV systems. Each system has circuits constructed overhead and underground. A geographic one-line diagram is displayed in Figure 3-1.

The LIPA electric system currently has two 345 kV and three 138 kV alternating current (AC) transmission interconnections and two high-voltage direct current (HVDC) interconnections to neighboring systems in operation. The load requirements

---

*National Grid manages LIPA's transmission and distribution (T&D) assets in Long Island, New York.

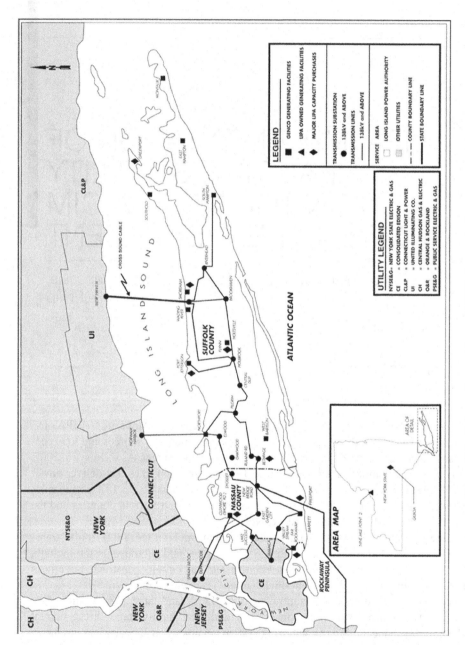

Figure 3-1. Schematic overview of LIPA's transmission system.

of LIPA, NYPA, and Long Island municipalities are served by: 5461 MW of National Grid generation, 195 MW from independent power producers and small industrial units, and 2062 MW that are imported from the remainder of New York State over the two 345 kV interconnections. In addition, 286 MW of power are wheeled to Con Edison's Jamaica substation over the jointly owned Shore Road–Dunwoodie interconnection.

These power exchanges take place across three transmission interfaces as follows:

1. NYISO Transmission interface, which is defined by the four interconnections to the Con Edison system
2. ISO-NE interface, which encompasses the Northport–Norwalk Harbor 138 kV cable and the HVDC Cross Sound Cable between Shoreham and East Shore in New Haven, Connecticut
3. PJM Interface, defined by the 660 MW HVDC interconnection from Raritan River, New Jersey to a converter station at Duffy Avenue, Hicksville, Long Island. From the Duffy Avenue station, the interconnection continues at 345 kV AC to Newbridge Road substation where it connects to the LIPA system. This interconnection was placed into commercial operation at the end of June 2007.

As shown in Section 3.2.1.1, special data acquisition and modeling provisions have been implemented to handle these interconnections, along with extended areas from the neighboring systems, in the real-time and study-mode static and dynamic security assessment calculations that support system operations at LIPA.

3.1.2 LIPA's CIM Compliant Environment

3.1.2.1 SCADA/EMS Overview

In 1997, the earlier generation SCADA facilities that provide LIPA the capability to monitor and control the power system operations were replaced with the distributed XA/21™ Network Solution (XA/21) from General Electric. The system includes the following software subsystems:

- Data Acquisition and Control (DAC), which acquires measurement data via remote terminal units and data links; validates data for reasonability and communication failures; performs checks on process variable limits and executes user-defined calculations; determines the quality and integrity of acquired data based on limits, normal/abnormal states, and so on; generates alarms for abnormal data conditions; and provides remote raise/lower, set point, and contact closure supervisory control capability.
- Generation Dispatch and Control (GDC), which controls the system frequency and the power interchanges with neighboring utilities and comprises automatic generation control (AGC), interchange scheduling, AGC performance monitoring, production costing, and reserve monitoring
- Power Network Applications (PNA), which provide standard real-time and study-mode network analysis functionality including parameter adaptation, con-

tingency selection, network configuration, state estimation, contingency analysis, and voltage VAr scheduling

The XA/21 architecture implemented at LIPA is depicted conceptually in Figure 3-2. All the servers and workstations use IBM RISC/6000 hardware running in the IBM AIX operating system. Not shown in this configuration diagram is the ICCP server that supports the connectivity with the PI OSIsoft-based Historical Information System (PI Historian), which is a key component of the CIM-compliant extension of the SCADA/EMS addressed in Section 3.1.2.3.

3.1.2.2 CIM, UIB, and GID Concepts

This section provides a brief introduction to the Common Information Model, Utility Information Bus, and Generic Interface Definition concepts, which form the foundation for understanding the implementation of the CIM-compliant environment at LIPA.

The Common Information Model (CIM) is an abstract model that represents all the major objects in an electric utility enterprise typically involved in operations [21]. It provides a standard way of representing power system resources as object classes and attributes, along with their relationships. From LIPA's perspective, CIM is an information object-oriented model representing real-world objects found in transmission and distribution operation and management. From other utilities' perspective, CIM may represent other real-world objects as well, for example, generation, protection, and so on. This is made possible by the partition of CIM into a set of *packages* that can be tailored for each user specific needs. A package "is a general purpose means of grouping related model elements" [21]. For example, the CIM Part 301 specification, which is governed by the IEC 61970 Part 301 standard, covers the Core, Domain, Generation, Generation Dynamics, LoadModel, Meas, Outage, Production, Protection, Topology and Wires packages, shown conceptually in Figure 3-3 and addressed in detail in [21].

It must be emphasized that CIM is not a database but a *model,* and in a CIM environment all databases are able to communicate with each other. In fact, CIM has been developed precisely for the purpose of facilitating the integration of SCADA/EMS applications developed by different vendors, between entire SCADA/EMS systems developed independently, and/or between a SCADA/EMS system and other systems concerned with different aspects of power system operations.

This important benefit provided by the CIM environment—the capability to share the data models available therein among multiple applications—relies on the Utility Information Bus (UIB). The UIB provides the tools needed to create a centralized maintainable enterprise view of data processing as well as a utility-specific way of maintaining the data model. It is essentially a software platform predicated on the IEC61970 and IEC61968 standards, and consists of relational tables and procedures in an Oracle database along with data import utilities that can be used to populate the UIB tables with CIM schema and object instance models. The UIB detailed specifications are provided in [22].

A parallel standard to the CIM, the Generic Interface Definition (GID), defines *how* to communicate CIM data across the enterprise environment supported by UIB. The GID specifications aim at minimizing the effort required to integrate utility applica-

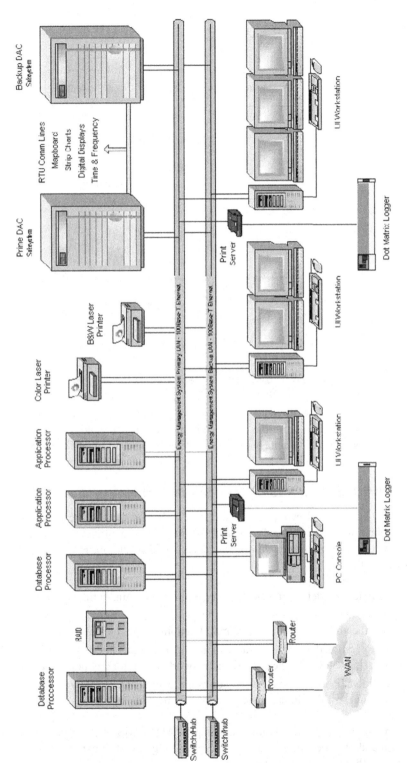

Figure 3-2. SCADA/EMS configuration overview.

49

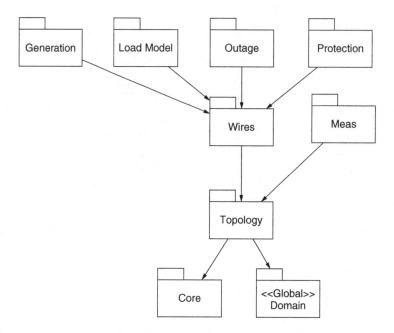

Figure 3-3. CIM Part 301 package diagram.

tions, isolating applications from the underlying middleware technology used to integrate them, and taking full advantage of CIM. A detailed description of GID is provided in [23].

3.1.2.3   CIM-Compliant PSS/ODMS—SCADA/EMS Integration

3.1.2.3.1 BACKGROUND. Between 2005 and 2007, LIPA undertook the implementation of a technologically advanced software solution to aid them in the complex and difficult task of power management, maintenance, and distribution. Four primary objectives were set:

1. To accurately model the transmission system for any point in time (current, past, and future) within minutes.
2. To have the ability to model future modifications to the transmission system.
3. To use operational SCADA data, both analog and status points stored in the PI Historian, to regenerate the operational network model for any point in time.
4. To provide LIPA with the capability of running a PSS/E load-flow study for any point in time within minutes using the specific network operations model.

It became immediately clear that the existing SCADA/EMS, which runs in a Unix environment and provides limited, if any, interoperability and data export capabilities, could not meet these demanding objectives.

Accordingly, the only solution at hand was to expand the existing SCADA/EMS with a CIM-compliant environment. In order to accomplish this goal and to meet the stated requirements, LIPA chose to use the EPRI CIM Installer software package as the foundation of their solution.

3.1.2.3.2 CIM-COMPLIANT PSS/ODMS—XA/21 INTEGRATION. At the core of this development was the integration of the existing XA/21 system, the Siemens Energy–Power Technologies International (PTI) Power System Simulator for Operations (PSS/ODMS) and the PI OSIsoft into a CIM compliant hardware and software solution. The basic architecture is depicted in Figure 3-4. The data flow across this architecture is as follows:

- Real-time metering data, such as analog values, status indications, and accumulator data, are gathered by the XA/21 DAC function and written in the real-time database.
- Snapshots are taken from the real-time database at 2 second intervals for status indications and 10 seconds for analog values, and written in ICCP format on the ICCP server and then sent to the PI Historian.
- The PI Historian stores the snapshots by using a data-compression algorithm that makes it possible to record large amounts of historical data over a significant period of time.
- The metering data are transferred from the PI Historian to the UIB via a CIM PI Adapter.
- PSS/ODMS takes over, executes the network analysis sequence and, if a valid state estimate has been obtained, triggers the real-time stability calculations as shown in Section 3.2.2.2.

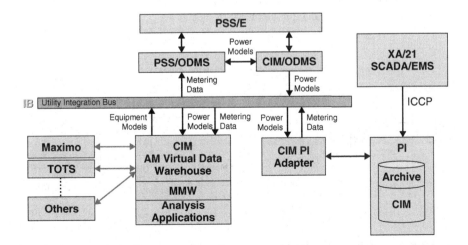

Figure 3-4. PSS/ODMS CIM-compliant integration with the existing SCADA/EMS. The ICCP server is indicated conceptually (not shown in Figure 3-2).

Phase One of the CIM-compliant integration of PSS/ODMS with the PI Historian and the XA/21 made it possible to:

- Track network changes to recreate historical model, which is made possible by the time stamping of every network model change recorded in PI Historian
- Retrieve metering data from PI Historian
- Run PSS/ODMS to support operations with real-time load-flow analyses
- Create PSS/E load-flow models for electric planning
- Establish access to asset condition data via the CIM integration bus.

Figure 3-5 shows how a solved state estimation case, originating either in the real-time system or in the historical archives, can be built and exported as a PSS/E solved load-flow case for subsequent use in real-time, study-mode and/or offline network analysis calculations.

Side (a) in Figure 3-5 illustrates the transfer of a real-time database snapshot into the PI Historian. In a nutshell, the PI Historian is the historical repository of all the information coming from the power system, generated under normal operating conditions or during disturbances. A data compression algorithm is used to store and maintain data from the real-time environment from any system point. All the data items retained in the PI Historian are time stamped. In order to provide for after-the-fact auditing of power system events and to ensure the credibility of the historical information retained therein, the data originally recorded in the PI Historian are not modifiable, but can be copied for analysis and studies, as shown symbolically at the time T1 in Figure 3-5, side (a).

Side (b) of Figure 3-5 illustrates how the PSS/ODMS retrieves the snapshot with the time stamp T1 from the PI Historian, merges it with the CIM Model, and executes the network analysis sequence that culminates with a state estimate and load-flow solution that are converted to an equivalent PSS/E case.

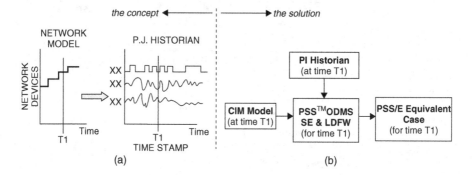

Figure 3-5. Converting a real-time or archival snapshot into a state estimate formatted as a PSS/E solved load-flow case.

The LIPA transmission model also includes substation distribution banks, as shown in Figure 3-6. This type of load-flow model, which is unusual in the industry, is very important to capture the total reactive losses.

Utilizing the CIM and GID on their network gives LIPA the ability not only to communicate with its neighbors and model its system, but to also integrate many additional applications such the maintenance management workstation or MMW and real-time stability software like QuickStab. All these applications help LIPA in assessing its daily needs more accurately.

3.1.2.3.3 SYSTEM PLANNING VISION OF CIM. One of the key future goals of the industry-wide CIM initiative is to extend the current CIM to include stability data. At the present time, the dynamic models are getting more complex, and there is a growing need for manufacturers to start providing models based on the CIM standard rather than the current practice of providing data in accordance with industry de-facto standards such as the proprietary Siemens-PTI PSS/E and General Electric International (GE) PSLF formats. Anticipating that this would eventually happen, it is expected that major software manufacturers such as Siemens-PTI and GE will take the first step and convert their existing models to CIM-compliant formats. The availability of CIM-compliant stability data would allow CIM-compliant applications to perform stability calculations *without* using external machine data files.

Another objective is to use the huge and continuously updated reservoir of historical data to:

- Perform event reconstruction. This is an important step in postfacto and post-mortem analyses that need to be performed in order to audit, and/or learn from, local and/or system-wide disturbances.
- Improve the results of similar-day load forecasting. The idea is to perform similar-day load forecasting in "tracking mode" to compare the current load demand as it unfolds with the most recent forecast and to update in near real time the model used to forecast the load.
- Improve the quality of loss analysis. Transmission loss studies are an important step toward optimizing the system operations and reducing the real and reactive losses. The availability of historical data would allow refining this process and placing the studies in a more accurate context.

Another useful CIM extension to support the planning activities is the so-called model on demand for operation (MOD) capability, which would allow the users of a CIM-compliant system to manage their model edits using projects. Changes to the network configuration could be saved as modeling projects and, after editing, review, and approval, would be stored as permanent model changes. Then, when the network changes are actually implemented and become operational, the CIM model would be automatically updated to correctly reflect the new system configuration. These and other potential future applications of CIM in planning are illustrated in Figure 3-7.

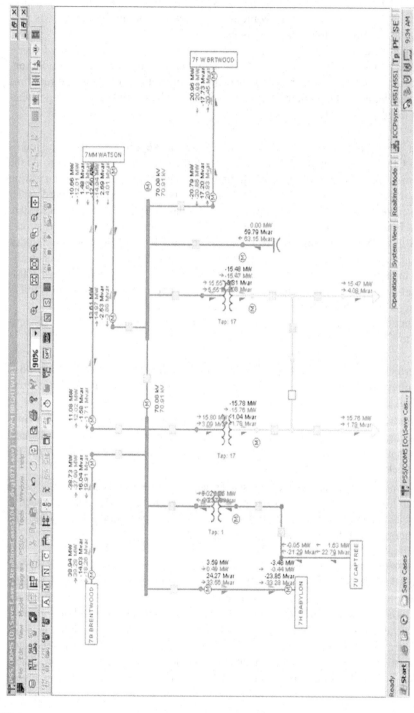

Figure 3-6. Example of real-time substation modeling; includes distribution transformers to capture losses.

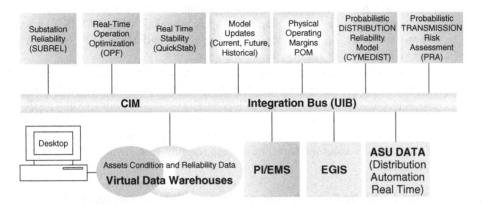

Figure 3-7. System planning vision of CIM.

3.2 STATIC AND DYNAMIC SECURITY ASSESSMENT AT LIPA

3.2.1 Overview

3.2.1.1 Static Security Assessment—Current Capabilities

Contingency analysis encompassing single and multiple contingencies is executed during the real-time network analysis sequence cycle at five-minute intervals for $N - 1$, single, and complex contingencies. The LIPA system has seven interconnecting cables to the outside world, two of which are oil filled and have dynamic ratings. These ratings are determined based on real-time heat-sensor readings along the cables and heat dissipation in the ground. A computer program determines the cable rating during hourly operation. The computer program also calculates the lowest rating based on the previous day; this rating is entered daily in the security analysis package.

3.2.1.1.1 THERMAL SECURITY ANALYSIS. The thermal security analysis function automatically analyzes outages for all generators, transformers, and single and complex contingencies in the LIPA system and produces a table that shows the element out of service, the MVA flow before and after the outage, and the percent thermal overload for normal (NOR), long-term emergency (LTE), and short-term emergency (STE). Complex contingencies are manually created for each substation and include bus faults and breaker faults. These contingencies are usually the worst since they take several lines out of service.

3.2.1.1.2 VOLTAGE SECURITY ANALYSIS. Based on LIPA operating criteria, the voltage at each bus should never go below 0.95 p.u. and should never exceed 1.05 p.u. during normal and contingency operation. The contingency analysis program can be set to monitor these limits and a similar table as the one for thermal analysis can be generated.

The key results of voltage as well as thermal security analysis are displayed into a spreadsheet format. Another graphical display provided by PSS/ODMS is contours. As shown in Figure 3-8, the pictorial image allows identification of thermal and/or voltage violations in the LIPA system at a glance.

The voltage security functionality is enhanced with real-time and offline voltage stability assessment. At LIPA, the system operator uses the QuickStab Voltage Stability application. In real time, this application runs every five minutes and produces stability margins for normal, contingency, and a selected load pocket in the system. These limits can be monitored in real time by using the trending capability of PI-Historian. Further details about the real-time voltage stability functionality are provided in Section 3.2.2.

3.2.1.1 Static Security Assessment—Future Enhancements

Because of the flexibility of PSS/ODMS, which is able to export a real-time RAW file of the current system every five minutes, LIPA was able to run QuickStab in real time. By using a product called POM developed by V&R Energy, LIPA will be able to run in real time optimum power mitigation (OPM), which will tell the system operator how to resolve some of the thermal and voltage problems by optimizing the power system. An additional application, also developed by V&R Energy, is POM-TS. This ap-

Figure 3-8. Graphical display showing "normal" and "trouble" spots of the system. The dark spot in the upper-right corner is shown in red in the actual display and indicates low voltages in that area.

plication will run transient stability for selected contingencies in near real time at a 30 minute intervals.

3.2.1.2 Need to Assess Voltage Stability in Real-Time

By using the static security assessment tools, the system operator at LIPA has a very good handle on thermal and voltage constraints. However the operator has less of an understanding of system stability during peak and off-peak periods. Evaluating off-peak stability is just as important because all the maintenance is done during off-peak conditions. Furthermore, the stability limits of the power system change constantly and, under certain operating conditions, the system state may become vulnerable to events that eventually may trigger instability. Simply stated, such degraded states may be caused by:

- *Topology changes.* Line and transformer outages reduce the maximum loadability of the transmission system; generator outages that are compensated by raising the generation elsewhere may also push the system closer to its stability limit.
- *Lower system voltages.* Insufficient reactive compensation and decaying system voltages are well documented reasons for approaching the point of voltage collapse.
- *Transmission congestion.* External MW flows wheeled across the transmission network increase the total network MW utilization and push the system to its maximum power transfer limit.

These considerations pointed to the need to perform stability analysis in real-time, immediately after a valid state estimation case becomes available and sufficiently fast so that the results could be used for online decision making. But since a real-time case may require further analysis in order to develop corrective action, if needed, the need to replicate the suite of real-time stability calculations in study-mode, by modifying the real-time case for the purpose of simulating "what-if" scenarios was also envisioned.

In addition, the need was identified to perform stability evaluation of single and multiple contingencies, execute day-ahead stability runs using the PSS/ODMS output (RAW file), and perform stability calculations to support electric system operations (ESO) clearance requests during major construction projects.

Accordingly, the following requirements have been established for the stability program that will be integrated with the CIM compliant PSS/ODMS computing environment:

- The stability application should demonstrably be capable of running truly in real-time* as well as in study mode.

*As shown in the Chapter 1 of this book, by "real-time" we understand that the input data reflect the most recent picture of the system conditions; processing is performed within very short delays, typically not exceeding a couple of seconds; and the output is usable almost instantly, that is, approximately one to two seconds.

- The solution technique must be fast and reasonably accurate.
- The results must include "indicators," such as "distance to instability," that are easy to interpret.
- The output must be presented graphically in simple format.
- The software must be able to evaluate single and multiple contingencies.

Several products were evaluated and the voltage and steady-state stability assessment software from Energy Consulting International, Inc. (ECI) called QuickStab was selected because of its speed, accuracy, and easy-to-interpret graphics.

3.2.2 Implementation of Real-Time Stability Assessment and Monitoring

3.2.2.1 Key Functional Characteristics

The maximum loadability of a transmission system is the state at which voltages may collapse and units may get out of synchronism, and is a severe constraint. The distance from a given operating point to the state of maximum loadability is called the *steady-state stability reserve;* it changes when the load, topology and voltages change, and may be quite different from values computed offline. Its recalculation after each state estimate and load-flow solution is required because operating the system near its stability limit may lead to blackouts. This problem is solved rapidly and with acceptable accuracy by the fast steady-state stability assessment application* that is used in real-time and study mode in ESO at LIPA. Given a load-flow solution or solved state estimate of a multiarea power system, the program:

- Executes fast voltage and steady-state stability computations to determine, for the base-case conditions of the study area, the maximum loadability and the safe system loading for a user-defined security margin. The "maximum loadability" is defined as the total MW network utilization of the study area, including both area MW generation and the MW imports, immediately before steady-state instability.
- Performs contingency evaluation for each single or multiple contingency specified in a contingency list. First, it determines the postcontingency state by running a full AC Newton–Raphson load-flow computation. Then it executes the complete suite of voltage and steady-state stability computations, and, upon completion, ranks the contingencies in the order of their stability margins.
- Identifies the generators, as well as the virtual machines that model tie-line imports, that may cause instability and ranks them[†] in order of their impact on system's stability, both for the base case and for contingencies.

---

*The program is known as QuickStab® Professional; however, in order to avoid commercial overtones, it will be referred to in this chapter as "fast steady-state stability assessment application."
[†]The ranking criterion is the value of the term $(Y_m E_m/\cos \delta_m)$ in equation 8 of Appendix A.

- Presents the results in graphical formats, in charts and diagrams that are easy to understand and to interpret, and in tabular displays.

These functions are executed both in real time and in study mode. The solution technique is based on the methodology described in [3, 4, 7, 13, 14, 20], and is presented in detail, along with numerical examples that illustrate the technique, in Appendix A. In a nutshell, the algorithm consist of alternating steady-state stability checks with "case-worsening" calculations that stress the system* from the base operating conditions toward a case in which the determinant of the dynamic Jacobian becomes singular, thus indicating steady-state instability.

The state-estimation solution or solved power-flow case is supplied in a comma-separated ASCII file in the PTI PSS/E "raw" format version 30. In addition, the input data include the:

- Transient, subtransient and synchronous reactances of the generators and synchronous condensers,
- Reactances of the step-up transformers (if the machines were shown on the high-voltage buses),
- P-Q capability curves of the generators,
- List of contingency scenarios, in which each contingency may consist of single or multiple transmission, generation, shunt, and/or load events, and
- Execution control parameters, such as specification of subareas to be retained for stability calculations, load-pocket area specification if load-pocket analysis needs to be performed, and so on.

The "load-pocket" calculation is a feature developed specifically for LIPA. In power systems that contain distribution networks, which is the case at LIPA, it is useful to know how much additional load can be increased at a user-defined group of load buses, identified as a "load-pocket area," before getting close to a state at which voltages may collapse. This is an inherently difficult computational proposition, but a simple method can easily be devised based on the assumption that when the total MW load in the system or subsystem area is increasing during the case-worsening process, the total load in the "load pocket" that is being analyzed increases conformingly, that is, in the same proportion as the total system load. For example, if the total system load has increased, say by 10%, the sum of the loads within a load pocket would also have increased by 10%.

Accordingly, once the case-worsening procedure has been completed, and if the "load pocket" option has been specified, the sum of the loads within the user-defined load pocket is projected from the actual case value to the critical case by using the same rate of increase of the total study-area MW load from the actual to the critical

---

*As shown in [3, 4, 13] and in Appendices A and C in this book, there are potentially many ways to stress the system conditions; accordingly, each worsening strategy may lead to a potentially different limit. The "case-worsening" approach deployed in this case is the method used by Dimo [3, 4, 13] and is illustrated in detail in Appendix A.

case. Of course, the results of "load-pocket" calculations only indicate what would be the MW load within the load pocket if the entire study-area load would have been pushed to the limit of instability, but nothing else. Furthermore, the accuracy of such calculations depends upon how conforming the loads are.

When evaluating contingencies, two operational modes are available:

1. Run the entire suite of full Newton–Raphson postcontingency load-flow and steady-state stability calculations for each contingency case at a time, then stop to allow the user to examine and assess the partial results.
2. Execute all the contingency cases nonstop until the list has been exhausted.

At the end of the calculations, the contingencies are ranked in the order of their steady-state stability reserves and the results corresponding to the base case and the worst contingency are saved for the final presentation of results. This process is illustrated in Figure 3-9.

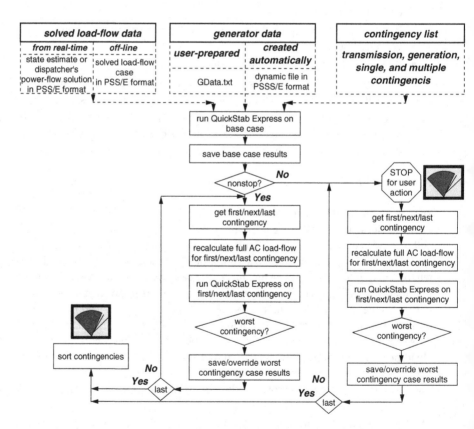

Figure 3-9. Conceptual flow chart illustrating the processing of base and contingency cases in step-by-step and nonstop execution modes, respectively. Early interruption and recovery traps for user and/or data errors are not shown.

3.2.2.2   *Integration of the Fast Stability Tool with PSS/ODMS*

The integration of the fast steady-state stability tool with LIPA's CIM-compliant SCADA/EMS was performed jointly by LIPA, PTI, and ECI. At the core of the solution is the PSS/ODMS software which:

- Collects analog and digital metering data in real-time, merges them with the transmission model, and then runs a state-estimation and a power-flow calculation,
- Exports the state-estimate solution into a RAW file,
- Executes the real-time interface that controls the fast steady-state stability computations and displays, and
- Repeats the cycle automatically every five minutes and on demand upon user request.

This process is illustrated in Figure 3-10. When executed, the real-time interface module (identified in Figure 3-10 by the name "Real-Time QuickStab Extensions") triggers the fast steady-state stability computational engine and refreshes the displays on the system operator workstation in accordance with user-defined commands stored in a text file.

Technically speaking, this is a "loosely integrated" approach, but in reality it provides the key advantage of a seamlessly integrated solution in the sense that the real-time stability process is initiated on the SCADA/EMS rather then from the stability program itself, thus triggering a new stability calculation each time the state estimator detects a change in the power system conditions. Other significant benefits of this implementation include:

- The stability calculation results are automatically displayed and continuously refreshed on the system operator workstation. If the time stamp posted by the stability application display engine is not current, the operator immediately knows that the results are "old" because, for whatever reason, the state estimator did not run or did not converge.
- The current value of the steady-state stability reserve is continuously transferred to the historical database, thus allowing it to be monitored in real time on a standard SCADA trending display.
- The solved real-time stability case can be examined on other PCs connected to the ESO LAN. This is made possible by a process that runs on the main server, saves the results for archival purposes, and then copies them in a directory that can be accessed by properly authorized users. Depending upon the jurisdiction level, certain users can simulate alternate scenarios by recalculating the real-time stability case with different computational options.

3.2.2.2   *User Interface and Performance*

Another benefit of the "semiloosely" integrated solution implemented by LIPA for assessing stability in real time is the availability of the full range of display options of the

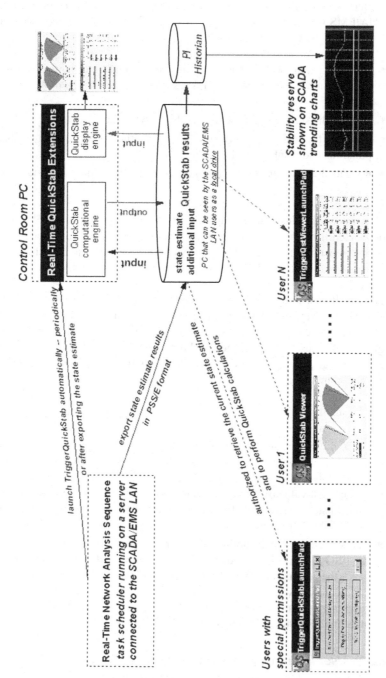

Figure 3-10. Integration of the fast stability assessment tool with PSS/ODMS.

fast steady-state stability assessment application, both on the system operator workstation and on the user PCs connected to the ESO LAN. Some of these graphical and tabular display capabilities are illustrated below:

- *Two MW speedometer* charts that display two one-needle speedometers side by
  side, one for the base case and the other one for the worst contingency case (Figure 3-11),
- *Linear speedometer* charts, which can be shown sorted or unsorted and visualize
  the impact of contingencies on the system's stability reserve (Figure 3-12),
- *Unit-ranking* bar charts that depict the impact of generators and tie-line injections on the steady-state stability conditions of the study area (Figure 3-13), and
- *Extended summary report,* which displays, in tabular format, some of the key
  base-case data and stability calculation results (Figure 3-14).

The fast steady-state stability software is extremely fast. The average solution time,
including both the post contingency load-flow, which uses a full AC Newton–Raphson
algorithm, and the steady-state stability computation times, is approximately 1.5 seconds per case (approximately 1000 buses); the steady-state stability calculations alone
are performed in less than 1 second. The most relevant aspect, however, is the fact that
the program runs after each successful state estimate and the results are displayed al-

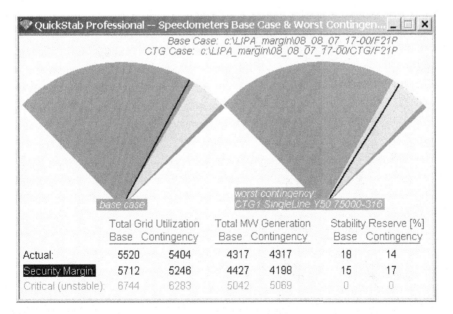

Figure 3-11. MW speedometers for the base case (left) and the worst contingency
(right) depicting the distance to instability (red sector, shown in dark gray) and to the
security margin (separation line between the two shades of gray).

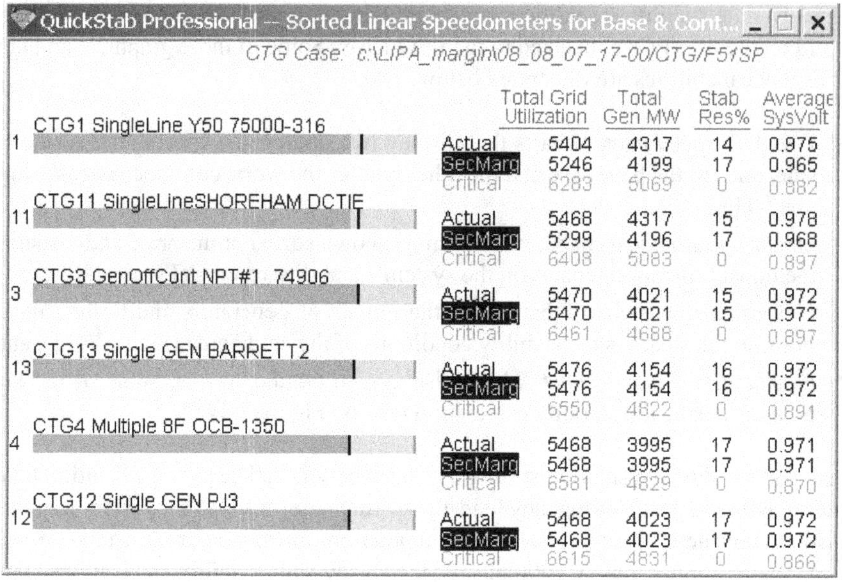

Figure 3-12. Sorted linear speedometers depicting the distance to instability (red sector, shown in dark gray) and to the security margin (separation line between the two shades of gray) for the six worst contingencies.

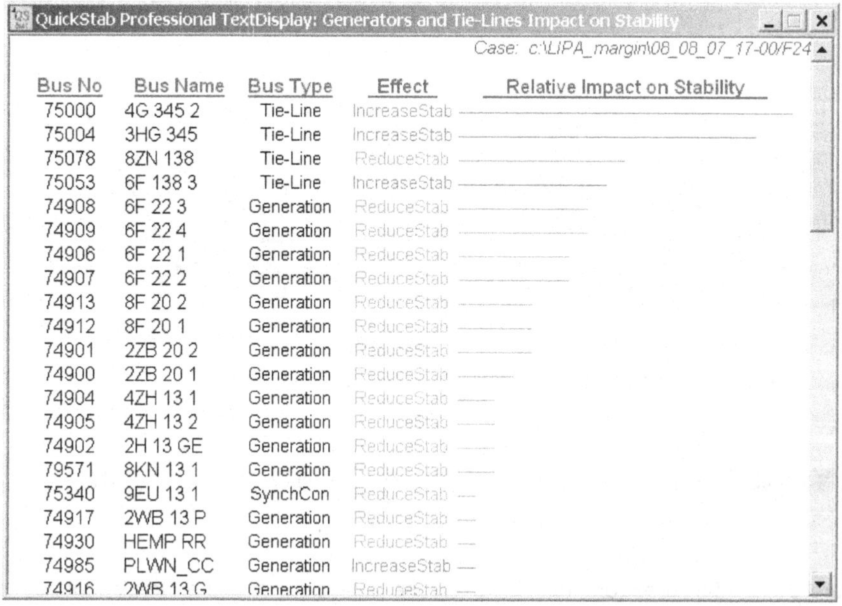

Figure 3-13. Unit-ranking bar chart showing the impact of generators and tie-line injections on system's stability.

QuickStab Professional TextDisplay: Extended Summary Report		
Case: c:\LIPA_margin\08_08_07_17-00/FX21P		
Extended System Summary	**Actual**	**Critical**
Total MW System (Area) Load	5255.62	6421.35
Total MW System (Area) Generation	4316.59	5041.51
Total MW System (Area) Grid Utilization	5519.65	6743.94
Total MW System (Area) Load+Exports	5416.86	6618.36
Total MW System (Area) Losses	102.79	125.58
Total MW System (Area) Imports	1203.06	1702.43
Total MW System (Area) Exports	161.24	197.00
Net MW System (Area) Interchange	1041.82	1505.43
Average System (Area) Bus Voltage [p.u.]	0.98	0.88
System (Area) Stability Reserve [%]	18.15	0.00
MW Generation read on input	4316.59	** n/a **
MW Generation converted to negative load	-0.00	** n/a **

Figure 3-14. Extended summary report showing the key base-case data and stability calculation results in tabular format.

most instantly. This computational speed provides sufficient time for the system operator to monitor the current stability conditions of the system as the operating conditions unfold.

3.3 BENCHMARKING THE REAL-TIME STABILITY APPLICATION

After the fast steady-state stability software was integrated with PSS/ODMS and the preliminary testing with actual real-time stability cases was completed, LIPA benchmarked the application before starting its production-grade use in daily system operations. The benchmarking was conducted in two stages. The first stage consisted of determining how accurate is the prediction of the "distance to instability" for the very important purpose of gaining confidence that states predicted to be unstable would, indeed, be unstable or closed to instability, and vice-versa. The accuracy testing procedure and results are discussed in Section 3.3.1.

The second part of the benchmarking effort aimed at fine-tuning the value of the "security margin," that is, the amount of "steady-state stability reserve"* that needs to be maintained in order to ensure that none of the critical contingencies that can reasonably be envisioned would cause the power system to become unstable. This goal was considered equally important because operating the power system with too large a security margin would translate into not using all the available transmission capability,

---

*The concepts of "steady-state stability reserve" and "security margin" have been addressed in [13, 17, 18] and related references, and are extensively discussed in this book in Section 2.2.4.3 of Chapter 2, Annex 4-1 of Chapter 4, and Section 6.2.1.1 of Chapter 6.

whereas, on the other and, using too small a security margin would be like walking on thin ice. This benchmarking step is addressed in Section 3.3.2

3.3.1 Accuracy Testing

3.3.1.1 Approach

The validation and accuracy testing of a computer application should encompass the following steps:

- Demonstrate that the application behaves as per the stated specifications. This is achieved by running the software both with good and with bad data, and comparing the outcome with the program's specifications. If there are no discrepancies, it can be inferred that the software fully meets the specifications.

- Validate the results, that is, demonstrate that the results produced by the software and those obtained by applying the same algorithm in a different way, for example, performing the calculations by hand, are practically identical. This is typically done by developing a small-scale model and performing the full array of calculations by hand. If the results produced by the software are close or identical to those calculated by hand it can be inferred that the underlying algorithm has been programmed correctly.

- Demonstrate the accuracy of the solution technique, that is, show that the software produces results that are similar to those produced by other technologies. The key assumption is that the method and its approximations, if any, are theoretically sound. On this basis, the numerical testing should demonstrate that the conclusions derived from the results produced by the application are consistent with those obtained by using an alternate method. When it comes to stability, this is inherently difficult because the assessment depends upon the type of stability evaluation that has to be conducted. For example, in order to avoid comparing apples with oranges, a steady-state stability program that performs "case worsening" must be benchmarked against another steady-state stability application that executes "case worsening" as well. As a minimum, it must be demonstrated that system states predicted to be unstable by the program being benchmarked are found to be unstable by using an alternate approach.

The first two steps are relatively straightforward but the third is not so trivial. One way to achieve it is to demonstrate that states predicted to be unstable by the stability tool are found to be at the limit of convergence by running a good load-flow program. This validation approach [17, pp. 65–92] relies on the observation that near the steady-state stability limit of a power system, voltages are low and load flows may diverge [16].

Of course, a nonconvergent load flow does not necessarily mean that the system is unstable. It was shown in [10, 11, 17] that the system load at which the load flow diverges is just an upper bound, for it is possible for one or several units to get out of synchronism before that point. Also, "for voltage collapse and voltage instability analysis, any conclusions based on the singularity of the load-flow Jacobian would apply only to the voltage behavior near the state of maximum power transfer. Such analysis would not detect any voltage instabilities associated with synchronous machines characteristics and their controls" [10, p. 1380]. However, running load flows at increasingly high load levels until they diverge is the only way to obtain a base case near the limit of stability. Then, starting from this base case, voltage and steady-state stability tools can be used to refine the assessment.

3.3.1.2 Benchmarking Procedure and Results
Initially, LIPA tested the fast steady-state stability software for conformance with the stated specifications for 120 days, which was successfully concluded with site acceptance testing. The numerical verification of the algorithm implementation was not performed; instead, the results obtained in the course of a similar exercise conducted at CAMMESA* were accepted and relied upon.

The accuracy benchmarking per se consisted of demonstrating that steady states predicted to be unstable by the stability tool are found to be at the limit of convergence by the PTI PSS/E load-flow program. The goal was to demonstrate that, given an output of LIPA's state estimator in solved PSS/E load-flow format, the critical state[†] predicted by the stability tool can also be found by running a sequence of load flows with PSS/E at increasingly higher MW levels, recalculating the stability limit for each new case, and stopping when the PSS/E load-flow computations diverge. Accordingly, the following procedure was implemented:

- Start with a base case retrieved from the historical database and run a preliminary stability calculation to determine the ΔMW increase of the total system load needed to push the network near instability. Since the base case was far below the stability limit (24.82% stability reserve), it was inferred that the initial prediction may be too conservative. Accordingly, the initial ΔMW was set at 8000.

- Run a new PSS/E load flow after increasing the system load in the original base case by 8000 MW. The bus loads and generation schedules were scaled by PSS/E.

- Compute the stability reserve of the new case. Since the total MW system grid utilization in the new case was much larger, it was expected that the stability reserve would be much smaller than in the base case. Indeed, as shown in Table 3-1, the total MW system grid utilization increased from 21,125.37 MW to 29,441.13 MW, whereas the stability reserve dropped to 1.14%, which is very close to instability.

*In a separate testing and validation effort conducted by the System and Market Operator of Argentina (CAMMESA) and ECI, it was demonstrated that the results computed with the fast steady-state stability tool on a small network taken directly from the Argentinean system are 100% consistent with the results of manual calculations performed independently by the two different project teams. The results are fully documented in the Appendix A of this book.
†The critical state is a hypothetical system state near the stability limit where voltages may collapse and units may loose synchronism.

Table 3-1. Summary results of benchmarking calculations

System load *increases* toward instability (↓)

Case ID	Computed by PSS/E	Predicted by the steady-state stability tool	
	Actual system grid utilization [MW]	Critical state [MW]	Stability reserve [%]
Base case Oct. 11, 2007 (recalculated by PSS/E) total load = 20,731.57 MW	21,125.37	28,098.46	24.82
Base case + 8000 total load = 28,591.17	29,441.13	29,780.84	1.14
Base case + 8100 total load = 28,691.18	29,554.94	29,784.97	0.77
Base case + 8200 total load = 28,791.17	29,704.86	29,704.86	0
Base case + 8300 total load = 28,891.17	29,791.80	29,791.80	0
Base case + 8400 total load = 28,991.18	29,918.21	29,918.21	0
Base case + 8500 total load = 29,091.18	load flow diverged	—	—

Stability reserve *decreases* toward instability (↓)

These steps were repeated by recalculating PSS/E load flows at continuously increased levels of load—base case + 8100 MW, then base case + 8200 MW, and so on. The results are summarized in Table 3-1.

As shown in Table 3-1, with each system load increase, the stability reserve computed by the stability tool became smaller and then reached 0%*; the PSS/E calculations diverged at base case + 8500 MW. It was thus inferred that for the system area† *and* the system conditions modeled in the test case the:

- Total MW system grid utilization at which voltages may collapse and units may lose synchronism is at some point between 28,991.18 MW and 29,091.18 MW, that is, approximately 29,000 MW.
- Total MW system grid utilization immediately before instability is 29,918.21 MW.

*The stability tool finds the stability limit by performing a case-worsening algorithm. Starting from an actual case, for example, base case + 8100, the generators' MW output is increased in small steps "coherently" (proportionally) and the stability criterion $d\Delta Q/dV$, which must be *negative* for stable states, is recalculated. The process stops when $d\Delta Q/dV > 0$, and the last state that was stable is labeled critical. If the first step of case worsening pushes the system beyond the stability limit, the system state is called critically stable, the $MW_{critical}$ is set equal to MW_{actual}, and the stability reserve is set to 0. If the MW steps of case worsening are larger than the MW steps used to raise the system load in PSS/E, it is possible to get a couple of load-flow solutions beyond a critically stable state. This is what actually happened in the LIPA experiment in which the last three cases before divergence were found to be critically stable.

†At LIPA, the state estimator model encompasses, in addition to the entire network under LIPA's jurisdiction, external areas from ConEdison, Connecticut, and upstate New York.

The results thus obtained are fully consistent with both the theoretical foundation*
of the underlying solution technique and the testing and validation results obtained by
other users,[†] and demonstrate that the:

- Stability limits computed with the stability tool depict reliably the maximum
  grid loadability, which is the hypothetical system state at which voltages may
  collapse and units may lose synchronism. The average error between the MW
  system grid utilization at instability computed with the stability tool and the
  value where PSS/E diverged was approximately 0.5%, which is excellent precision.

- The predictions are more conservative when the current system conditions (base
  case) are far from instability, but become accurate and precise at high system-
  load levels.

- Continuous stability monitoring by using this technique offers the significant
  benefit of getting a reliable picture of the distance to instability at any moment
  and, in particular, near and during the peak, when it matters most.

3.3.2 Tuning the Security Margin

3.3.2.1 *Background and Approach*

Just as the steady-state stability limit (SSSL) of the power system changes as the system topology and operating conditions change, the MW value of the "safe" loadability
changes as well. Empirically, a *security margin* expressed as a percentage of the SSSL
can be identified such that, for any system state with a steady-state stability reserve
higher than this value, no contingency, no matter how severe, would cause transient instability [5,13,17,18]. In other words, all the states with a total MW loading smaller
than the MW value of the security margin are inferred to be safe, even if a "safe" system state with higher MW loading might possibly be found. The intrinsic technical
complexity and the large number of credible contingencies render the exact solution of
this problem practically impossible, so heuristics and practical experience must be
used instead. For example, the 15% security margin recommended in references [4],
[5], and [9] is predicated on many years of practical experience with Dimo's method;
these values are still used in daily operations in Romania, as shown in the Chapter 5 of
this book, but are not necessarily applicable to power systems with significantly different topologies, voltage profile, and MW transfer patterns. A heuristic, however, has
been described in [17] and consists of the following steps:

- Step 1. Start with a base-case load flow for a typical system topology under peak
  load conditions. Assume an initial percent value for the security margin, say
  15% below SSSL, and compute the SSSL and related MW value of the security

---

*Please refer to Appendix A in this book.
[†]In January 2005, the steady-state stability assessment software described herein was tested and validated by
ETESA in Panama, as described in detail in [17].

margin, which, in this case, would be equal to 0.85 × SSSL. As part of this step, the MW outputs of the generators, referred to as security margin MW generation schedules, should also be computed.

- Step 2. Run an extensive suite of transient stability cases. If no instability has been detected, go to Step 5. If at least one contingency (fault) case was found to be unstable, go to Step 3.

- Step 3. Use the security margin MW generation schedules from Step 1 to calculate a new base-case load flow.

- Step 4. For the load flow computed in Step 3, run again the same suite of transient stability cases that were evaluated in Step 2.

- Step 5. If no instability has been detected, repeat Step 4 for successively increased MW levels until at least one contingency (fault) causes transient instability. The percent value of the steady-state stability reserve for the immediately precedent state is the percent value of the security margin of the system under evaluation.

- If the stability calculations in Step 4 detected at least one contingency (fault) that would cause instability, build a new load-flow case for a slightly *reduced* load level and repeat the transient stability checks. If no instability has been detected, recalculate the SSSL and the steady-state stability reserve, which correspond to the security margin of the system under evaluation.

- Repeat the entire process for alternate, yet credible system topologies that differ significantly from the base scenario and select the security margin that yields the most conservative results.

3.3.2.2 Security Margin for LIPA System

It should be clear by now that if the percent value of the security margin is set too high, the evaluation of the current operating conditions becomes too conservative, and vice versa, to small a percent value of the security margin could convey a false sense of security. Therefore, in order to assist the system operator to issue quickly a reliable judgment regarding how far from, or how close to, the safe margin of operation is the transmission system, the percent value of the security must be fine tuned.

Due to the importance of knowing, even if just approximately, the reasonable amount of security margin that should be observed in order to ensure that no blackout due to instability may occur, LIPA set up a project team responsible for executing the calculations described above. At the time when the text of this chapter went to press, the results of the simulations were not yet available.

3.4 PRACTICAL EXPERIENCE AND OUTLOOK

The Control Center Project and CIM is one way LIPA was able to bridge the gap between planning and operations. It was a very successful project and, as far as we know, it was the first of its kind. When presented to the system operator, the first suggestion they had was to add a real-time voltage stability assessment capability that was easy to interpret. After evaluating several application packages, QuickStab was selected both because of its computational speed and for its unique speedometer graphics, which were very well accepted by LIPA's system operator.

Another request from the system operator was to perform maximum loadability analysis both for the entire system and for a particular area, called East End and referred to as the "load pocket," which is very sensitive to voltage collapse. This functionality was implemented and, at the present time, critical contingencies are being evaluated in real-time for voltage stability and continuously monitored both system wide and for the East End load pocket.

In addition, the system operator suggested the development of a module within the voltage stability application that would prompt the operator to carry out the recommended remedial action if stability conditions need to be improved. Future enhancement of the software may include this request.

3.5 REFERENCES

[1] Anderson, P. M., and Fouad A. A., *Power System Control and Stability,* Iowa University Press, Ames, Iowa, 1990.

[2] Crarý, S. B., 1945, *Power System Stability,* General Electric Series, Schenectady, New York, Copyright 1945, Third Printing October 1955.

[3] Dimo, P., "Etude de la Stabilité Statique et du Réglage de Tension," *R.G.E.,* Paris, 1961, Vol. 70, No. 11, pp. 552–556.

[4] Dimo, P., *Nodal Analysis of Power Systems,* Abacus Press, Kent, England, 1975.

[5] Dimo, P., Manolescu, G., Iordanescu, I., Groza, L., Ionescu, S., Albert, H., Moraite, G., and Ungureanu, B., *Computation and Design of Electrical Energy Systems* (Romanian edition), Editura Tehnica, Bucharest, Romania, 1971.

[6] Dobson, I., and L. Liu, "Immediate Change in Stability and Voltage Collapse when Generator Reactive Power Limits are Encountered," in *Proceedings of International Seminar on Bulk Power System Voltage Phenomena II,* edited by L. H. Fink, pp. 65–74, 1993.

[7] Erwin, S. R., Oatts, M. L., Savulescu, S. C., "Predicting Steady-State Instability," *IEEE Computer Applications in Power,* July 1994, pp. 15–22.

[8] Ionescu, S., and Ungureanu, B., "The Dual Power States and Voltage Collapse Phenomena," *Rev. Roum. Sc. Tech., Série Electrotechnique et Energétique,* Vol. 26, No. 4, pp. 545–562.

[9] Magnien, M., Rapport spécial du Groupe 32 Conception et Fonctionnement des Réseaux, Conférence Internationale des Grands Réseaux Electriques à Haute Tension, CIGRE Session 1964.

[10] Sauer, W. P., and Pai, M. A., "Power System Steady-State Stability and the Load-Flow Jacobian," *IEEE Transactions in Power Systems,* Vol. 5 T-PWRS, No. 4, pp. 1374–1381.

[11] Sauer, W. P., and Pai, M. A., "Power System Steady-State Stability and the Load-Flow Jacobian," *IEEE Transactions in Power Systems,* Vol. 5 T-PWRS, No. 4, pp. 1374–1381.

[12] Sauer, W. P., Pai, M. A., "Relationships between Power System Dynamic Equilibrium, Load-Flow, and Operating Point Stability," in *Real-Time Stability in Power Systems,* pp. 1–30, Springer-Verlag, Norwell, MA, 2006.

[13]   Savulescu, S. C., "Fast Assessment of the Distance to Instability. Theory and Implementation," in *Real-Time Stability in Power Systems,* pp. 31–64, Springer Verlag, Norwell, MA, 2006.

[14]   Savulescu, S. C., Oatts, M. L., Pruitt, J. G., Williamson, F., and Adapa, R., "Fast Steady-State Stability Assessment for Real-Time and Operations Planning," *IEEE Trans. Pow. Sys.,* Vol. 8 T-PWRS, No. 4, Nov. 1993, pp. 1557–1569.

[15]   Venikov, V. A., "Transient Processes in Electrical Power Systems," Edited by V. A. Stroyev, English Translation, MIR Publishers, Moscow, 1977.

[16]   Venikov, V. A., Stroev, V. A., Idelchick, V. I., and Tarasov, V. I., "Estimation of Electrical Power System Steady-State Stability," *IEEE Trans. on PAS,* Vol. PAS-94, No. 3, May/June 1975, pp. 1034–1041.

[17]   Vergara, J. S., Thai, T. A., Cuong, N. D., Nam, N. T., Campeanu, H. S., and Savulescu, S. C., "Accuracy Testing and Real-Time Implementation of Dimo's Stability Analysis Technique," in *Real-Time Stability in Power Systems,* pp. 65–92, Springer Verlag, Norwell, MA.

[18]   Vickovic, D., Eichler, R., and Savulescu, S. C., "Real-Time System Stability Monitoring in the Transmission Network of Bosnia and Herzegovina," paper presented at the Power-Grid Europe Conference and Exhibition, Feria de Madrid, Madrid, Spain, 26–28 June 2007.

[19]   Vournas, C. D., Sauer, P. W., and Pai, M. A., "Relationships between Voltage and Angle Stability of Power Systems," *Electrical Power and Energy Systems,* Vol. 18, No. 8, pp. 493–500, 1996.

[20]   EPRI, "Power System Steady-State Stability Monitor Prototype," Final Report EPRI TR-100799, July 1992; and "Power System Steady-State Stability Monitor," Final Report EPRI TR-103169, December 1993.

[21]   EPRI, "Common Information Model (CIM): CIM 10 Version," EPRI Final Report 1001976, November 2001.

[22]   CISCO, "Utility Integration Bus Toolkit User's Guide, Revision 10," © SISCO, Inc. 2002–2005.

[23]   EPRI, "Generic Interface Definition (GID) Design Specification. Control Center Application Program Interface (CCAPIU), Recommended Component Interface Specification," EPRI Report.

[24]   http://www.quickstab.com.

4

REAL-TIME STABILITY MONITORING AT THE INDEPENDENT SYSTEM OPERATOR IN BOSNIA AND HERZEGOVINA

Dusko Vickovic and Roland Eichler

4.1 INTRODUCTION

4.1.1 Bosnia and Herzegovina Transmission System in the Southeastern European Context

The Independent System Operator (NOS) of Bosnia and Herzegovina (BiH) has full dispatch responsibility and authority for the operation of the BiH power system and electricity market, which comprises the electrical generation and transmission facilities in the entire country and is operated as a UCTE (Union for the Coordination of Transmission of Electricity) Single Control Area within the UCTE Control Block encompassing Bosnia and Herzegovina, Croatia, and Slovenia. The BiH transmission system is schematically depicted in Figure 4-1. As further illustrated in Figure 4-2, the BiH transmission system is located at the center of significant MW transfers between the market participants in Southeastern Europe, which may significantly affect the operating reliability of its transmission system.

Therefore, in addition to its main responsibility to maintain the quality and continuity of the electric service within the country, NOS BiH must also facilitate the significant power transfers across the BiH transmission network caused by energy exchanges

Real-Time Stability Assessment in Modern Power System Control Centers. Edited by Savu C. Savulescu
Copyright © 2009 The Institute of Electrical and Electronics Engineers, Inc.

Figure 4-1. BiH transmission system and its regional interconnections.

in the region while protecting the BiH customers against blackouts and unwanted disturbances.

4.1.2   Need to Monitor the Risk of Instability

Just like many other transmission grids, the BiH transmission system must sustain MW transfers that can be quite different from those for which it was planned. This is because energy transactions across the Southeastern European interconnection may impact the BiH transmission system and cause parallel flows, significant network loadings, and low bus voltages. Such system states may further deteriorate and become unstable, for once the system gets in the neighborhood of the mathematical condition that characterizes the state wherein voltages may collapse and units may loose synchronism, instability develops instantly and leaves no time to react. But in order to be *prevented,* the risk of blackout due to instability must first be *predicted,* so that adequate corrective actions could be quickly enacted if and when needed.

Figure 4-2. Bosnia and Herzegovina geographical position within Southeastern Europe.

The ability to predict the risk of blackout rests on the capability to quantify and compute the *distance* to the stability limit that corresponds both to the current operating state and to the strategy deployed to mathematically approach it, that is, the trajectory followed when stressing the system. Therefore, the distance between the current conditions and a hypothetical state wherein voltages may collapse and units may loose synchronism must be reevaluated after each state estimate and after each load flow.

In other words, the detection of the risk of instability must be envisioned as a continuous process; that is, the distance to instability must be *monitored*. "Monitoring" is a real-time process and points to the need to perform stability calculations:

- In the data environment offered by SCADA/EMS facilities
- With a minimum amount of information
- Within continuously running real-time network analysis sequences
- Quickly enough so that the results would be available immediately after the execution of the state estimator, thus allowing the user to examine them *before* the next real-time network sequence run.

To further increase the complexity of such an already difficult proposition, the results must be presented in a user-friendly format as required for fast and reliable online decision making.

This problem was solved effectively and efficiently by NOS BiH by blending its SCADA/EMS facilities with a very fast, reasonably accurate and field-proven steady-state stability assessment tool as shown in the following.

4.1.3   Overview of NOS BiH Approach to Real-Time Stability Monitoring

In order to fulfill its operational mission, NOS BiH has undertaken the implementation of control, information, and communications support consisting of a state-of-the-art SCADA/EMS, extended backbone fiber network, and data acquisition and substation automation facilities. The project has been contracted with Siemens Power Transmission and Distribution (Siemens) and is currently in its final stages. The new system is expected to become fully operational in the middle of the year 2008.

However, due to the long duration of this extremely complex effort, NOS BiH also developed an interim solution aimed at providing at an early stage the key tools needed to maintain and enhance the operating reliability of the grid. Accordingly, it first acquired off-line network analysis software, consisting of Power Technologies Inc. PSS/E load-flow and transient stability programs and, in 2001, it implemented basic SCADA functionality predicated on Siemens SINAUT Spectrum® technology.

Then, in 2005, the interim SCADA system was enhanced with a SINAUT Spectrum state estimator and the fast steady-state stability program known as QuickStab® Professional [32], which is used:

- In real time, to perform maximum loadability analysis and determine on a continuous basis the distance to instability of the current system state obtained from the latest run of the state estimator.
- Offline, in conjunction with study cases developed with the PSS/E software.

The tools implemented in the interim phase fulfilled their purpose and made it possible to continuously and reliably assess the actual utilization of the BiH transmission network, support online decision making, and help maintain and enhance the operating reliability of the grid.

In preparation for the field installation of the new SINAUT Spectrum SCADA/EMS, an upgraded version of the fast steady-state stability application used in the interim phase has already been tightly integrated by Siemens with the baseline real-time and study-mode network analysis applications. In addition, the solution was enhanced with the ability to display the key stability calculation results and to monitor the distance to instability in real time *directly in the native SCADA/EMS user interface.*

The implementation approach, practical experience, and lessons learned from the interim real-time stability assessment solution, as well as the salient features of the upgraded implementation are documented in the subsequent sections.

4.2   INTERIM IMPLEMENTATION OF REAL-TIME STABILITY ASSESSMENT AT NOS BiH

4.2.1   Real-Time Stability Tool

The implementation of real-time stability assessment at NOS BiH was preceded by a competitive bidding process. The terms of reference specified a "fast maximum transfer capability analyzer" that can be used in real time, near real time, and offline. In addition to speed, an important requirement was the ability to determine the "risk of blackout," where the "risk of blackout" is computed by performing steady-state stability calculations and quantifying the "distance" between the current power system state and the steady-state stability limit rather than just characterizing it as stable or unstable. Additional requirements included the ability to recommend a transmission reliability margin within a user-defined security margin, and the presentation of results in easily interpreted user-friendly charts and graphics.

The fast steady-state stability application that was selected uses the methodology developed by Paul Dimo in Europe in the early 1960s [3,4,5]. This technique was introduced in the United States in the early 1990s [8,17,29] and has been subsequently deployed in several real-time installations in Europe, Latin America, and the United States [16,22,23,24,32]. A detailed description of this technology goes beyond the scope of this chapter but is provided for reference in Appendix A.

4.2.2   Implementation Overview

NOS BiH implemented fast steady-state stability assessment in two computing environments:

1. Offline, to support the short- and mid-term operations-scheduling processes
2. Real time, in which the stability assessment capability is used in system dispatching

In study mode, the fast stability application runs on offline PCs and uses offline load-flow and generator data in PSS/E format prepared on the PSS/E platform.

In real time, the fast steady-state stability assessment software is installed on a separate PC connected to the SCADA/EMS LAN. On the SCADA/EMS side, after each successful execution of the state estimator, the output is saved in PSS/E format on the data administration server. On the PC side, a control program developed specifically for this purpose:

- Runs at fixed time intervals that are user definable, for example, 5, 10, 15 minutes
- Retrieves the most recent state estimate file from the SCADA/EMS server via FTP and copies it on the PC
- Automatically triggers the computational engine of the fast steady-state stability assessment program

The calculations are extremely fast and take less than one second to complete. Then, the control program triggers the display engine of the fast stability assessment program and presents the results on the PC. In addition, both the computational results and the input data are copied in a special directory for archival purposes. The archived input data and computational results can subsequently be exported to an external data storage device. This process is schematically depicted in Figure 4-3. The configuration diagram depicting the loose integration of the interim SCADA/EMS system with the PC on which the fast steady-state stability application is executed is shown in Figure 4-4.

4.2.3   Experience with the Interim Implementation

The ability to compute the steady-state stability reserve of the BiH power system in real time and to display the results in a graphical format that can be easily interpreted and understood by the system dispatcher has proven useful, and the experience acquired to date has been positive.

One of the key displays that are continuously updated on a Windows operator console is the so-called two-speedometer chart for the base case, that is, the current system state as computed during the most recent successful run of the state estimator (Figure 4-5).

The right-hand speedometer displays the distance to instability on a linear MW scale. The needle corresponds to the total MW system grid utilization in the base (current) state. The left edge of the red sector (shown in black in the figure) depicts the steady-state stability limit (SSSL, please refer to Annex 4-1 for the meaning and definition of this concept). The distance between the black needle (base case) and the red area (critical state) is quantified by the stability reserve in percent below the SSSL. The width of the light-gray sector (yellow on the actual display) is proportional with

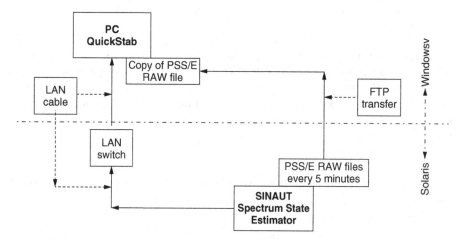

Figure 4-3. Schematic description of the interim implementation of real-time stability assessment at NOS BiH.

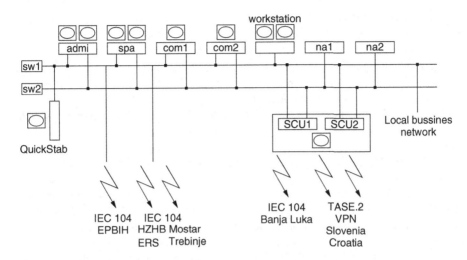

<u>Figure 4-4.</u> Interim phase—loose integration of the SCADA/EMS with the fast steady-state stability assessment program.

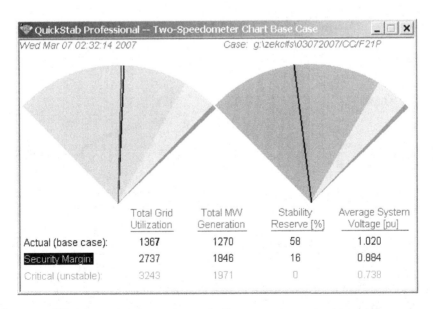

<u>Figure 4-5.</u> Two-speedometer chart depicting the distance to instability (red sector, shown in dark gray) and to the security margin (separation line between the two shades of gray).

the percent value of the security margin. This representation can be related to the "stability envelope" concept illustrated in Annex 4-1, where the "safe" operating region corresponds to total system MW loadings smaller than the MW security margin.

The left-hand speedometer shows the distance to instability on a nonlinear scale, where the values of the $d\Delta Q/dV$ reactive power steady-state (voltage) stability criterion are mapped for the base case (black needle), and for a hypothetical state (blue needle, shown in black in the figure, but always at the left of the "black" needle) in which the generators and the virtual units representing the tie-line injections would have been redispatched to maximize the distance to instability. In the case illustrated in Figure 4-5, the blue and the black needles are very close to each other, thus suggesting that even if the units were redispatched to achieve maximum stability, the gain would be minimal. Incidentally, since the stability reserve is already very large, an attempt to maximize stability would not be necessary. Let us also note that the distance between the black needle and the red sector on the left (blue) speedometer is smaller than the distance between the black needle and the red sector on the right (green) speedometer. This suggests that during the case-worsening process, the system approaches "slowly" the critical state, that is, many small MW increases are needed for the $d\Delta Q/dV$ derivative to increase from its initial negative value up to the point where it changes sign from minus to plus. This information is corroborated by the P–V curve shown in Figure 4-6, which shows many successively degraded states from the base case through the state of instability.

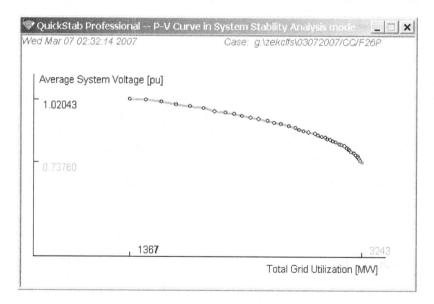

Figure 4-6. P–V curve depicting the case-worsening process—the average system voltage decays while the total MW system grid utilization increases up to the point of voltage collapse.

Another important output of the stability tool is the bar chart that ranks the units, generators, and tie-line injections, in decreasing order of their impact on the system's stability (Figure 4-7).

A relatively small increase in the MW output of the units labeled "IncreaseStab" (shown in green in the actual display) causes the stability conditions to improve, whereas the machines labeled "ReduceStab" (represented in red in the actual display) need to reduce their output in order to improve the stability conditions. Let us also note that the lengths of the bar charts in Figure 4-7 are proportional with the moduli of the short-circuit currents of the generators and indicate how "large" or "small" appear the machines when "seen" from the system. A detailed discussion of this concept goes beyond the scope of this chapter. The interested reader is directed to the reference [16] where all the mathematical aspects are presented in detail.

To summarize, the computational speed of the steady-state stability tool and the user-friendly format of the displays used to present the results are significant benefits and fully justify its real-time implementation. But the interim solution also revealed some areas that needed further improvement.

In terms of functionality, the contingency evaluation capability had not yet been implemented in the fast steady-state stability program version that was available at the time when the interim solution was devised, and since neither contingency analysis nor dispatcher's power-flow functions were available on the interim SCADA either, the only way to evaluate contingencies for stability violations would have been to run the fast stability assessment program in conjunction with PSS/E, using the real-time case as a base case. In reality, this was not possible, on the one hand because of the different bus naming and bus numbering conventions of the real-time and offline databases, and, on the other hand, because the real-time model used by the state estimator encompassed only the BiH transmission system and, since exter-

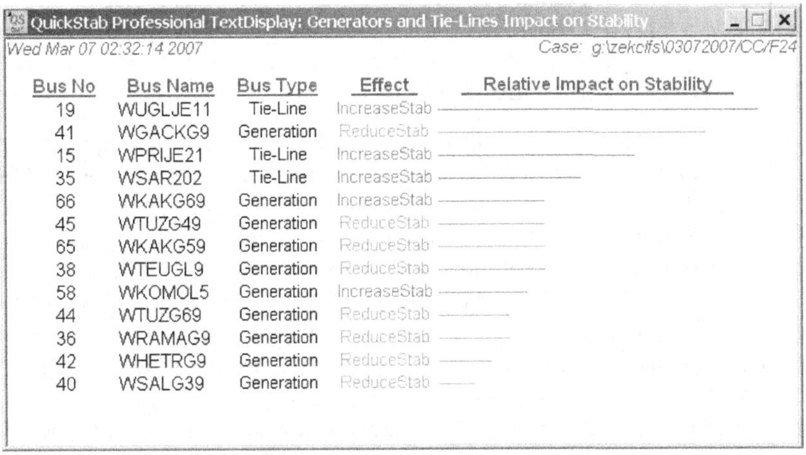

Bus No	Bus Name	Bus Type	Effect	Relative Impact on Stability
19	WUGLJE11	Tie-Line	IncreaseStab	
41	WGACKG9	Generation	ReduceStab	
15	WPRIJE21	Tie-Line	IncreaseStab	
35	WSAR202	Tie-Line	IncreaseStab	
66	WKAKG69	Generation	IncreaseStab	
45	WTUZG49	Generation	ReduceStab	
65	WKAKG59	Generation	ReduceStab	
38	WTEUGL9	Generation	ReduceStab	
58	WKOMOL5	Generation	IncreaseStab	
44	WTUZG69	Generation	ReduceStab	
36	WRAMAG9	Generation	ReduceStab	
42	WHETRG9	Generation	ReduceStab	
40	WSALG39	Generation	ReduceStab	

QuickStab Professional TextDisplay: Generators and Tie-Lines Impact on Stability
Wed Mar 07 02:32:14 2007 *Case: g:\zekciffs\03072007/CC/F24*

Figure 4-7. Unit ranking bar chart.

nal areas were not included, contingencies would not have been simulated realistically.

QuickStab has since been upgraded with a contingency analysis feature that scans the list of contingencies and then, for each contingency, performs a full Newton–Raphson AC load-flow calculation followed by the complete suite of steady-state stability calculations. This new functionality is now supported by the new SCADA/EMS, in addition to a more complex contingency evaluation process as described later in this chapter.

Another limitation of the interim solution stems from the fact that the fast stability assessment program uses a Windows PC user interface that is very different from the SCADA/EMS user interface. This forces the operators to navigate between two display environments, one on the SCADA/EMS console and the other one on a Windows PC, which is neither easy nor comfortable.

Also missing in the loosely integrated interim implementation was the ability to monitor in real time the distance to instability. Although the data needed to implement stability monitoring are calculated by the fast stability assessment program, they do not get transferred to the real-time database. A significant benefit of the seamlessly integrated approach is that after each calculation cycle, the steady-state stability reserve is written in the real-time database and then it can be trended by using the standard trending facilities of the new SCADA/EMS. This, in turn, will allow the operator to see the evolution of the distance to instability over a user-selectable period of time. Furthermore, the new implementation would allow the activation of the fast stability assessment program after state estimator solutions outside the fixed time grid chosen at the PC, for example, after switching-event-triggered activations. These and other substantial benefits of the seamlessly integrated approach are briefly discussed in the following section.

4.3 REAL-TIME STABILITY ASSESSMENT IN THE NEW SCADA/EMS ENVIRONMENT

4.3.1 Overview of the New System

The system was contracted with Siemens and is based on the SINAUT Spectrum platform. The overall project encompasses the single large SCADA/EMS Control Center in NOS BiH, which has overall authority for power system operations, four SCADA systems located in the transmission company regional control centers, an extensive telecommunication network based on fiber-optic cables and SDH/PDH communication equipment, as well as RTUs and intelligent substation automation systems (ISAS) in 105 substations. Additional RTUs and ISASs that have been implemented through other projects are also connected, which brings the total number of RTU and ISAS to 135.

The capabilities provided by the SCADA/EMS include: system and support software such as operating systems, compilers, communications software, database management, and user interface support; and various applications software subsystems, including SCADA, historical information system (HIS), generation control, short-term load forecasting, transmission network applications, and a dispatcher training simulator envisaged primarily for operators training but also capable of simulating the modeling and operation of the power system.

The SCADA subsystem provides a classic array of functions such as data acquisition and processing, supervisory control, alarm processing, trending, tagging, report generation, operator logbook, and so on. The HIS capability is implemented on a separate computer system and supports data processing, storage and retrieval, calculation and presentation capabilities, long-term historical data storage, and access to the historical data. The generation control subsystem encompasses load frequency control, reserve monitoring, interchange scheduling, and energy accounting, and is fully coordinated with the short-term load forecasting function.

The fast steady-state stability assessment program is part of, and fully integrated with, the transmission network applications subsystem, which consists of topology and network processor, state estimator, contingency analysis, dispatcher power flow, optimal power flow, bus load forecast, and penalty factor calculation, in addition to general functions and user interface.

The conceptual system architecture is illustrated in Figure 4-8.

Shown in Figure 4-9 is the configuration diagram that depicts the hardware platform consisting of components connected via a dual-redundant LAN. In addition, there is a complete system for uninterruptible power supply (UPS) consisting of rectifier, inverter and diesel generator with UPS installed power of 24 kVA.

The telecommunication subsystem, as a part of the overall BiH power system communication infrastructure and equipment, has the following functional parts and equipment:

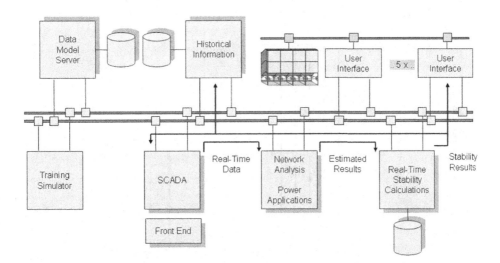

Figure 4-8. Conceptual architecture of the NOS BiH SCADA/EMS.

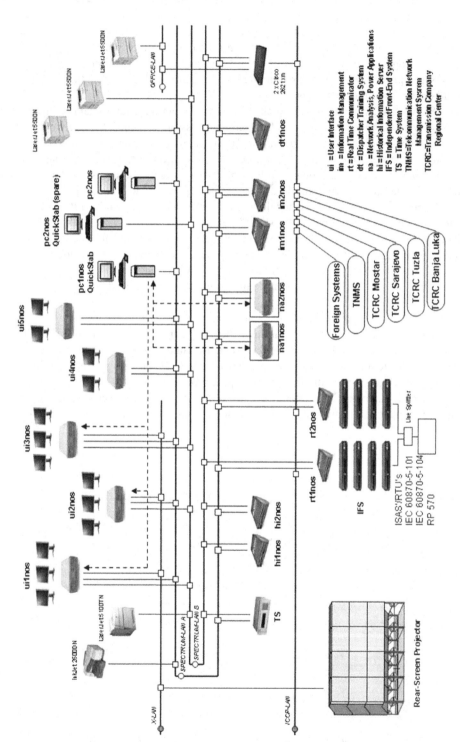

Figure 4-9. NOS BiH SCADA/EMS configuration diagram.

ui = User Interface
im = Information Management
rt = Real Time Communicator
dt = Dispatcher Training System
na = Network Analysis, Power Applications
hi = Historical Information Server
IFS = Independent Front-End System
TS = Time System
TNMS = Telecommunication Network
         Management System
TCRC = Transmission Company
         Regional Center

- Optical cables (two cables with 48 fibers each) for the connection of the NOS BiH location with the telecommunication network in a ring configuration,
- SDH equipment,
- PDH/TDM equipment,
- Synchronization equipment,
- Telecommunication network management system (TNMS) for the monitoring, control and reconfiguration of the overall telecommunication network, which is also an NOS BiH responsibility,
- PABX, and
- Power supply equipment.

4.3.2   Implementation of Real-Time Stability Monitoring

At the time when Siemens offered the SINAUT Spectrum SCADA/EMS upgrade, the standard transmission network analysis (TNA) package already included a voltage stability application (see Annex 4-2). However, the specifications issued by NOS-BiH for real-time steady-state stability evaluation and monitoring were better matched by the steady-state stability program described earlier in this chapter. It was thus decided to integrate this solution with the SINAUT Spectrum SCADA/EMS platform.

When designing the integration of the steady-state stability program with SINAUT Spectrum, emphasis was put on the ability to access it easily by the existing users of the TNA package as well. Further considerations governing the real-time implementation of QuickStab into the SCADA/EMS were derived from the major drawbacks experienced from the interim solution, as outlined in the previous section. Hence the major goals were:

- Immediate update of the operator on the distance from instability periodically as well as after each significant system change, as opposed to periodical updates only,
- Display of the decisive results in the user interface of the SCADA/EMS, as opposed to a separate and unfamiliar user interface,
- Capability to evaluate system instability for perceived situations by means of the familiar SCADA/EMS study-case management and in the same convenient way as in real-time, as opposed to using a separate and unfamiliar offline study environment.

The design shown in Figure 4-10 was chosen.

For the following explanations, please keep in mind that the SCADA/EMS is Unix-based, whereas the PC runs under the Windows operating system. As opposed to the interim implementation outlined in Section 4.2, the fast steady-state stability calculations are triggered *from within* the SCADA/EMS rather than from the PC. For that purpose, the Network Analysis Sequence Control (NASC) program of the SCADA/EMS has been extended so that after producing a PSS/E data file by means of the existing

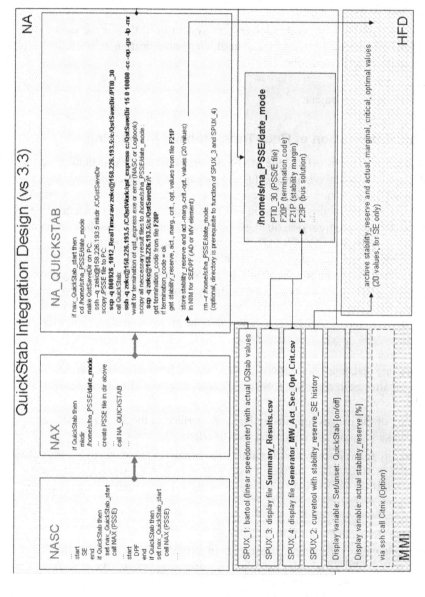

Figure 4-10. Schematic description of the final implementation of real-time stability assessment at NOS BiH.

Network Analysis Data Export program (NAX), a dedicated script named NA_Quick-Stab is activated. This activation takes place after each run of the State Estimator (SE), in real-time mode, as well as after each run of the Dispatcher Power Flow (DPF), in study mode. In order to separate SE-based activations from several simultaneously running DPF-based activations, their respective working directories are named "Real-time" versus "Study Case Alpha," "Study Case Beta," and so on.

The script NA_QuickStab uses "secure shell commands" as well "secure copy commands" to create directories on the PC simultaneously with those created on the SCA-DA/EMS and to populate them with the respective PSS/E file. The fast steady-state stability tool is then invoked by the script to be executed on the PC using the respective working directory; after completion, the results are stored in the respective working directory on the PC. Here they reside until they are deleted. This means that at any time the PC user can:

- Access the complete input and output data of any calculation performed with data received from the SCADA/EMS by means of the QuickStab detailed visualization tools.
- Use the data for performing additional steady-state stability analysis in study mode.

The script NA_QuickStab waits for the termination code, then it copies selected results from the PC into the respective working directory on the SCADA/EMS. The current *stability reserve* values are written into the SCADA/EMS database in either the real-time area or the study-case areas, as appropriate. Finally, the script NA_Quick-Stab terminates.

At this point the regular SCADA/EMS features for alarming and displaying take over. As the most prominent result, the current stability reserve value is displayed in the execution control display of SE or DPF (please refer to the lower-right corner in Figure 4-11). The display is spontaneously updated as soon as new calculation results are written by NA_QuickStab into the real-time database. If limits are violated, an alarm is triggered in the SCADA/EMS in the same way as for any other analog value.

By means of buttons QstDisplay 1 and QstDisplay 2 on the execution control display of the SE or DPF application, the user can call tabular displays providing more detailed results out of the SCADA/EMS real-time database. Figure 4-11 shows the Critical Bus Data Table as an example (middle area on left side).

Last but not least, since the current stability reserve values are treated by the SCA-DA/EMS like regular analogs, they can be archived and displayed in trend curves, as shown in the lower-left area in Figure 4-11. This allows the operator to *monitor* the evolution of the distance to instability over a user-selectable period of time.

In case the operator is actually interested in seeing more results of the most recent stability calculation he/she can press a button on the execution control display of the SE or DPF application that, by the means of VNC Viewer, opens a window on the Unix-based console of the SCADA/EMS actually displaying the desktop of the PC

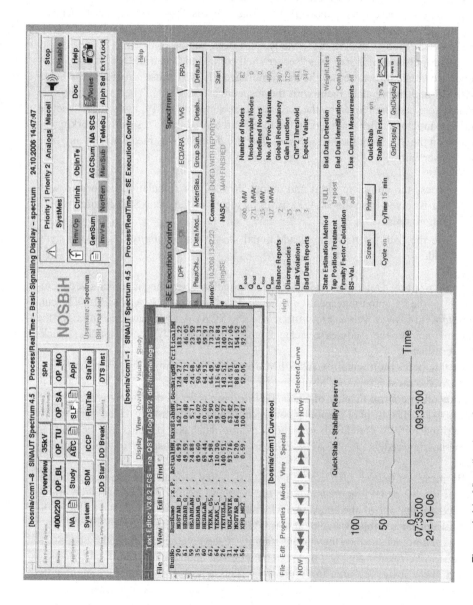

Figure 4-11. Visualization of real-time stability analysis results in the SCADA/EMS user interface.

running QuickStab. By these means, the operator can use all the displaying facilities of the genuine QuickStab program without actually moving to another monitor.

4.4 CONCLUSIONS AND RECOMMENDATIONS

Power system stability analysis during system operation is very important and, especially over the last few years, even more transmission system operators have realized the pressing need for such analysis after energy markets have been opened and large cross-border energy trades started to impact the stability of power systems. Proper stability analysis requires continuous application to track the continuous changes of the power system loading and to assess the impact of discontinuous events such as switching-state changes. Such real-time monitoring of system stability, however, is hampered by the computational burden imposed. Therefore, in order to be applicable, the solution approach needs to be based on acceptable approximations. The Independent System Operator of Bosnia and Herzegovina has elected to implement a technology based on Dimo's steady-state stability analysis method. Using this technique allows extremely fast calculation of the stability reserve of the power system while reasonably accurate generator models are retained, thus yielding results with excellent accuracy around the operating point.

It is important for the results produced by sophisticated tools such as real-time stability analysis to be presented to the operator in a way that makes it easy and convenient to use them. Therefore, the integration of such a tool in the real-time mode of the SCADA/EMS of the Independent System Operator in Bosnia and Herzegovina was done in a way that results in immediate update of system stability reserves not only after a cycle time such as 5 minutes has elapsed, but also after each significant system change. It was particularly important to meet the operators' requirement for displaying the key results of the stability analysis directly in the user interface of the SCADA/EMS and to provide the capability to evaluate system stability reserves for perceived situations by means of the familiar SCADA/EMS study-case management (study mode). Thus, the need to get into a user interface environment other than SCADA/EMS was entirely abolished.

The thorough evaluation of power system stability conditions requires the consideration not only of the base case but also of contingency cases. With the approach of system stability analysis integration in SCADA/EMS as described in the chapter, it is possible to perform such contingency case studies using the separate QuickStab environment on a separate PC whose user interface, however, is fully accessible from the operators' well-known Unix user interface of the SCADA/EMS. The next step of integration will enable contingency-case stability analysis based on the contingency lists of the SCADA/EMS entirely in the user environment of the latter.

4.5 REFERENCES

[1] Barbier, C., and Barret, J. P., "An Analysis of Phenomena of Voltage Collapse on a Transmission System," *Revue Générale de l'Electricité,* Vol. 89, No. 7, pp. 3–21, 1980.

[2]   Crary, S. B., *Power System Stability*, General Electric Series, Schenectady, New York, Copyright 1945, Third Printing October 1955.

[3]   Dimo, P., *Nodal Analysis of Power Systems*, Abacus Press, Kent, England, 1975.

[4]   Dimo, P., "Etude de la Stabilité Statique et du Réglage de Tension," *Revue Générale de l'Electricité*, Paris, Vol. 70, No. 11, 552–556, 1961.

[5]   Dimo, P., Manolescu, G., Iordanescu, I., Groza, L., Ionescu, S., Albert, H., Moraite, G., and Ungureanu, B., *Computation and Design of Electrical Energy Systems* (Romanian edition), Editura Tehnica, Bucharest, Romania, 1971.

[6]   Dobson, I., and L. Liu, "Immediate Change in Stability and Voltage Collapse when Generator Reactive Power Limits are Encountered," in *Proceedings of International Seminar on Bulk Power System Voltage Phenomena II*, edited by L. H. Fink, pp. 65–74, 1993.

[7]   Ejebe, G. C., Irisarri, G. D., Mokhtari, S., Obadina, O., Ristanovic, P., and Tong, J., "Methods for Contingency Screening and Ranking for Voltage Stability Analysis of Power Systems," *IEEE Transactions on Power Systems*, Vol. 11, No. 1, February 1996, pp. 350–356.

[8]   Erwin, S. R., Oatts, M. L., and Savulescu, S. C., "Predicting Steady-State Instability," *IEEE Computer Applications in Power*, Vol. 4, No. 7, pp. 15–22, 1994.

[9]   Ionescu, S., and Ungureanu, B., "The Dual Power States and Voltage Collapse Phenomena," *Revue Roumaine de Sciences Techniques, Série Electrotechnique et Energétique*, Vol. 26, No. 4, pp. 545–562, 1981.

[10] Kundur, P., Introduction to *Techniques for Power System Stability Search*, A special publication of the Power System Dynamic Performance Committee of the IEEE PES, TP-138-0, pp. 1–3, 1999.

[11] Magnien, M., "Rapport Spécial du Groupe 32 Conception et Fonctionnement des Réseaux," *Conférence Internationale des Grands Réseaux Electriques à Haute Tension*, CIGRE Session 1964.

[12] Moraite, G., Ionescu, S., Feldmann, S., and Chenzbraun, I., "Problèmes Soulevés par la Stabilité Statique des Réseaux Bouclés," *Conférence Internationale des Grands Réseaux Electriques à Haute Tension*, CIGRE Session 1966).

[13] Navarro-Perez, R., and Prada, R. B., "Voltage Collapse or Steady-State Stability Limit," in *Proceedings of International Seminar on Bulk Power System Voltage Phenomena II*, edited by L. H. Fink, pp. 75–84, 1993.

[14] Sauer, P. W., and Pai, M. A., "Power System Steady-State Stability and the Load-Flow Jacobian," *IEEE Transactions in Power Systems* Vol. 5 T-PWRS, No. 4, pp. 1374–1381, 1990.

[15] Sauer, W. P., and Pai, M. A., "Relationships between Power System Dynamic Equilibrium, Load-Flow, and Operating Point Stability," in *Real Time Stability in Power Systems*, pp. 1–30, Springer Verlag, Norwell, MA, 2005.

[16] Savulescu, S. C., "Fast Assessment of the Distance to Instability. Theory and implementation," in *Real Time Stability in Power Systems*, Springer, Norwell, MA, 2005.

[17] Savulescu, S. C., Oatts, M. L., Pruitt, J. G., Williamson, F., and Adapa, R., "Fast Steady-State Stability Assessment for Real-Time and Operations Planning," *IEEE Transactions in Power Systems*, Vol. 8 T-PWRS, No. 4, pp. 1557–1569, 1993.

[18] Tweedy, J., "Assessing Stability Conditions to Evaluate the Risk Of Blackout. A Real-Time Indicator Based on Field-Proven Techniques," *Transmission and Distribution*, June 2004, pp. 622–628, 2004.

[19]   Venikov, V. A. *Transient Processes in Electrical Power Systems,* Edited by V. A. Stroyev, English Translation, MIR Publishers, Moscow, 1977.

[20]   Venikov, V. A., Stroev, V. A., Idelchick, V. I., and Tarasov, V. I., "Estimation of Electrical Power System Steady-State Stability," *IEEE Transactions on PAS,* Vol. PAS-94, No. 3, pp. 1034–1041, 1975.

[21]   Vergara, J. S., Thai, T. A., Cuong, N. D., Nam, N. T., Campeanu, H. S., and Savulescu, S. C., "Accuracy Testing and Real-Time Implementation of Dimo's Stability Analysis Technique," in *Real Time Stability in Power Systems,* Springer Verlag, Norwell, MA, 2005.

[22]   Vickovic, D., "Concept of the Fast Maximum Transfer Capability Analyzer FMTCA Application in the Bosnia and Herzegovina ISO (NOS BiH) System," paper R.C2./D2.06, CIGRE Meeting in Neum, Bosnia and Herzegovina, 25–29 September, 2005.

[23]   Vickovic, D., Eichler, R., and Savulescu, S. C., ""Real-Time System Stability Monitoring in the Transmission Network of Bosnia and Herzegovina," paper presented at the Power-Grid Europe Conference and Exhibition, Feria de Madrid, Madrid, Spain, 26–28 June 2007.

[24]   Virmani, S., Vickovic, D., and Savulescu, S. C, "Real-Time Calculation of Power System Loadability Limit," Paper No. 576 presented at the Powertech 2007 Conference, July 1–5, 2007, Lausanne, Switzerland.

[25]   Vournas, C. D., Sauer, P. W., and Pai, M. A., "Relationships between Voltage and Angle Stability of Power Systems," *Electrical Power and Energy Systems,* Vol. 18, No. 8, pp. 493–500, 1996.

[26]   Wu, F. F., and Narasimhamurti, N., "Necessary Conditions for REI Reduction to be Exact," *IEEE PES Winter Meeting 1979,* Paper A 79 065-4, 1979.

[27]   IEEE PES, "Techniques for Power System Stability Search," A Special Publication of the Power System Dynamic Performance Committee of the IEEE PES, TP-138-0, 1999.

[28]   IEEE, PES Task Force on Terms and Definitions, "Proposed Terms and Definitions for Power System Stability," *IEEE Transactions on PAS,* Vol. PAS-101, No. 7, 1982.

[29]   EPRI, *Power System Steady-State Stability Monitor Prototype,* Final Report EPRI TR-100799, July 1992; and *Power System Steady-State Stability Monitor,* Final Report EPRI TR-103169, December 1993.

[30]   EPRI, "Factors Related to the Series of Outages on August 14, 2003," a White Paper, Product ID 1009317, November 20, 2003.

[31]   NERC, 1996, *Available Transfer Capability Definitions and Determination,* North American Electric Reliability Council, June 1996.

[32]   http://www.quickstab.com.

ANNEX 4-1. TSL, TTC, AND THE STABILITY ENVELOPE

Just as for each system state there is a steady-state stability limit (SSSL), a transient stability limit (TSL) can also be thought to exist. However, as opposed to the SSSL, and because of the computational procedures used to detect transient instability, the TSL is not quantifiable through a specific formula. This is because in order to find the TSL, transient stability simulations would have to be performed for each potential disturbance starting with a base-case scenario and continuing with a sequence of successively degraded operating states until the first unstable state has been identified. The intrinsic technical complexity and the large number of credible contingencies render such a problem practically unsolvable. However, intuition suggests that:

- For a given set of relay settings, TSL depends, just like the SSSL, upon topology, voltage levels, and system loading.
- For any system state, SSSL and TSL are interrelated and move in the same direction; if SSSL is high, TSL is also high, and vice-versa.

In the past, empirical values approximating the TSL/SSSL ratio have been used to compute a "safe" system loading, expressed as a percentage of the SSSL and referred to as the *security margin,* such that, for any system state with a steady-state stability reserve higher than this value, no contingency, no matter how severe, would cause transient instability [12]. Accordingly, the security margin can be regarded as a stability envelope: all the states with a total MW loading smaller than the MW value of the security margin are safe, even if a "safe" system state with higher MW loading might possibly be found. This concept is illustrated in Figure 4-12.

The security margin depends upon the specific combination of topology, loads, generators, and reactive compensation, and must be determined, and periodically reassessed, for each particular transmission system. For strong and highly meshed net-

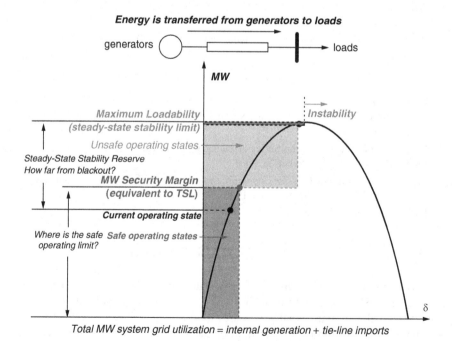

Figure 4-12. Stability envelope concept.

works in which the postdisturbance configuration is relatively close to the predisturbance state, the ratio $\sigma = $ TSL/SSSL can be assumed to be approximately constant.

If this ratio σ is known, the system loading at TSL can be determined by first computing SSSL and then identifying a new system state with total MW loading equal to $\sigma \times$ SSSL. We are not aware of a mathematical formula relating TSL and SSSL, nor do we know whether such a relationship can be developed analytically, but the empirical approach described in [21] can be expanded to build the following heuristic:

- Step 1. Start with a base-case load flow for peak load conditions, assume an initial percent value for the security margin, say 15% below SSSL, and compute the SSSL and related MW value of the security margin, which, in this case, would be equal to $0.85 \times$ SSSL; as part of this step, the MW outputs of the generators, referred to as *security margin MW generation schedules,* should also be computed.
- Step 2. Run an extensive suite of transient stability cases. If no instability has been detected, go to Step 5. If at least one contingency (fault) case was found to be unstable, go to Step 3.
- Step 3. Use the security margin MW generation schedules from Step 1 to calculate a new base case load flow.
- Step 4. For the load flow computed in Step 3, run again the same suite of transient stability cases that were evaluated in Step 2.
- Step 5. If no instability has been detected, repeat Step 4 for successively increased MW levels until at least one contingency (fault) causes transient instability. The percent value of the steady-state stability reserve for the immediately precedent state is the percent value of the security margin of the system under evaluation.
- If the stability calculations in Step 4 detected at least one contingency (fault) that would cause instability, build a new load-flow case for a slightly *reduced* load level and repeat the transient stability checks. If no instability has been detected, recalculate the SSSL and the steady-state stability reserve, which correspond to the security margin of the system under evaluation.

Once a percent value $x\%$ of the security margin has been determined, the stability envelope associated with any given system state is obtained as follows:

- First, starting from a state estimate or solved load flow, determine the steady-state stability reserve, that is, the distance to SSSL.
- Then, for the $x\%$ value of the security margin, determine the corresponding safe system MW loading below the SSSL.

Each system has its own stability envelope. As we already pointed out, it may be difficult, or even impossible, to reach the exact value of σ, but operating experience provides invaluable hints. For example, reference [5] recommended a 20% security margin* for the Romanian power system as it was in the 1970s. Reference [21] describes the procedure used by ETESA of Panama to validate the value of the security margin (15%) that is currently used in conjunction with its real-time stability assessment application. A 15% value of σ is used in the NOS BiH system, too.

ANNEX 4-2. SIEMENS IMPLEMENTATION OF THE CONTINUATION POWER FLOW

The security margin to the point of voltage instability is defined as the difference between the initial MW load and the collapse point MW load minus a MW back-off value equivalent to the security margin. One way of obtaining those margins is running Continuation Power Flow (CPF) as the system is stressed by increasing the load in a sink area and generation in the source area. CPF traces PV-curves at the monitored buses and determines the critical equilibrium point on this curve, and thus computes the stability margins.

CPF is a power-flow program enhanced with continuation methods. Its input consists of the state of the power system as determined by a state estimator or conventional power flow solution typically stored in the real-time database of the SCADA/EMS. In the CPF method, the power system approaching voltage instability is calculated as a series of solutions to power flow equations that have been extended by a parameter λ used to represent the change in demand at all buses of the power system as compared to the operating point. For slow changes of λ, these solutions involve computing the equilibrium points defined by the well-known steady-state power flow equations given as

$$f(x, \lambda) = 0 \tag{1}$$

where x is the n-vector of state variables (voltage magnitudes and angles at all buses). The above equation may be written as

$$P_{Gi}(\lambda) - P_{Li}(\lambda) = \sum_{j \varepsilon i} V_i \times V_j \times (G_{ij} \times \cos \theta_{ij} + B_{ij} \times \sin \theta_{ij})$$

$$Q_{Gi}(\lambda) - Q_{Li}(\lambda) = \sum_{j \varepsilon i} V_i \times V_j \times (G_{ij} \times \sin \theta_{ij} - B_{ij} \times \cos \theta_{ij})$$

$$\tag{2}$$

*The Grid Operation Code in Romania specifies the minimum steady-state stability reserves that must be observed in normal (20%) and contingency (8%) operating states. In Romania, however, the reference for computing the stability reserve is the total MW in the base case, as shown by formula (1) in the Section 5.2.1.1 of Chapter 5 in this book. Therefore, a value of 20% stability reserve (for normal operating conditions) calculated this way corresponds to approximately 15% stability reserve computed by taking for reference the total MW grid utilization at the SSSL.

where

$$P_{Gi}(\lambda) = P_{Gi0} \times (1 + \lambda \times K_{Gi})$$
$$P_{Li}(\lambda) = P_{Li0} \times (1 + \lambda \times K_{Li}) \qquad (3)$$
$$Q_{Li}(\lambda) = Q_{Li0} \times (1 + \lambda \times K_{Li})$$

P_{Li0} and Q_{Li0} are the active and reactive load at bus i, and P_{Gi0} is the active generation at bus i at the operating point (base case).

The above power-flow equations may be more compactly written as

$$f(x, \lambda) = F(x) + \lambda \times b = 0 \qquad (4)$$

The direction vector b represents the changes in real and reactive power demand and the changes in real power generation.

Solutions of this set of equations are used to trace both the stable and unstable branches of the voltage versus power (or λ) curve for any particular bus voltage magnitude in the power system. Any power-flow method can potentially be used to solve for the state variables in the above equations given a particular load change. To trace the complete branches of the V-versus-λ curve, however, the use of a continuation method is required. A continuation-based method consists of two steps: a predictor step, which produces an approximate solution to be used as initial condition to the second step, the corrector step. Figure 4-13 shows these steps. Differences between continuation methods are usually due to how these steps are implemented.

For the purposes of this chapter it is sufficient to know that by the successive application of these steps it is possible to trace the stable branch of the V-versus-P (or λ)

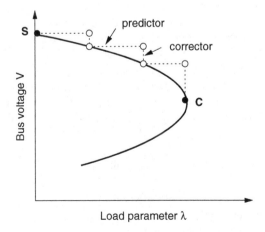

Figure 4-13. Steps of the continuation power-flow method.

curve to obtain the point C in Figure 4-13. Given C, the distance to collapse is then obtained as the difference in the power at C and the power at the initial point S.

Often, CPF is applied as part of a voltage stability analysis (VSA) package. VSA is used in a SCADA/EMS in conjunction with static voltage or thermal security analysis to provide a more complete picture of system security by determining how close the system is to voltage collapse. Integrating VSA with the real-time transmission network analysis applications of the SCADA/EMS eliminates the need to perform off-line voltage stability studies and to avoid the associated effort required for setting up off-line studies. As a result, voltage security analysis results are available in a timely fashion.

The objective of VSA is to determine the distance of the base case from system collapse as well as to determine those contingencies from a large set of potential contingencies that may lead to voltage stability problems. For describing the change of the system loading situation from the current system load level to the collapse point, the user can set up and study several distinct combinations of sources and sinks along with selected monitored buses and contingencies. Source and sink combinations are selected on a company or zone basis. Sinks can be defined as load increase or generation decrease. Generation in the source areas is adjusted to match changes in the sink areas.

The user interface of a real VSA implementation from which a new combination of source and sink can be added to the list or an existing combination can be edited is shown in Figure 4-14. In the middle of the window area the user can specify the location of source and sink as well as the sequence in which several participating companies/zones share the changes. In the upper window area, some general parameters of the combination are defined, for example, concerning the way generation increase is distributed to generators or contingencies to be considered. Finally, the lower window area allows the specification of buses to be monitored and contingencies to be included.

The definition of contingencies to be analyzed starts from a large list of selected contingencies. For optimal integration with the SCADA/EMS, VSA utilizes the same contingency list and associated outaged equipment as the security analysis function of the SCADA/EMS. A flag is associated with each contingency, allowing the user to select which contingencies are to be considered by VSA. Each contingency also has a priority assigned for use by VSA. Specifying the desired range of contingency priorities prior to execution can control VSA contingency selection.

The initial large list of contingencies is then screened and ranked using fast ranking algorithms [7]. Finally, contingencies flagged as potentially harmful during the screening and ranking phases are studied in detail using the CPF method. The output of this process is limits to collapse in terms of MW load changes for each fully analyzed contingency case.

Thus, the VSA algorithm consists of the following steps:

- Step 1. Operating point stability assessment,
- Step 2. Contingency selection,
- Step 3. Contingency screening and ranking, and
- Step 4. Contingency evaluation.

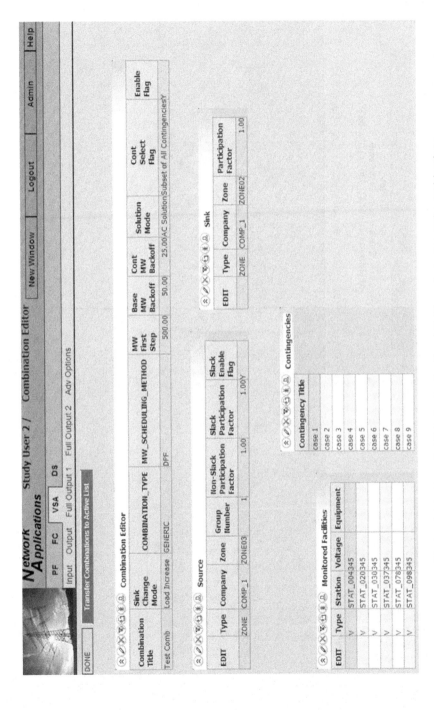

Figure 4-14. Sample VSA display for the definition of source–sink combinations.

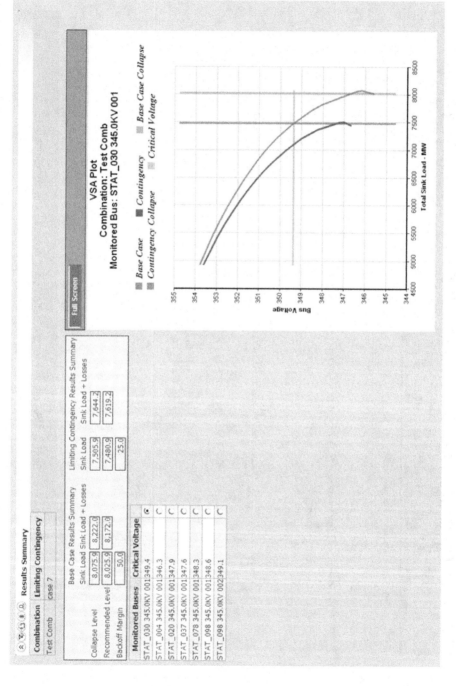

Figure 4-15. Typical result of VSA.

PF | FC | **VSA** | DS

Input | Output | Full Output 1 | Full Output 2 | Adv Options

Execution Date/Time: 14:11 10/23/2006
Combination For Full Output: Test Comb
Contingency For Full Output: case 1
STEP_NUMBER_1_FOR_FULL_OUTPUT 1
STEP_NUMBER_2_FOR_FULL_OUTPUT 5

Critical Buses
1 - 25 / 54

Message
COMBINATION = Test Comb
BASE CASE CALCULATION

Eigenvalue Analysis				Tangent Vector Analysis			
Critical Buses:			Value:	Critical Buses:			Value:
STAT_246	23.9KV	005	1.0000	STAT_215	18.0KV	002	1.0000
STAT_246	23.9KV	006	.9997	STAT_215	18.0KV	003	.9982
STAT_246	23.9KV	003	.9996	STAT_224	23.9KV	005	.9490
STAT_246	23.9KV	004	.9989	STAT_224	23.9KV	003	.9487
STAT_249	1.0KV	005	.9844	STAT_215	16.5KV	005	.9484
STAT_249	1.0KV	004	.9842	STAT_215	16.5KV	006	.9484
STAT_249	1.0KV	006	.9838	STAT_224	23.9KV	004	.9484
STAT_249	1.0KV	003	.9835	STAT_215	16.5KV	004	.9473
STAT_249	69.0KV	002	.9828	STAT_215	16.5KV	007	.9450
STAT_224	23.9KV	005	.9821	STAT_217	1.0KV	005	.9277
STAT_224	23.9KV	003	.9818	STAT_217	1.0KV	003	.9276
STAT_224	23.9KV	004	.9814	STAT_217	1.0KV	004	.9275
STAT_259	23.9KV	002	.9726	STAT_217	69.0KV	002	.9266
STAT_247	23.9KV	003	.9706	STAT_224	1.0KV	006	.9197
STAT_262	1.0KV	004	.9655	STAT_224	1.0KV	007	.9194
STAT_262	1.0KV	003	.9647	STAT_224	1.0KV	008	.9194
STAT_262	69.0KV	002	.9636	STAT_224	69.0KV	002	.9188
STAT_252	23.9KV	003	.9602	STAT_215	161.0KV	001	.9183
				STAT_221	1.0KV	006	.9151

Figure 4-16. Sample VSA display for critical bus analysis

Typical output results of VSA comprise, for each combination of source and sink:

- Collapse MW level,
- The most limiting contingency,
- The most limiting contingency collapse MW level,
- Critical voltage at monitored buses,
- Transfer flows on critical transmission lines, between areas, and between zones,
- The base case and the limiting contingency case are shown in Figure 4-15 with their associated MW back-off value (indicated by the two vertical lines). The critical voltage for the monitored bus is determined by projecting the contingency-case collapse point onto the base-case P–V curve.

Besides the CPF method for the determination of the distance from voltage collapse, VSA contains modal and sensitivity analysis methods to determine the nature of the equilibrium point. This additional information is provided to the user for the base case and the worst contingency case. The display in Figure 4-16 shows as an example the bus ranking for the base case. Two ranked lists are provided for each case: the first based on eigenvalue-analysis of the power-flow Jacobian matrix and the second based on the continuation power-flow tangent vector analysis. Both eigenvalue analysis and tangent vector analysis are done at the point of voltage collapse. The first bus listed is the weakest bus based on that particular ranking algorithm.

5

EXPERIENCE WITH REAL-TIME STABILITY ASSESSMENT AT TRANSELECTRICA

Horia S. Campeanu, Cornel Erbasu, and Cornel Aldea

5.1 INTRODUCTION

5.1.1 Overview of the Romanian Transmission System

Transelectrica S.A. (Transelectrica) is the owner, administrator, and operator of the Romanian transmission system and, as a member of UCTE (European Union for the Coordination and Transmission of Electricity) and ETSO (European Association of Transmission System Operators), facilitates the energy exchanges between the Southeastern European countries. In this capacity, Transelectrica supervises the transmission of electric energy, performs system dispatching and market operations, and is responsible for maintaining and enhancing the operating reliability of the Romanian National Electric System (SEN).

The company started its operations on July 31, 2000 as a result of the privatization and deregulation process in Romania when the vertically integrated utility CONEL (National Electricity Company) was split in four independent companies: Transelectrica; Electrica S.A., which is responsible for the distribution of electric energy; Hidroelectrica S.A., which handles the hydro generation in Romania; and Termoelectrica, S.A., which is responsible for the generation of electric energy in conventional fuel thermal plants (nuclear generation not included). Therefore, the transmission, system and market services have been completely separated from generation and distribution.

Transelectrica's activities are regulated by the National Regulatory Energy Agency (ANRE) consistently with the Energy Law, as well as with the secondary legislation, and in full agreement with the Transmission and System Operator License, Electric Transmission Network Code, Commercial Code, and the Metering Code.

The system and market operator functions are performed by an administrative unit known as UNO-DEN (National Dispatch Center) which monitors, supervises, and controls:

- 77 substations, including one 750 kV substation, 32 400 kV substations and 44 220 kV substations, as well as approximately 125 transformers with a total installed power of 34,525 MVA,
- 8,950 km of transmission lines encompassing 155 km of 750 kV lines, 4630 km of 400 kV lines, 4132 km of 220 kV lines, and 38 km of 110 kV tie lines between Romania and the neighboring countries.

Consistently with the operating rules and guidelines issued by UCTE, and except for the few 110 kV lines that belong to the "stability constrained transmission corridors" identified in Section 5.2.1.2, the 110 kV network is under the jurisdiction of the distribution company. A simplified diagram of the Romanian transmission system, which is currently interconnected with the power systems of Ukraine, Moldova, Hungary, Serbia and Bulgaria, is shown in Figure 5-1.

In order to meet the operating reliability requirements, the SEN network model that is used to perform static and dynamic security assessment also includes both the high-voltage transmission system and the subtransmission and distribution networks that operate at 110 kV; approximately 50–60% of the total system generation is connected. As a result, the electric model of the Romanian transmission system encompasses over 660 substations, 1350 overhead transmission lines, and 475 transformers and auto-transformers.

In addition to its duties in capacity of system operator, which are supported by a state-of-the-art SCADA/EMS system, UNO-DEN is also responsible for the "green certificates" and for the allocation of transmission capacities across interconnections.

The market operation functions of UNO-DEN are handled by the

- Balancing Market Operator, known by the acronym OPE,
- Commercial Market Operator, or OPCOM,
- Wholesale Market Metering Operator, or OMEPA.

The information and telecommunications services in support of the Romanian Electricity Market are provided by the Information Technology (IT) and Telecommunications Operator, which is known by the acronym TELETRANS S.A. and is a wholly owned subsidiary of Transelectrica. The overall picture of the Electricity Market in Romania is illustrated conceptually in Figure 5-2.

Romanian Transmission System in the South-Eastern European Context

Figure 5-1. Romanian transmission system overview.

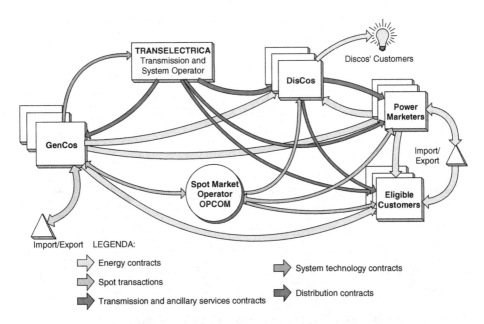

Figure 5-2. Overview of the Romanian electricity market.

5.1.2 The SCADA/EMS

5.1.2.1 Hierarchy and Configuration

The dispatching of the electric power system and the operation of the electricity market in Romania are supported by an integrated information system that creates and maintains an extensive repository of raw and processed data that can be accessed, based on appropriate access jurisdiction, by all the players in the Romanian electricity market. The information architecture encompasses a hierarchical SCADA/EMS and a balancing market system. Both systems are operated by Transelectrica and are located at the National Dispatch Center (DEN). The hierarchical SCADA/EMS structure is shown in Figure 5-3.

Transelectrica's SCADA/EMS is responsible for the real-time acquisition and processing of all the data pertaining to the operation of the SEN and provides for the monitoring and telecontrol of approximately 350 generators and 660 substations under the operational jurisdiction of the National Dispatch Center (DEN) and five Territorial Dispatch Centers (DET). One of the DETs—the Bucharest Dispatch

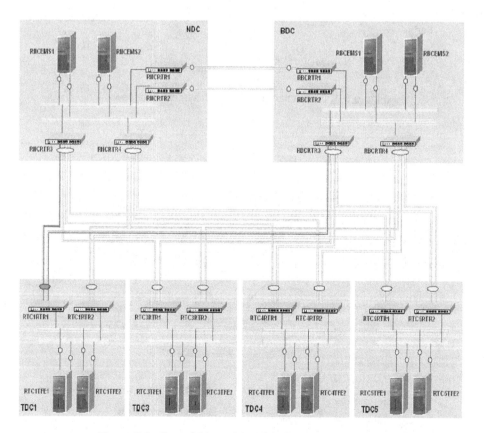

Figure 5-3. Overall hierarchical SCADA/EMS structure.

Center or BDC—is situated in the same building as the DEN. The system was designed, manufactured, and installed by AREVA T&D (AREVA). The data collection in substations is performed by RTUs provided by AREVA and General Electric (GE) and by substation automation systems provided by AREVA, Siemens, ABB, and GE. Figure 5-4 depicts the central SCADA/EMS system at the National Dispatch Center DEN.

5.1.2.2 *System Dispatching Support Applications*
Transelectrica's SCADA/EMS applications available at DEN encompass

- SCADA, including analog, status and accumulator data processing, limit checking, value replacement, calculations, historical data recording, intersite data processing, tagging, controls, and load shedding,
- Generation control and scheduling, including AGC, reserve monitoring, AGC performance monitoring, historical loss model update, transaction scheduling, transaction evaluation, load forecasting, and market interface,
- Network analysis, including topology processing, state estimation, real-time steady-state stability assessment and monitoring (QuickStab, [25]), monitored element processing, bus load model update, loss sensitivities, contingency analysis, security enhancement, power flow, optimal power flow/voltage VAr dispatch, offline short-circuit analysis, and off-line stability analysis (SAMI, EuroStag, and offline QuickStab).

This is illustrated in Figure 5-5. The real-time applications, such as contingency analysis and security enhancement, can be executed in study-mode, too. In addition, a number of pure study-mode applications are also available, such as hourly transmission capacity assessment based on the market clearing schedules received from OP-COM.

5.1.2.3 *Ancillary Services*
In addition to the standard and nonstandard real-time and study-mode applications identified in the previous section, DEN performs the following ancillary services that are needed in the electricity market:

- Primary regulation reserve. In 2006, the minimum amount required by UCTE for SEN was 60 MW; this target was met by maintaining at each generator a reserve in the amount of ±1% of the nominal power.
- Secondary regulation reserve. In 2006, the internal procedures specified 460 MW for this reserve, out of which 50 MW have been allocated to thermal power plants.
- One-minute reserve. In 2006, the internal procedures specified that 1020 to 1175 MW be available at any moment in order to compensate for the unscheduled outage of the largest unit installed in the SEN.

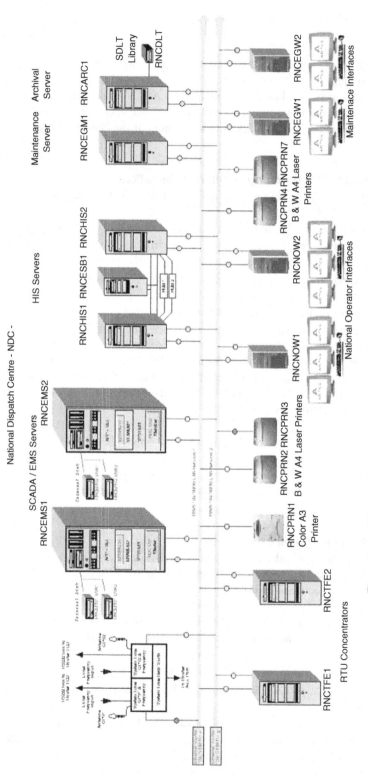

Figure 5-4. SCADA/EMS configuration at the National Dispatch Center DEN.

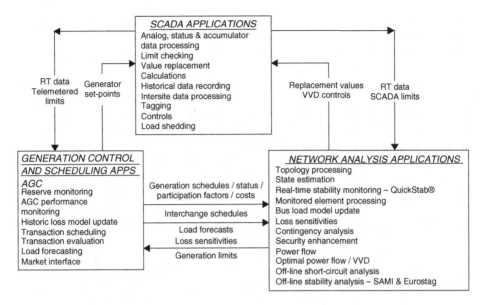

Figure 5-5. SCADA/EMS applications implemented at DEN.

5.2   SECURITY ASSESSMENT PHILOSOPHY AND CRITERIA

Transelectrica has deployed, and is currently using, a broad range of state-of-the-art software tools for assessing, maintaining, and enhancing the operating reliability of its transmission system. Static security assessment is performed with load-flow and contingency evaluation tools that are used both in real time and study mode, with data from the most recent state estimate, and offline, on postulated scenarios. Dynamic Security Assessment (DSA) is also performed in real time and offline.

In real-time, the DSA calculations consist of computing, after each automatic or on-demand execution of the state estimator, the distance to the steady-state stability limit system-wide, and for each of the electric areas connected through "stability-constrained transmission corridors"* by using the fast steady-state stability assessment application described in [25]. Ass shown in Section 2.1.1, the distance to steady-state instability is called steady-state stability reserve; it is an important indicator of how "healthy" the current operating conditions are, and its computation is required by the official rules and guidelines that regulate the power system operations in Romania. The current values of the steady-state stability reserves thus determined are stored in the real-time database as calculated points, and then are tracked on standard SCADA/EMS trending charts both on operator workstations and on the video projection system.

In the off-line environment, the DSA studies encompass an extensive set of contingency and system loading scenarios that are evaluated in detail with a variety of

*The concept of "stability constrained transmission corridor" is addressed in Section 5.2.1.2.

steady-state and transient stability analysis programs. The calculations are performed both system-wide and for the stability-constrained transmission corridors by using input retrieved form the real-time system and postulated data as well.

5.2.1   Steady-State Stability Assessment

The importance of steady-state stability assessment has been recognized in Romania for a long time, from the early days, when the power system was operated by a vertically integrated utility, until now, when the generation, transmission, and distribution of electricity take place in the electricity market context.

Steady-state stability analysis goes way beyond just evaluating the risk of instability when the maximum loadability limit is approached via gradual load changes. Rather, the key objective of the steady-state stability analysis procedures deployed at Transelectrica, regardless of the underlying solution technique, is determining the *distance* to a system state at which voltages may collapse and units may loose synchronism, and to *quantify* such distance by means of a simple indicator known as the *steady-state stability reserve*. The assessment is performed both system-wide and for stability-constrained transmission corridors. These important aspects of the real-time and offline steady-state stability assessment at Transelectrica are discussed in the following section.

5.2.1.1   Power and Voltage Steady-State Stability Reserve Indicators

The concept of "steady-state stability reserve" provides a very simple way of saying "how far" the current, or actual, system state is from the "critical" state in which a small change of the operating parameters may cause steady-state instability. In accordance with the system operating reliability guidelines developed in the 1960s for SEN [10, pp. 200–201], two types of stability reserve indicators are defined as follows:

$$stability\ reserve_{\text{power}} = \frac{P_{\text{max}} - P_{\text{base case}}}{P_{\text{base case}}} \times 100\ [\%] \qquad (1)$$

$$stability\ reserve_{\text{voltage}} = \frac{V_{\text{base case}} - V_{\text{critical}}}{V_{\text{base case}}} \times 100\ [\%] \qquad (2)$$

where:

$stability\ reserve_{\text{power}}$ quantifies the stability reserve in terms of MW power

$stability\ reserve_{\text{voltage}}$ quantifies the stability reserve in terms of voltage

P_{max} is the total MW network utilization, which includes generation and imports, in the critical state corresponding to steady-state instability

$P_{\text{base case}}$ is the MW power in the actual or base case

V_{critical} is the average system voltage in the critical case

$V_{\text{base case}}$ is the voltage in the base case

The state of maximum MW network utilization, or maximum MW loadability, that is, the system operating conditions immediately before the state of voltage collapse, is determined by alternating steady-state stability calculations that provide the verdict as to whether the system is stable or unstable, with system-stressing procedures that are aimed at "moving" the current operating point toward the steady-state stability limit in accordance with some prespecified scenarios.

These concepts were developed half a century ago and have successfully been used ever since. Initially, the recommended values for the *stability reserve*$_{power}$ indicator were 20% or more for base cases and 10% or more for postcontingency cases; the recommended values for the *stability reserve*$_{voltage}$ indicator were at least 10% for base cases and 5% for postcontingency cases. Subsequent operational experience with real-life situations suggested that the recommended minimum value, or *security margin,* for the *stability reserve*$_{power}$ can safely be relaxed to 15% for normal operating states and 8% for postcontingency states. At the present time, the official operating guidelines in Romania call for the *stability reserve*$_{power}$ to be higher or at least equal to 20% for normal operating states and 8% for postcontingency states.

Most of the steady-state stability assessment software developed in Romania since early 1960s was based on the REI methodology introduced by Paul Dimo [8,9,10]; this technique has been extensively described in the literature [18,19,22] and is addressed in detail in Appendix A of this book. Other steady-state stability analysis methods have been used as well, for example, evaluating the free term of the characteristic equation of the dynamic Jacobian in conjunction with load-flow algorithms that stress the system until it becomes steady-state unstable [1,3,5,6].

In the past, such studies were performed with off-line data that were periodically updated in order to build scenarios capable of simulating the actual operations as closely as possible. Today, as shown in Section 5.3, the steady-state stability reserve* of the SEN is computed at the DEN in real-time and is monitored continuously on standard SCADA/EMS trending charts by using the fast steady-state stability assessment software described in [2], [18], and [25]. In addition, off-line steady-state stability assessment checks with data retrieved from the state estimator are also performed by using the in-house developed program SAMI [3,4,5,6], which is briefly addressed in Section 5.4.

5.2.1.2 *Stability-Constrained Transmission Corridors*

BACKGROUND. A common scenario in the operation of interconnected power systems consists of compensating load increases and/or generation outages in a system area by raising the generation elsewhere. As shown in the following, and further discussed in Appendix C, if the system areas that participate in such power transfers are con-

---

*Currently, the steady-state stability reserve is calculated as $[(P_{critical} - P_{actual})/P_{critical}] \times 100\%]$, where $P_{critical}$ corresponds to the critical MW network utilization before voltage collapse and P_{actual} is the total MW network utilization in the base case. A 15% security margin calculated this way is approximately equal to a 20% security margin computed with formula (1).

nected through transmission corridors ("links") which, if removed, would split the network in two areas, one on each side of the link, the maximum power that can be transferred is restricted both by thermal limits and by stability constraints.

The stability limit of a transmission corridor can be quantified by the *additional* amount of power that can be sent from one side of the link to the other side, without causing instability. This indicator can be computed by means of steady-state stability analysis referred to as the *stability reserve* of the link and can be expressed either in MW or in percentage of the maximum link loading [15,16]. Under certain circumstances, the MW transfers between system areas that are interconnected this way may exceed the stability limits of the links; hence, the terminology *stability-constrained transmission corridor* or *weak link*. In order to ensure that the grid does not get too close to its stability limits, it is therefore important to evaluate the maximum transfer capability across the "links" that interconnect the areas involved in such transactions.

STABILITY-CONSTRAINED TRANSMISSION CORRIDORS IN ROMANIA. The Romanian transmission system encompasses several tightly meshed electrical areas that are interconnected through clearly identifiable transmission corridors which, if removed, would cause the system to become islanded. These electrical areas are not necessarily disjoint; some of them overlap. They are not fixed, either, and change depending upon the pattern of load, size and location of the generation reserves, scheduled and unscheduled generation and transmission outages, power flows in transmission equipment, bus voltage levels, and size and location of the reactive compensation resources. Under certain circumstances, the MW transfers between two areas may exceed the stability limits of the link; hence, the terminology stability-constrained transmission corridors or weak links, and since by removing them the network would become disconnected, they may also be referred to as "cut-sets."*

In order to better understand this concept, let us note that the populated areas and the industrial zones of Romania are aggregated in concentric areas divided by the Carpathian Mountain chain. The center area is surrounded by mountains and encompasses a dense 110 kV network sustained by a 220–400 kV backbone. Around it, there is an outer ring of major power plants that inject power into a strong 220–400–750 kV transmission system. The power flows essentially from the outer ring into the center area. The main directions of the power transfers are from south-southwest toward the center, from south-southeast toward the northeastern part of the outer ring, and from the northern part of the central area toward the northeastern part of the outer ring. The existence of stability-constrained transmission corridors in the Romanian power system is a direct consequence of this particular pattern of the MW transfers across the network.

Five stability-constrained transmission corridors or "sections" have been identified as follows and are depicted in Figure 5-6:

---

*The word used in Romanian language for this concept is *"secţiune,"* which designates, with some approximation, the mathematical concept of cut-set. Due to the difficulty inherent in the translation, we will use in the following the word "section," with the understanding that it identifies a "stability-constraint transmission corridor" or a "weak link."

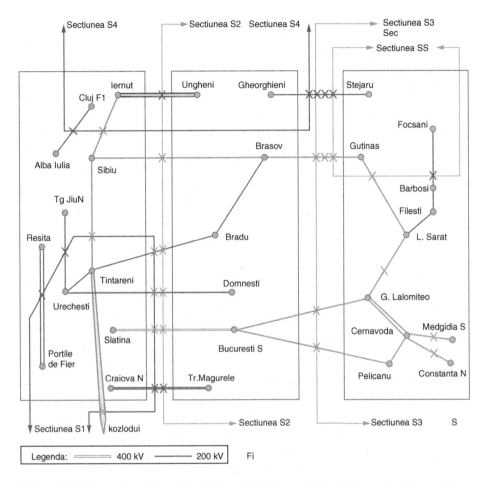

Figure 5-6. Current configuration of the stability-constrained transmission corridors ("sections") in the transmission system of Romania.

- Section S1, in Oltenia, which encompasses six 400 kV lines, including the international tie line Iron Gates–Djerdap, and three major 220 kV lines,
- Section S2, consisting of five 400 kV lines, including the international tie line Isaccea–Dobrudja, as well as two 220 kV and two major 110 kV lines,
- Section S3, across Moldova, Dobrogea, and part of Muntenia, consisting of four 400 kV, one 220 kV, and one 110 kV lines,
- Section S4, in northern Transilvania, which consist of thee 400 kV lines, including the international interconnection Oradea–Bekecsaba, and two major 220 kV lines,
- Section S5, in Moldova, encompassing four lines, two at 400 kV and two at 220 kV.

In accordance with the official operating guide PEO26/92, the minimum steady-state stability reserve requirements, or security margins, that should be met for the entire SEN and for each one of the sections are 20% with all the N power system components in service and 8% for $N - 1$ states. For the purpose of the operating guide PEO26/92, the minimum security margins are computed with formula (1).

A key operational objective is to maintain the MW transfers across each "section" below, or at most equal to, the stability reserve of the link. For this purpose, operations planning studies are performed on a daily basis to verify that these limits would not be violated. In the course of the actual system operations, the dispatcher monitors the MW loading of the "sections" and implements whatever actions are necessary to ensure that no "section" would become critical. Since the *topological configuration* of the "sections" may change due to the factors mentioned in the earlier paragraphs, the actual definition of the "sections" is reassessed twice an year, but only via offline studies.

In real life, the actual stability limits across these "sections" may differ substantially from those computed offline and need to be evaluated continuously based on the system conditions determined by the state estimator. As shown in Section 5.3, this is achieved by recalculating the stability reserve for each one of the system areas separated by "sections" after each automatic and/or on-demand execution of the state estimator. Then, the current values of the stability reserves across sections are saved in the real-time database and subsequently monitored on SCADA/EMS trending charts.

5.2.2   Transient Stability Studies

In addition to steady-state stability assessment, which is conducted both in real-time and offline, extensive transient stability studies are also performed, but only offline, to ensure that the transmission system would withstand a full range of postulated contingencies. Particular emphasis is placed on verifying the transient stability conditions in the Iron Gates area, on the Danube, and in the northeastern area of the SEN, including the international interconnections. The transient stability studies are governed by the following computational assumptions:

- For the full system configuration (N elements in service), three-phase and single-phase short circuits are simulated for each one of the transmission lines in the system.

- For degraded system configurations ($N - 1$ elements in service), three-phase short-circuit calculations are performed both for the branch (line or transformer) that was disconnected and for those branches that are sensitive to such outages. The evaluation criteria for $N - 1$ transient stability analysis include the maximum MW loading criterion; the critical "section" criterion, where the branch outages are simulated for each line that belongs in a "section"; the critical generators criterion, whereby the ability to transfer power from the main power plants into the system is evaluated during short circuits simulated for configurations with all the network elements in service.

At the present time, the transient stability calculations are not executed online, but the capability is provided at the DEN to use the offline software tools currently available with input retrieved from the real-time system, as shown in Section 5.4.2.

5.3 REAL-TIME STEADY-STATE STABILITY ASSESSMENT AND MONITORING

5.3.1 Implementation Overview

5.3.1.1 Needs and Justification
A primary concern at Transelectrica is the ability to facilitate MW transfers between parties situated beyond the geographical borders of its transmission system while meeting a broad range of operating reliability constraints. A further complication comes from the fact that the network consists of electrical areas interconnected through stability-constrained transmission corridors. Under certain conditions, the energy transactions that typically take place in this region may cause parallel flows, excessive network loadings, and low bus voltages. Accordingly, the system operation becomes quite complex and, in order to maximize the use of the transmission system while avoiding the risk of blackout, the dispatchers must meet conflicting requirements.

The analysis of blackouts due to instability that occurred in the past in various parts of the world revealed that most of them followed a similar pattern:

- Large MW blocks get transferred from areas with inexpensively priced energy toward areas where the load demand has increased due to an actual increase in load, or perhaps because one or several local generating units are scheduled for maintenance, or simply because the local generation is too expensive.
- As a result, certain transmission corridors get loaded closer and closer to their stability limits, whereas their stability reserves get smaller and smaller.
- At this moment, a generation or transmission outage takes place. Typically, such incidents evolve into cascading outages.
- Since the transmission corridor was already operating within a shrinking stability reserve, the widespread disturbance becomes unavoidable.

Therefore, in addition to conventional security monitoring based on classic network analysis applications, there is also a need to perform *split-second stability checks* and to *monitor continuously* the:

- Distance to the voltage and steady-state stability limit and the safe operating margin of the system as a whole,
- Stability reserve across the interfaces between system areas separated by stability-constrained transmission corridors.

At Transelectrica, this assessment is performed in real-time and is reinforced by monitoring the steady-state stability reserves of the entire system and of the five electric areas separated by "sections." The following sections describe the state-of-the-art solution implemented by AREVA on the DEN's SCADA/EMS to perform real-time stability calculations and to provide the results quickly and in a user-friendly format as required for fast and reliable online decision making.

5.3.1.2 Key Functional Characteristics

Given a state estimate or a solved load-flow solution of multiarea power system, the fast steady-state stability assessment tool:

- Executes steady-state stability calculations, system-wide and for each area, to determine the maximum MW network utilization and the safe system MW loading for a user-defined percent value of the stability reserve that corresponds to a safe operating state, or security margin,
- Identifies generators that may cause instability and ranks the machines and tie-line injections in order of their impact on stability, system-wide and for "sections,"
- Presents the results in graphical formats, on charts and diagrams that are easy to understand and to interpret, and on tabular displays.

These functions are executed both in real time and in study mode. The solution technique is based on the REI–Dimo methodology [8,9,18,19,22] and is described in detail in Appendix A. The application also performs the full range of fast steady-state stability computations for single and multiple contingencies, both system-wide and for areas separated by "sections." At Transelectrica, however, this capability is available only in the offline environment. The overall architecture of the real-time and off-line stability calculations supported by the SCADA/EMS is depicted in Figure 5-7.

The state estimation solution or solved power-flow case must be supplied in a comma-separated ASCII file in an industry-standard format such as Siemens Power Technologies, Inc. PSS/E; General Electric International, Inc. PSLF; IEEE Common Exchange Format; and so on. In AREVA's implementation at Transelectrica, the state estimate is provided in PSS/E format version 26. In addition, the input data include the transient, subtransient, and synchronous reactances of the generators and synchronous condensers; reactances of the step-up transformers (if the machines are shown on the high-voltage buses); and $P–Q$ Capability Curves of the generators. For the off-line implementation, the input also includes an optional list of contingency scenarios, in which each contingency may consist of single or multiple transmission, generation, shunt, and/or load events.

5.3.2 Tight Integration of the Stability Tool with the SCADA/EMS

AREVA has tightly integrated the computational engine of the steady-state stability assessment program with the real-time network analysis sequence of Transelectrica's

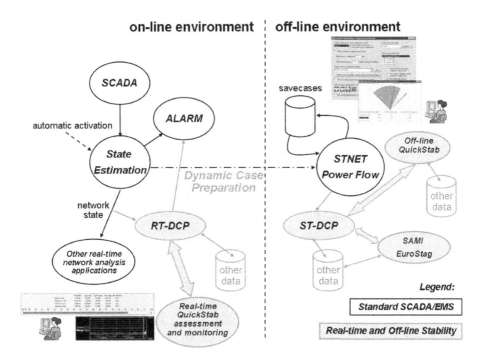

Figure 5-7. Real-time and off-line steady-state stability tools at the DEN.

SCADA/EMS. After a successful state estimation run, a snapshot file in PSS/E format is created and, together with a dynamic-generator data file, is written in six separate working directories—one directory for the entire system and one directory each for the five "sections." Due to the special bus-naming and numbering conventions used by AREVA in the SCADA/EMS solution at the DEN, a special algorithm was implemented to match the generators identified in the dynamic generator data file with the electrical nodes created automatically at each execution of the real-time network analysis sequence. This procedure, which also recognizes the fact that the boundaries of the electrical areas separated by "sections" are not fixed, provides for assigning correctly the generators to the electrical nodes and in the areas to which they actually belong.

Immediately after the real-time data have been exported in the working directories, six steady-state stability calculations are triggered simultaneously, one for the entire system and one for each "section" (Figure 5-8). The results are stored in the same working directories in which the input had been placed and are made available for display. If any of the stability reserves thus calculated is smaller than a user-defined threshold, a "QuickStab alarm" is triggered. This is shown in Figure 5-8.

In addition, the values of the stability reserves, for the system and for each "section", respectively, are written in the real-time database as calculated values. This makes it possible to track the distance to instability in real-time and to monitor it on

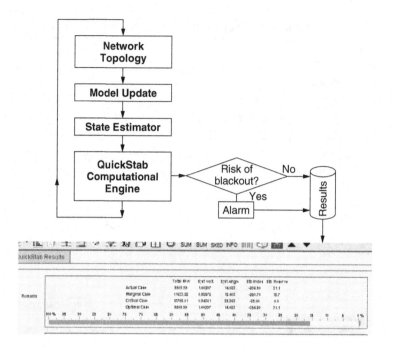

<u>Figure 5-8.</u> Seamless integration of the fast steady-state stability computational en-
gine with the SCADA/EMS real-time network analysis sequence.

standard SCADA/EMS trending charts. This and other aspects of the user interface are
discussed in the next section.

5.3.3 User Interface, Performance, and Operational Experience

The relevant information is summarized both in a results table and on linear speedometers that provide a graphical indication of the current stability reserves system-wide and
for each "section" (Figure 5-9). Furthermore, after each analysis an alarm is sounded if
the stability reserve percentage is below a user-specified danger level. It must be emphasized that the real-time displays of the fast steady-state stability assessment application have been designed and implemented by AREVA directly in the native
SCADA/EMS user interface. This approach offers the following significant benefits:

- The dispatcher can use the application without having to access a different workstation and/or a different user-interface environment.
- Standard, vendor-provided SCADA trending facilities can easily be used to
  monitor the evaluation of the stability reserves side by side with the linear
  speedometers, as shown in Figure 5-10.
- The key output results of the real-time stability calculations can be shown on the
  video projection system, as illustrated in Figure 5-11.

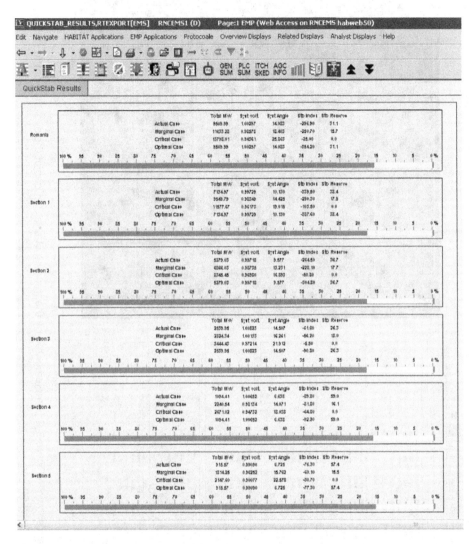

<u>Figure 5-9.</u> Results of real-time steady-state stability calculations displayed on linear speedometers for the entire system (top) and for each "section."

This, in addition to the linear speedometers that depict the stability reserves and the trending charts that allow monitoring the distance to instability as it changes in real-time, allows the dispatchers to anticipate critical system conditions that may lead to instability.

The fast steady-state stability software is very fast. The application performance was tested offline on a standard laptop with a 3382 bus, 4064 line Brazilian system by running nonstop a suite of base and contingency cases. The average solution time, including *both* the postcontingency load flow, which uses a full AC Newton–Raphson

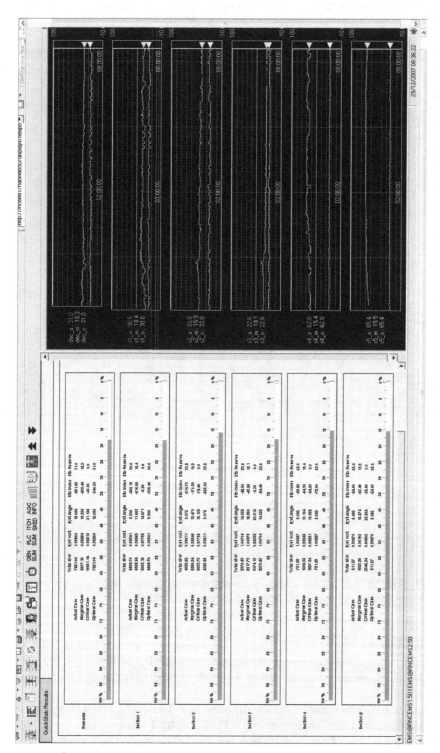

Figure 5-10. Linear speedometers shown in Figure 5-9 complemented by the trending charts that depict the evolution of the stability reserves system-wide and for each "section."

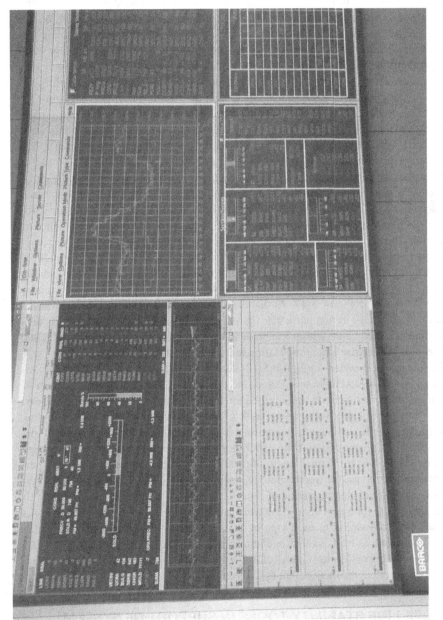

Figure 5-11. Results of real-time steady-state stability calculations displayed on linear speedometer charts for the entire Romanian system (top speedometer) and for each one of the five "sections."

algorithm, *and* the steady-state stability computation times, was approximately 1.5 seconds per case. Off-line runs conducted on a 1500 bus system representing a subset of the Southeastern European interconnection converged in less than a second. The program performance was not benchmarked on the real-time system at the DEN, but the execution time was estimated at approximately one second for performing six sets of steady-state stability calculations, one for the entire system (approximately 1000 buses) and one for each "section."

The most relevant aspect, however, is the fact that the fast steady-state stability application is executed after each state estimate and the results are displayed almost instantly. This computational speed provides sufficient time for the dispatcher to implement the most appropriate remedial action if the operating conditions have deteriorated. An interesting example is offered by the unscheduled outage of a 710 MW major generating unit that occurred in March 2006.

When the 710 MW machine tripped, representing approximately 10% of the entire installed MW capacity in Romania, the system entered a prealarm state, but the capacity reserve existing at that time and the additional help from interconnections were sufficient to compensate for such a significant loss of production. Although the system state changed drastically, the classical procedure that controls the transfers across "sections" within prespecified MW limits did not catch the event, but the shape of the stability monitoring trending charts instantly sensed the reduction of the steady-state stability limits both for the system and across "Section 4," which dropped below the security margins, as shown in Figure 5-12. At that time, the system dispatcher concluded that a second $(N-2)$ outage at that particular moment would be damaging, temporarily suspended the "market," and, within less than fifteen minutes, manually requested a 400 MW generation increase in a hydro station located in a "sink" area of the system.

The linear speedometers in Figure 5-12 correspond to the moment immediately after the additional 400 MW were injected into the system. The system and "Section 4" pointers in the trending charts (highlighted on the right-hand side of the picture) clearly show how the stability was getting back to normal. The time span of the trending charts in Figure 5-12 was set at fifteen minutes (1/4 of an hour) for the purpose of displaying the most recent system state which, for practical purposes, depicts the current operating conditions. If the time length of these curves would have been twelve or perhaps twenty-four hours, the trend would have allowed reviewing the quality of the services over a relatively longer period of time.

The time cycle for a new state estimation is three minutes, but the state estimator is automatically activated after each topology change, new AGC state, and manual dispatcher command, as well as other operational criteria. Due to this granularity, the stability-trending curves follow very closely the system operations and the linear speedometers reflect the most recent system state.

5.4   OFF-LINE STABILITY TOOLS IN SUPPORT OF SYSTEM OPERATIONS

The fast steady-state stability assessment application that runs in real-time as described in Section 5.3 is also available for off-line studies. In addition, the DEN deployed var-

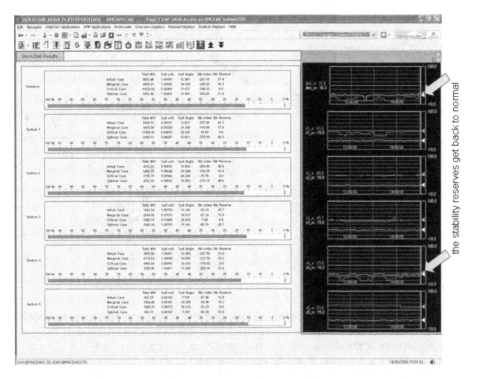

<u>Figure 5-12.</u> The trending charts (right side of the picture) show that the stability re-
serves for "Section 4" and for the entire system have were *smaller* than the prescribed
values when the 710 MW unit tripped unexpectedly. As soon as the dispatcher implemented remedial action, the stability reserves went back to normal, as shown on the
linear speedometers.

ious other stability tools such as EuroStag, from Tractebel, and SAMI, which is an in-
house developed steady-state stability program.

The off-line stability-assessment process in support of operations planning at the
DEN has been shown schematically in Figure 5-7. The data flow between the real-time
and the offline environments is illustrated in Figure 5-13.

5.4.1 SAMI

5.4.1.1 Application Overview

SOLUTION TECHNIQUE OVERVIEW. The software SAMI, which is the Romanian
acronym for Aperiodic Stability by Integration Method, uses an Integral Steady-State
Stability Margin (ISSM) indicator [3,5,6] to check the steady-state stability of a power
system that is subjected to gradual changes of the parameters that define the operating
state. The method belongs to the family of techniques that evaluate the sign of the last
term of the characteristic equation, which is the dynamic Jacobian determinant. A

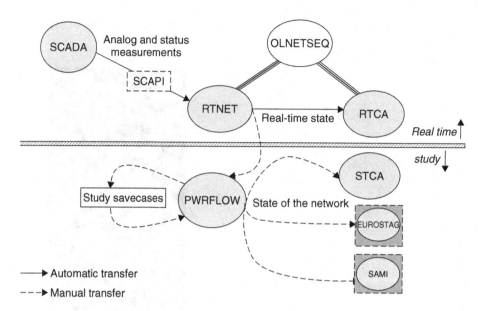

<u>Figure 5-13.</u> Data flow between the real-time and the offline computational environments at the DEN. Abbreviations: OLNETSEQ = on line network sequence, RTNET = real-time network applications, RTCA = real-time contingency analysis, STCA = study-mode contingency analysis, EUROSTAG = transient stability analysis (by Tractbel), SAMI = steady-state stability analysis, as described in Section 5.4.1.

change of sign from positive to negative with further loading of the system indicates aperiodic instability [20]. Although the instability in the form of self-oscillations remains unrevealed by this method, the hypothesis of aperiodic displacement of the power system state variables when exceeding the stability limit proves to be realistic unless negative damping is introduced by generator excitation systems that are inadequately tuned.

METHODOLOGY. The methodology consists of alternating the computation of the ISSM indicator with load-flow calculations that aim at stressing the system until it becomes unstable. The power flow routine was specifically designed to circumvent numerical instability when the system is approaching a critical state. As illustrated in Section 5.4.1.3, the procedure for stressing the operating conditions consists of either raising the generation in areas with large MW generation reserves and, simultaneously, reducing the generation in areas where there is a MW generation capacity deficit; or gradually increasing the nodal injected power; or modifying the network topology. By performing stability checks on a sequence of successively degraded system states, the technique provides for determining the distance between the current state and a hypothetical state in which the voltages may collapse or units may loose synchronism.

The following simplifying assumptions have been adopted in order to reduce the computation burden and the data requirements:

- Neither the network parameters nor the loads vary with the system frequency. This hypothesis is conservative because, in real life, the imbalance caused by a disturbance is compensated by the variation of the system frequency.
- The internal resistances of the synchronous machines are neglected. The generators are represented in terms of steady-state power equations with which the automatic voltage regulators (AVR) control the voltage until the unit reaches the excitation limit; beyond that point, the machine model is replaced with a constant electromotive force behind the synchronous reactance.

One way to deteriorate the system state and move the operating conditions from the initial steady state toward an unstable, or critical, state is to redispatch the MW generation to:

- Further increase the MW deficit on one side of the "section" and, simultaneously, raise the MW generation on the other side (please refer to Section 5.4.1.3 for additional details).
- Gradually increase the real and reactive generated powers and the real and reactive loads.

During this system-stressing procedure, the reactive generated powers are modified by changing the emfs of the machines. The real power changes are compensated by increasing the output of a number of prespecified generators. The reactive power changes are compensated for both by the intervention of the system generators and by the static characteristics of the loads. The power mismatches due to load reductions caused by lower voltages are compensated for at the swing bus.

The other way to stress the system operating conditions is to outage certain transmission elements, such as major lines and transformers, while maintaining unchanged the structure of the loads. These topology changes result in modifications of the power flows and the bus voltages in the network and, also, may produce load reductions due to the static characteristics of the loads.

5.4.1.2 Mode of Execution
The SAMI calculations are executed:

- Periodically (twice a year), as part of the operational planning studies performed at DEN when the normal operating conditions are defined for the summer and winter seasons as well as for the morning and evening peaks. The computations are performed for N and $N-1$ scenarios for the "sections" and the areas separated by "sections" and, in some cases, for $N-2$ scenarios.
- Occasionally, when permanent scheduled outages become imminent, for example, due to removal of transmission elements as a result of technical refurbishing.

In both cases, the steady-state stability calculations performed with SAMI are backed up with transient stability simulations.

5.4.1.3   Using Calculation Results Obtained with SAMI

If on one side of the "section" there is an area with a MW deficit, this deficit is amplified by disconnecting additional generators and by increasing proportionally the bus loads until the steady-state stability limit has been reached, thus identifying the maximum MW transfer across that "section" that is needed to compensate for the area's MW deficit.

It must be emphasized that the critical states in areas separated by "sections" identify the maximum MW grid utilization of those areas, rather than maximum transfer capacity across "sections." Accordingly, the transfer capacities across "sections" are computed for operating conditions *within those areas* with 20% steady-state stability reserve when N transmission components are in service and 8% stability reserve for $N-1$ sates.

Then the MW transfers across sections are determined such that: (a) the bus voltages in the entire interconnected system (SEN) are within the limits required by the Grid Operation Code, and (b) there are no thermal violations. Since the MW transfers across "sections" have already been determined for the 8% (contingency states) and 20% (normal states) scenarios, the next step is to determine the minimum MW transfers that accommodate both the required steady-state stability reserves and the voltage and thermal limits. The same procedure is used to identify the most conservative limits for states corresponding to permanent scheduled outages.

Finally, the limits thus computed are introduced in the SCADA/EMS and are monitored online by the system dispatchers, who then are responsible to implement operational decisions, for example, network switching operations and generation rescheduling, which may be needed to maintain and enhance the operating reliability. These procedures are followed both for areas where there are MW deficits (sinks) and for those where there is excess of generation (sources).

In all these cases, the network model used in computations takes into account the international interconnections by maintaining constant the MW generation schedules outside the Romanian borders and by allowing the MVAr generation to vary within the Q_{max} and Q_{min} limits of the machines in order to provide reactive support across the tie lines if needed.

5.4.2   Other Tools

In addition to SAMI and the off-line version of the fast steady-state assessment application that runs in real-time, DEN also uses the EuroStag software to conduct off-line stability studies both with input originating in the SCADA/EMS and with postulated data. The loose integration of EuroStag with the SCADA/EMS is based on a standard solution developed by AREVA, as shown schematically in Figure 5-14.

EuroStag uses three main input files as follows:

1. *Initial State of the Electrical Network.* This file contains the state of the network (voltage, branch flows, topology, etc.) as well as the main parameters of the

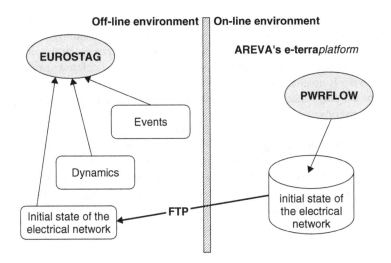

Figure 5-14. Loose integration of EuroStag's data environment with the SCADA/EMS.

transmission system equipments (line and transformer resistances, reactances, etc.).

2. *Dynamics.* This file contains the dynamic data of the machines, such as generator and turbine models, voltage regulators, protections, and so on.

3. *Events.* This file describes the simulation scenarios.

The Initial State of the Electrical Network file is created automatically, as well as upon operator request, from the results of the State Estimator. The other two files, Dynamics and Events, are created and modified via the EuroStag native interface.

5.5 CONCLUSIONS AND OUTLOOK

This paper described the architecture, functionality and implementation of the real-time stability-assessment and monitoring software that is currently used in the National Dispatch Center of Transelectrica. The application has been seamlessly integrated by AREVA with the Real-Time Network Analysis sequence and runs automatically after each successful state estimate.

The real-time stability software tool is extremely fast and allows performing six sets of steady-state stability calculations within the state-estimation cycle, for the entire interconnected system of Romania and for the five electrical areas separated by stability-constrained transmission corridors or "sections." The "sections" are identified via off-line studies and are dynamically reconfigurable. The key results are posted on intuitive displays, including industry-unique real-time stability-trending charts, which were developed by AREVA directly in the native SCADA/EMS user interface.

This approach to visualizing the computational results in a user-friendly manner allows the operator to continuously monitor the evolution of the stability reserve of the interconnected system as well as across critical network interfaces. However, the real-time stability assessment and monitoring is just the first line of defense against dangerous states in the Romanian transmission system. In addition, detailed off-line stability calculations are performed at DEN on a daily basis by using both real-time input and postulated scenarios.

The experimental results obtained to date have been very good. The real-time stability tool described in this chapter is used twenty-four hours a day, seven days a week in system dispatching and has also been adopted in short-term operations planning, together with the other off-line stability analysis programs deployed at the DEN. Plans are currently under way to enhance the balancing market system with a fast steady-state and voltage stability assessment capability that would quickly identify market-clearing MW schedules that may pose a risk of blackout due to instability.

5.6 REFERENCES

[1]   Arie, E., Pomarleanu, M., Bejuscu, L, and Botgros, M., "Determinarea Rezervelor de Stabilitate Statică a Sistemelor Electroenergetice Complexe," *St. Cerc. Energ. Electr.*, Vol. 23, No. 3, pp. 709–727, 1973.

[2]   Campeanu, H. S., L'Helguen, E., Assef, Y., and Vidal, N., "Real-Time Stability Monitoring at Transelectrica," Paper No. 1288 presented at the Real-Time Stability Applications in Modern SCADA/EMS Panel, IEEE Power Systems Conference and Exposition 2006, Atlanta, GA, October 29–November 2, 2006.

[3]   Constantinescu, J., "Practical Assessment of the Power System Stability Margins," *Rev. Roum. Sc. Tech., Série Elect. et Energ.*, Vol. 39, No. 2, pp. 217–222, 1994.

[4]   Constantinescu, J., "Study of the Transient Processes in Large-Scale Power Systems," *Rev. Roum. Sc. Tech., Série Elect. et Energ.*, Vol. 27, No. 2, pp. 211–227, 1982.

[5]   Constantinescu, J., and Aldea, C., "Sistem Operational pentru Supravegherea in Timp Real a Sigurantei Functionării Regimului SEN," *Energetica*, Vol. 43, No. 4, pp. 175–180, 1995.

[6]   Constantinescu, J., and Homos, M., "Metodˇa Practicˇa de Calcul Pentru Studiul Stabilităţii Statice a Sistemelor Electroenergetice Complexe," *Energetica*, Vol. 25, No. 9, pp. 328–336, 1977.

[7]   Crary, S. B., *Power System Stability*, General Electric Series, Schenectady, New York, 1945, Third Printing October 1955.

[8]   Dimo, Paul, "Etude de la Stabilité Statique et du Réglage de Tension," *R.G.E.*, Paris, Vol. 70, No. 11, pp. 552–556, 1961.

[9]   Dimo, Paul, *Nodal Analysis of Power Systems*, Abacus Press, Kent, England, 1975.

[10]   Dimo, P., Manolescu, G., Iordanesscu, I., Groza, L., Ionescu, S., Albert, H., Moraite, G., and Ungureanu, B., *Computation and Design of Electrical Energy Systems* (Romanian edition), Editura Tehnica, Bucharest, Romania, pp. 200–201, 1971.

[11]   Erwin, S. R., Oatts, M. L., and Savulescu, S. C., "Predicting Steady-State Instability," *IEEE Computer Applications in Power*, July, pp. 15–22, 1994.

[12]   Ionescu, S., and Ungureanu, B., "The Dual Power States and Voltage Collapse Phenomena," *Rev. Roum. Sc. Tech., Série Elect. et Energ.,* Vol. 26, No. 4, pp. 545–562.

[13]   Ionescu, S., "Theoretical Analysis of Algebraic Criteria for Voltage and Steady-State Stability Assessment," personal communication to Savu C. Savulescu, manuscript in Romanian language, Bucharest, Romania, February 9, 1994.

[14]   Kundur, P., *Power System Stability and Control,* McGraw-Hill, New York, 1994.

[15]   Pomarleanu, M., "The Selection and Ranking of Power System Dangerous Cuts," in *Symposium on Bulk Power System Dynamics and Control, IV–Restructuring,* Santorini, pp. 24–28, 1998.

[16]   Pomarleanu, M., and Bejuscu, L., "Selectarea Operativă a Sec,tiunilor Unui Sistem Electroenergetic cu Rezervele de Stabilitate Static̆a Cele Mai Mici," *Energetica,* Vol. 47, No. 2, 1999.

[17]   Sauer, W. P., and Pai, M. A., "Relationships between Power System Dynamic Equilibrium, Load-Flow, and Operating Point Stability," in *Real Time Stability in Power Systems,* pp. 1–30, Springer Verlag, Norwell, MA, 2006.

[18]   Savulescu, S. C., "Fast Assessment of the Distance to Instability. Theory and Implementation," in *Real Time Stability in Power Systems,* pp. 31–64, Springer Verlag, Norwell, MA, 2006.

[19]   Savulescu, S. C., Oatts, M. L., Pruitt, J. G., Williamson, F., and Adapa, R., "Fast Steady-State Stability Assessment for Real-Time and Operations Planning," *IEEE Transactions on Power Systems,* Vol. 8 T-PWRS, No. 4, pp. 1557–1569, 1993.

[20]   Venikov, V. A. Transient Processes in Electrical Power Systems, Edited by V. A. Stroyev, English Translation, MIR Publishers, Moscow, 1977.

[21]   Vergara, J. S., Thai, T. A., Cuong, N. D., Nam, N. T., Campeanu, H. S., and Savulescu, S. C., "Accuracy Testing and Real-Time Implementation of Dimo's Stability Analysis Technique," in *Real Time Stability in Power Systems,* pp. 65–92, Springer Verlag, Norwell, MA, 2006.

[22]   EPRI, "Power System Steady-State Stability Monitor Prototype," Final Report EPRI TR-100799, July 1992; and "Power System Steady-State Stability Monitor," Final Report EPRI TR-103169, December 1993.

[23]   IEEE PES, *Techniques for Power System Stability Search,* A Special Publication of the Power System Dynamic Performance Committee of the IEEE PES, TP-138-0, 1999.

[24]   IEEE PES Task Force on Terms and Definitions, "Proposed Terms and Definitions for Power System Stability," *IEEE Transactions on PAS,* Vol. PAS-101, No. 7, 1982.

[25]   http://www.quickstab.com.

6

IMPLEMENTATION OF ONLINE DYNAMIC SECURITY ASSESSMENT AT SOUTHERN COMPANY

Kip Morison, Lei Wang, Fred Howell, James Viikinsalo, and Alan Martin

6.1 INTRODUCTION

6.1.1 Problem Statement

All power system operators strive to operate their systems with a high degree of reliability. Reliability refers to the probability of a power system's satisfactory operation over the long run. It denotes the ability to supply adequate electricity service on a nearly continuous basis, with few interruptions over an extended time period [1]. This adequacy may be measured by the frequency, duration, and magnitude of adverse effects on the electricity supply or interruptions of service to customers. Although power systems may suffer occasional disruptions and still be considered reliable (provided the impacts of such disruptions are not too great), the key to reliable operation is to maintain satisfactory security at all times.

Unlike reliability, which is measured as performance over a period of time, security refers to the degree of risk in a power system's ability to survive imminent disturbances (contingencies) without interruption to customer service at any instant of time. It relates to robustness of the system to imminent disturbances and, hence, depends on the system operating condition as well as the contingent probability of disturbances. For a given system operating state, the security of a power system can be assessed by simulating potential disturbances, such as faults and/or loss of facilities, and determining if the disturbances will result in any adverse impacts that could cause unsafe conditions, customer interruptions, or equipment damage. These adverse effects may be the result of thermal overloading of elements, unacceptable voltage or frequency excursions, or loss of stability.

It is well known, and illustrated by the number of blackouts that have occurred in recent years, that modern power systems are often less secure than systems of the past. The reasons are numerous and include the reduced investment in transmission infrastructure over the last two decades, the proliferation of "economically sited" independent power producers, unusual power transfers driven by market activities, the use of complex controls and special protection systems (SPS),* and a general lack of system-wide oversight regarding reliable planning and operation. These problems cannot be solved quickly or without significant economic investments and as a result, operators facing increasing security challenges are looking for innovative solutions to improve system operations in a timely and affordable manner.

One such solution is the use of online dynamic security assessment (DSA) technology, which takes a snapshot of the system condition, performs comprehensive security assessment in near real time, and provides the operators with warnings of abnormal situations as well as remedial measure recommendations. The pressing need for such tools combined with the ever-increasing power of low-cost computers has, in recent years, lead to a significant number of implementations of online DSA applications in power systems worldwide. Online DSA systems are in use or in the implementation phases in countries including Canada, the United States, Brazil, Ireland, Greece, Italy, Portugal, Finland, Australia, New Zealand, Malaysia, Japan, and China [2,3].

The conceptual requirements and components of practical online DSA systems have been described in detail in the literature [4,6] but little has been written about the practical issues of actual implementation of such systems. Now that significant experience has been gained in online DSA installations, it is possible to detail the steps required and challenges faced in the specification, construction, installation, testing, and operation of these state-of-the-art software tools. It is intended that this overview, illustrated with the example of on-line transient security assessment at the control center of the Southern Company Services, Inc. (Southern Company) can assist power system operators and operation support engineers in the deployment of these advanced technologies, which are expected to become critical real-time applications in power system control centers to ensure system security and reliability.

6.1.2   Need for Online Dynamic Security Assessment at Southern Company

Southern Company is a superregional energy company with more than 40,000 megawatts of electric generating capacity in the southeastern United States (Figure 6-1). Southern Company supplies energy to a 120,000-square-mile service territory

---

*Although it may be true that special protection schemes (SPS) have prevented blackouts in other parts of the world, the Eastern Interconnection in North America has suffered some blackouts from failure of SPS.

Figure 6-1. Southern Company's service territory.

spanning most of Georgia and Alabama, southeastern Mississippi, and the panhandle region of Florida. Approximately 53 tie lines interconnect the Southern Company grid to neighboring utilities in the southeastern United States.

The desire to implement transient stability assessment on a real-time basis at Southern Company is driven by several factors. Reliability concerns arise when an increasing number of merchant generating plants interconnect to the transmission system, and opt not to procure the long-term transmission service necessary to solve various thermal and/or stability concerns which may exist from time to time as a result of this new generation. Transmission service may then be requested and offered on a shorter-term basis, usually as a result of a daily evaluation. Given the ever-changing system conditions encountered in real-time system operations, it simply is not feasible to study every conceivable system snapshot in the planning environment. Therefore, the risk exists that transmission service may be granted without realizing that a potential stability limit could be violated. Conversely, there may be occasions when the planning models and any resultant limits identified may be too conservative in the determination of a stability limit. Whatever the case, a greater degree of certainty should be derived from a transient security assessment based upon an accurate real-time state estimator power flow model. For all the reasons stated above, a goal was established to implement real-time transient stability assessment on at least an hourly study cycle in Southern Company's system control center.

Major challenges encountered along the way toward implementation of this real-time stability assessment included software procurement and system modeling issues. Southern Company needed to evaluate, procure, and install a transient stability package that would be capable of simulating critical contingencies and determining limits on an hourly basis. On a parallel track, Southern Company had to develop a process to build reduced dynamics-ready models that would provide accurate results, yet achieve substantial computer run-time savings. At the time this project was first initiated at Southern Company in 2003, very few utilities were actively pursuing the implementation of a real-time stability application. With limited commercial software availability because of a small user base, evaluation of different software packages was often times an "apples to oranges" comparison, since different packages tend to focus on different features or functions. After considerable evaluation, Southern Company ultimately purchased the Transient Security Assessment Tool (TSAT) application from Powertech Labs Inc. (Powertech) in December of 2003.

The primary objective of this chapter is to summarize the experience acquired by Southern Company throughout this complex yet successfully conducted project. In order to place this effort in the proper context, an in-depth discussion of many of the online DSA implementation concerns are also identified, covering the application scope, solution methods, main functionalities, and benefits. The chapter concludes with a brief overview of the full array of the DSA capabilities provided by Powertech.

6.2   DSA IMPLEMENTATION FUNDAMENTALS

6.2.1   Background

Power systems, and the manner in which they are operated, vary widely in nature depending on factors such as

- System size,
- Topological structure, for example, either predominately radial or highly meshed networks,
- Types and distributions of generation and load,
- Market activities,
- Interconnections,
- Operating criteria, including stability, security, and reliability requirements
- Types of controls and SPS,
- Exposure to disturbances due to weather or other factors, and
- Operation experience, especially high probability of occurrences of certain types of security problems.

As a result, an online DSA system must be specified and implemented in a manner that will ensure the specific requirements of a given system are met. The factors listed above dictate specifications such as the phenomena that need to be studied, the time in which a DSA cycle must be completed, the size of the models required, and the output provided to the operators. The basic implementation process can be partitioned into several steps as shown in Figure 6-2. Each step is described in some detail in the sections that follow.

6.2.2 DSA Objectives, Requirements, and Project Steps

The starting point for an online DSA project is to define its application objectives. This step is perhaps the most important as it effectively determines the overall capability and functionality of the system. The application objectives involve several aspects as described below.

6.2.2.1 Target Problems
Depending on the factors mentioned earlier a power system may exhibit different types of security problems in different operation modes. It is, therefore, ultimately important to clearly identify the problems that are security concerns and to address them by application of an appropriate online DSA system. For example, any of the following problems may be included in an online DSA project:

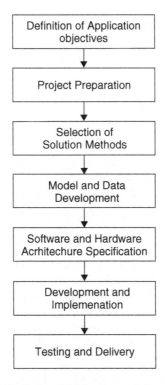

Figure 6-2. Online DSA implementation process.

- Thermal overloading,
- Voltage decline/rise,
- Frequency decline/rise,
- Voltage stability,
- Transient stability,
- Low-frequency oscillations,
- Stability limits,
- Determination of preventative and corrective control measures, and
- Determination of SPS settings

In addition to the online analysis capabilities, it is often as important for an online DSA system to provide adequate options for offline studies using data captured online as well as archived data. Offline analysis may include studies related to:

- Operational planning,
- Model calibration and validation,
- System restoration, and
- Postmortem analysis.

Consistency between the online analysis and offline studies in terms of modeling, analysis techniques, and results is considered essential to ensure success with the on-line applications; such consistency is necessary for operators to gain confidence in the online analyses.

6.2.2.2 Performance Requirements
One of the key differences between online and offline applications is the performance requirements. This difference is primarily related to:

- *Model size and detail.* The question of what size and detail of an online model is required to ensure acceptable results is critical. Consideration must be given to the availability and quality of the model. Experience has shown that model and data assembly is no doubt the most challenging task in the implementation of an online DSA and often represents the greatest hurdle to success. This modeling issue will be further addressed later in this chapter.
- *Speed.* The analysis speed of online DSA systems has been a research focus for many years. While pursuing the fastest analysis speed is a natural motivation, it is often necessary to balance analysis speed with other requirements. Comple-tion of an entire analysis cycle within 10 to 30 minutes is widely considered as a reasonable target. In some situations, analysis speed will have to be coordinated with other requirements such as market operation schedule or SPS arming and operation requirements.
- *Processing power.* The question is how much computer processing power is re-quired for a given online model. This power is a function of the scope of the

problems that need to be addressed in each DSA analysis cycle. The scope is defined, for example, in terms of the number of contingencies to be examined, the number of power transfers to be scheduled, the number of SPS settings to be determined, the types of phenomena to be analyzed, and so on.

6.2.2.3 System Reliability Requirements

With DSA to be implemented as an online system, reliability of the software/hardware system is an important consideration that impacts many aspects of the system design, such as hardware and software architecture, redundancy, failover, and system administrative security. The reliability requirements of an online DSA system are established based on the proposed use of the system, which can be categorized into the following classes ranked from the lowest to the highest reliability:

- *Occasional-Use Application.* A real-time tool primarily used by operational support engineers to provide information to system operators.
- *Continuous-Use Application.* A real-time tool available 27 hours a day, seven days a week, providing an important continuous function to system operators for security monitoring of the system. The online DSA system is required to operate continuously with minimum downtime.
- *Mission-Critical Application.* A real-time tool that is an essential component of the SCADA/EMS system; used in such a manner that unavailability would be considered as exposure of the power system to serious potential risks. This would, for example, be the case for systems with known stability problems or systems that depend on the online DSA to provide remedial action recommendations or even direct arming of SPS.

Most of the online DSA systems installed to date are considered occasional-use or continuous-use applications, mainly because the system operators previously had no similar function to assess security. However, as the technology matures, more online DSA systems are likely to provide integral control functions as part of the EMS, and therefore, become mission-critical applications.

An issue related to reliability is maintainability of an online DSA system. Such a system usually includes complex analytical tools, large data volumes, and significant computer network and hardware structures. Software and hardware components must be readily maintainable, by available personnel, and with minimum downtime of the DSA system.

6.2.2.4 Users of the System

A clear understanding of who will use the online DSA system can help in the system design and implementation. The term "user" includes those who directly control and operate the DSA applications, those who access and utilize the results, and those who may have authorization to view limited output. Users may include:

- System dispatchers/operators,
- EMS support engineers,

- Operational engineers,
- Planning engineers,
- Engineers in neighboring utilities or within the regional security council, and
- Others, including research engineers, senior management staff, and so on.

Another type of use of an online DSA system is a direct connection to control devices in the system, such as an SPS. In this capacity, online DSA forms a closed-loop control function to provide real-time computed settings to arm and trigger an SPS operation.

The major consideration for the users of the DSA system is its interface and custom functionality. The interface must be comprehensive to meet all security assessment and visualization requirements, yet simple enough to use and understand (particularly for users not specializing in power system dynamics). Custom functionality is critical if an online DSA system is to be interfaced directly with SPS.

6.2.2.5 Project Preparation

As in all commercial-grade software projects, an online DSA implementation requires careful planning and preparation. Main considerations in this stage are discussed here.

STATE ESTIMATOR. Operation of an online DSA system is based on the real-time system condition as acquired by the SCADA system and processed into a system power flow model by the State Estimator (SE). Therefore, a working and reliable state estimator solution is a must before an online DSA system can be implemented. Obviously, quality of the power flow solution provided by the SE is of primary importance; other factors also play significant roles, as will be discussed later.

NETWORK AND DYNAMIC MODELS. Often, the power flow provided by the SE may not include sufficient network details required for DSA. A typical situation is the detail in the distribution network modeling, which could be important, for example, for voltage stability analysis. Another challenging issue is the modeling of the external system not observable by the SE. Inclusion of these models in the real-time system models may require the development of adequate offline equivalent models, which can then be merged with the real-time SE models.

Dynamic models are a unique requirement for performing dynamic analysis in DSA and include models for generators and their associated controls, loads, HVDC links, static VAr compensators (SVC), special protection schemes, and any other devices with relevant dynamic characteristics that may impact the type of phenomena being analyzed. Although these models are normally available for off-line planning studies, the differences between offline and online analysis usually requires significant effort to incorporate these models for real-time applications. These models must be maintained properly to reflect the characteristics of the actual devices in the current system (unlike planning models, which may reflect some future state) and they must be properly matched to the network power flow model produced by the state estimator.

OTHER NECESSARY DATA AND INFORMATION. An online DSA system may require other data and information for various types of analysis, including:

- *Contingencies.* This data is required for all types of security analysis. Contingencies may be defined in SCADA/EMS for static contingency analysis, but their format is in most cases insufficient for DSA. For example, a contingency for transient-security assessment usually includes a fault applied in the network followed by the tripping of transmission elements due to relay actions to clear the fault. Since most SCADA/EMS advanced application software would not include dynamic assessment, information regarding the fault type, fault impedance, and clearing times would not be available. In more advanced online DSA applications, remedial actions, such as generator, load, and shunt tripping, may be applied to mimic a real-time switching sequence and these protective actions require comprehensive modeling of contingencies. All relevant contingency scenarios need to be developed for use in the online DSA.

- *Power transfers.* One of the key applications of an online DSA is to determine stability limits. Generally, a stability limit is found by maneuvering the system through an increasing (or decreasing) power transfer, applying contingencies at each power transfer level to determine the point at which the system reaches instability (or stability). The power transfer can be defined in terms of a source and sink combination; for example, generation increased in the source and load increased in the sink. The stability limit can be expressed in a number of ways, in terms of the transfer level, the interface flow, or the output of a key power plant. The proper definition of the stability limits should be consistent with operation planning definitions and criteria. To find the stability limits, a power-flow dispatch scheme must be developed for each stability limit problem so as to allow the software to iteratively find the point at which stability is lost. Often, a set of relevant contingencies (as a subset of the contingencies defined for the entire system) may be assigned to a particular stability-limit problem.

- *Monitored quantities.* The results from online DSA need to be properly displayed for the operators and engineers to understand the security status of the system. The results may range from simple swing curves, P–V curves, interface flows at limiting points, to more aggregated security indices. This range of available information leads to a need to carefully define the output results for display. In some situations, rule-based monitoring can be used; a quantity is displayed only when certain system performance criteria are met.

6.2.2.6 *Knowledge and Experience from Off-line Studies*
Knowledge and experience gained from off-line studies play a major role in ensuring a successful online implementation. After all, the objective is essentially to move the full or core capability of off-line studies into the near-real-time environment using selective and innovative computational methods. Knowledge and experience of the system often play an irreplaceable role in the entire online DSA project and operational planners should contribute significantly to the specification and implementation details of the project.

6.2.2.7 Knowledgeable and Dedicated Technical Staff

Last but not least, a team of knowledgeable and dedicated technical staff at the utility side must be assembled to kick start an online DSA project. The team should:

- Understand the project objectives completely,
- Have expertise with the existing SCADA/EMS functions,
- Be familiar with the related models,
- Have extensive offline study experience, and
- Be knowledgeable on other aspects related to the online DSA implementation.

6.2.3 Selection of Solution Methods

The selection of solution methods for the identified security problems is in most cases straightforward, based on technical understanding of the phenomena of interest and knowledge of available methods. In general, the following issues should be considered:

- As a universal rule for solving problems, the simpler a method, the better it usually performs. One should resist being overly intrigued by fancy mathematical formulations and long lists of academic references associated with many complex, but practically unproven, methods. On the other hand, use of overly simplified methods may produce inaccurate and unreliable results, and, therefore, should be avoided.
- The method selected must be capable of achieving the application objectives. For example, if contingency analysis is required, a measurement-based method would be insufficient.
- If possible, use same or similar methods for both online and offline analysis. This selection provides obvious advantages when cross checking online results with offline tools.
- The method must have manageable data requirements.

Other questions to be raised when selecting the solution method include:

- Is the method straightforward to implement?
- Is the method sufficiently flexible to handle new device models or other possible future additions?
- Will the method be easy to maintain?
- Does the method provide all required features?
- Is the method easy to understand and use?
- What are the training requirements?

6.2.4 Model and Data Development

When detailed security analyses are to be performed for online DSA, high-quality models of the interconnected system are needed. In fact, since all analyses are dependent on the quality of the system model, model development determines, to a large degree, the usefulness of the online DSA.

6.2.4.1 Data Quality

For obvious reasons, data quality is important to ensure the validity of online DSA results. In addition to the conventional data quality (such as accuracy of the SE results, reliability of SE solution, etc.), there are some unique issues that need to be looked into for an online DSA system. From experience, these tend to be mostly related to the generator data in the power flow model, such as:

- Reactive power output and the associated reactive power capability,
- Active power capability (maximum and minimum power ratings),
- Generator power factor,
- Generator MVA base,
- Handling of the step-up transformer,
- Source impedance, and
- Merging or splitting of generators in the same plant.

Other common data problems include incorrect switched shunt models and incorrect circuit/transformer thermal ratings.

6.2.4.2 External System Equivalencing

With the exception of small islanded power systems, SCADA systems typically provide limited observability of an entire power system. The resulting model produced by the state estimator will, therefore, have a detailed representation of a specific portion of the interconnected system (usually corresponding to a control area), and no model or a heavily equivalenced model for the "external system." In general, the external system, or an acceptable equivalent thereof, has to be included in the system model for online DSA. This external system is required since stability of the study area is often significantly impacted by the dynamics of the neighboring systems. Inclusion of the external system in the real-time models may involve use of special techniques such as dynamic reduction [5], in which equivalent models for the external system are created offline and merged with the online model produced by the state estimator. Several external system models can be created, each representing a specific system condition, and the appropriate equivalent merged with the SE model as required. At Southern Company, one base model of the entire Eastern Interconnection is used for the initial dynamic reduction. This base model is a lighter-load case, which produces more conservative results when used for transient stability evaluations. The resulting reference equivalent is then attached to the real-time state estimator case and balanced to achieve actual system tie flows, as described in Section 6.3.1.2. Ultimately, as data communications improve, it is possible that ever-expanding state estimator models can be developed in near real-time, thereby reducing the need for external equivalents. However, it should be noted that for performance considerations, dynamic reductions, if they can be readily determined, serve to obtain accurate security computations.

6.2.4.3 Data Matching

Data matching is another unique issue in an online DSA implementation. It refers to matching the real-time power flow with other data (dynamics, contingencies, transfers, etc.). This matching is often a problem because the nodal power flow (which has been converted from a bus-breaker model) may not have fixed bus numbers (or names). The challenge of data matching is twofold:

- Dynamic devices in power flow should be matched with correct dynamic models. This process is required for all dynamic studies. It is essential that all major dynamic devices be properly represented in time-domain simulations, otherwise the results may be erroneous, or numerical difficulties may be encountered in the simulations.
- Network elements in all real-time data should match each other (in terms of bus numbers and IDs, for example). Matching these elements may be achieved by using one of two possible approaches:
  1. All network elements (buses, circuits, generators, etc.) that may appear in real-time data are given unique identifications in term of a specific bus numbering scheme. Then, when a real-time power flow is obtained, the buses in the system are renumbered according to this numbering scheme.
  2. The bus numbers in the real-time power flow are used as the "masters" and bus numbers in all other data are mapped to this master through the use of a translation process.

6.2.4.4 Data Exchange

This refers to the exchange of real-time data from SCADA/EMS to the online DSA server and vice versa. It is essentially an implementation issue, but certain details are worth some extra attention here:

- *Data format.* In general, simple and common text formats are sufficient and preferred for speed and easy implementation. An added benefit of using such an approach is that it facilitates model and data verification, since most of the models and data used for online DSA analysis can be traced to those used for off-line studies in which text formats are the standard. More sophisticated methods involving databases or universal formats such the Common Information Model (CIM) have yet to prove their value in such an application due mainly to the fact that CIM definitions for dynamic devices are not mature.
- *Data exchange protocol.* This method is used to ensure that the data can be exchanged securely and reliably. Possible methods range from simple file copy (FTP) to the use of a dedicated message bus. Because of the relatively small amount of data to be transferred (even for very large systems, a complete set of data for one real-time case is on the order of several megabytes), experience has shown that there is no particular approach that is significantly superior to others.
- *Coordination between SCADA/EMS cycle and DSA cycle.* The issue here is twofold:

1. It is necessary to ensure that the DSA cycle starts with the latest available system snapshot.
2. It is necessary to ensure that if a critical system snapshot is available and needs to be assessed, DSA can abort the current case and start the critical case immediately.

6.2.4.5 Bad Data Detection and Correction

No matter how much validation, debugging, and testing is done for the real-time models and data, there will be unusual data ("bad data") passed to the online DSA engine. Unlike off-line analysis, human intervention cannot be used in online systems to detect and correct data errors. Therefore, when setting up an online system, data and models must be heavily scrutinized, and automatic bad data detection and correction functions put in place, to ensure that the system can be initialized and the analysis runs robustly and reliably. Some of the common issues include:

- *Power flow data correction.* Example: handling of dynamic SVC model interfaced with a generator that has nonzero active power
- *Dynamic model correction.* Example: elimination of mismatches in turbine rating and generator active power rating, resulting in governor limit violations
- *Other.* Example: handling of unstable small (trivial) islands isolated by circuit tripping in contingencies

6.2.5 Software and Hardware Architecture Specification

There are many issues related to the hardware and software architecture that need to be resolved. The following outlines some of the key ones.

6.2.5.1 Reliability

Reliability in this context refers to software and hardware reliability, which is important for any online systems. In addition to developing reliable code for software, the hardware (including task management workstation, computational servers, data server, and communication network) reliability is also of concern. One of the ways to address this concern is to use redundancy. Recovery from hardware failure requires software enhancements to provide automatic failover switching when necessary.

6.2.5.2 Portability

Portability of the various DSA software components is largely a maintenance issue for both developers and end users. Since operation of online DSA can be loosely coupled with SCADA/EMS (mainly through a simple data connection), a "plug-and-play" type of application is a natural result. All that is required is an appropriate data interface with any specific EMS.

6.2.5.3 Scalability

It is conceivable that a power system expands constantly, and the computational tasks required for an online DSA system may also be increased, depending on user demand. For example, more critical contingencies and/or transfers need to be assessed as the system expands or the system conditions change. This expansion requires that the computational capabilities of the online DSA system be scalable, without major changes to the software and hardware architecture. One practical approach is to use a multiserver architecture with the computational servers configurable by software. Using this approach, increasing computation power of the DSA system can be done easily by simply adding computational servers and performing appropriate configuration.

6.2.5.4   Security
Security in this context refers to software security, which has become increasingly important for software system deployment. For online DSA systems, software security is mainly concerned with two issues:

1. *Authorization to configure the system.* The ability to modify the DSA configurations (such as data exchange protocol definition, computation cycling, computation options and parameters, result broadcasting, etc.) should be restricted only to the authorized system administrator.
2. *Authorization to access analysis results.* The DSA results should be made available only to those who are authorized. This process is usually managed by the system administrator.

6.2.5.5   Redundancy
Redundancy is mainly used to improve the DSA system reliability. It also has the role of providing additional computation power when needed, for example, during the stressed system conditions, for which the amount of computation tends to increase within an analysis cycle.

6.2.5.6   Accessibility
The main users of an online DSA system are the system dispatchers and operators; therefore, accessing DSA results in the control room is usually the first priority. It is common that other users, such as operational planning engineers, planning engineers, or even authorized staff in remote offices, would like to have access to the DSA results. To accommodate such requirements, DSA results must be made available to remote users, using approaches such as secured Web access, LAN/WAN broadcasting, and so on.

6.2.5.7   Data Storage and Case Archiving
One of the common requirements for an online DSA system is input data storage and computation case archiving. Archiving capability often turns out to be valuable when analyzing a historical event or recent incident. The capacity to store at least one month of data is considered as a reasonable threshold. The key here is the convenience to browse and use the archived data, which must be organized in a way that is easy to understand and easy to use (for example, for loading into the offline study tools).

6.2.5.8 Presentation of Results

One of the most important aspects of online DSA software is the manner in which results are displayed. Although an endless variety of display types are possible (some examples were given in [4]), experience has shown that operators wish to have a very clear and simple high-level display to present the critical results of a given DSA cycle. System operators are constantly presented with a great deal of information from a variety of SCADA/EMS applications and, therefore, it is important not to overwhelm the user with excessively detailed or nonsalient information. The results must be filtered to ensure that all critical information reaches the operator and that the facilities are included for the operator to readily access details by "drilling down" through the main display. An example of a main display for an online VSA and TSA application is shown in Figure 6-3.

This display shows:

- The voltage and transient security ("secure" or "insecure") of the current system condition.
- The stability limits for two specified power transfers. The display also shows which criterion is restricting the limit (i.e., voltage decline, voltage stability, transient stability, etc.),

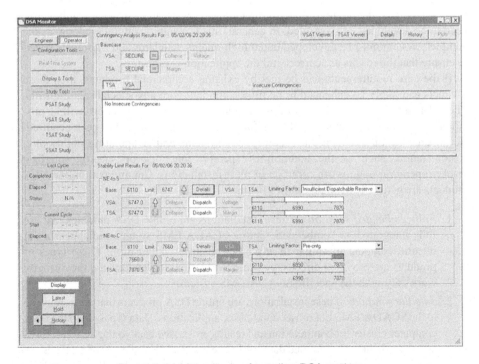

Figure 6-3. Main display for online DSA system.

- Critical contingencies,
- Time elapsed for the current DSA computation cycle, and
- Elapsed time and status for the last DSA computation cycle.

The operator also can click to view details such as plots or histograms. Access is also provided to all off-line study tools through this console.

6.2.5.9 Offline Study Capability
Offline study capability is an important part of an online DSA system. It is primarily used to:

- Examine the system conditions for a real-time case,
- Verify real-time models and data,
- Perform what-if type scenario analysis for critical system conditions,
- Perform postmortem analysis, and
- Troubleshoot the problems shown in online operation.

6.2.6   Development and Implementation

Once the preceding design tasks have been completed, the online DSA system can be developed and implemented. If the core requirements cannot be met by available commercial products, the required systems, including software and hardware components, need to be developed or assembled. In general, this is a significant undertaking, as the complete system must be developed under high-quality standards and rigorously tested to ensure that the results are correct and the application is robust.

If the core requirements can be met by existing commercial products, then these products can likely be enhanced to be fully compliant with any special requirements. The risks associated with this approach are quite low as the base functions will be provided by the validated commercial tools for which field experience has been acquired. Because of the unique nature of many power systems and/or operating philosophy, customization, such as the addition of specific functions or specific interface display/operation options are common in most installations. Commercial online DSA products are generally available in two forms:

1. *Fully embedded solutions.* These installations are online DSA modules that are fully integrated, or "embedded," into a SCADA/EMS system and appear as an integral component similar to any other advanced application. These types of products are provided directly by some SCADA/EMS vendors and they usually work only with the specific SCADA/EMS system.
2. *Satellite solutions.* These installations are online DSA products that are connected to the SCADA/EMS but run as stand-alone applications. Data is passed to the DSA computer cluster for computation and results are passed back to the SCADA/EMS as required, through a defined data-exchange protocol. These types of products are provided by third parties other than SCADA/EMS vendors. Unlike fully embed-

ded systems, satellite solutions have the "plug-and-play" portability advantage. They have minimum connections to the SCADA/EMS and, therefore, can be integrated with any SCADA/EMS, provided that the required data is available.

6.2.7 Testing and Delivery

Testing and delivery is the last step in the implementation of a real-time system. Successful completion of this step involves several stages:

- A quality system must be adopted, most commonly with quality controls based on ISO 9000 standards. Central to the quality system is the compilation of a detailed and workable testing plan and the associated testing procedure.
- It is desirable to use effective testing tools. In addition to the conventional testing tools, specialized software such as a real-time data emulator can be very useful. Features can be included in such software to generate either correct or faulty data so as to complete various testing tasks. This testing is conducted in the developer's shop and is commonly referred to as factory acceptance testing or FAT. The software must be shown to meet the necessary functional requirements and operate with acceptable reliability and accuracy.
- After the system has been successfully installed and tested in the control center (a process commonly referred to as site acceptance testing or SAT), an additional reasonable trial operation period of a few months is usually planned to have the online DSA system exposed to the true operation environment prior to using it in the production environment. This trial period helps identify unexpected operation situations and data problems that cannot be otherwise discovered in the factory or site acceptance testing.
- It is essential that users of the online DSA system (operators/dispatchers, SCADA/EMS support engineers, planning engineers, etc.) get involved in the testing and delivery process, so that they can contribute to the final tuning of the system so as to fully meet the application requirements.

6.3 TRANSIENT SECURITY ASSESSMENT IMPLEMENTATION AT SOUTHERN COMPANY

This section describes the practical implementation of online transient security assessment at Southern Company as well as some operational experiences encountered.

6.3.1 Model Development

6.3.1.1 Modeling the Internal System

The Southern Company state estimator network model used for steady-state power flow analysis applications has insufficient internal machine detail required for transient stability analysis. Because of limited measurement availability or other reasons, combined cycle units and hydro units are often modeled as a single generator. In order to provide sufficient detail for transient analysis, expanded internal generator representation is accomplished by mapping any aggregated machine output to an assumed explicit machine output as derived from more detailed planning models. The first step in creating a transient stability base case, therefore, is to internally expand the latest state estimate and to solve the expanded case.

6.3.1.2 Modeling the External System

The external equivalent portion of the Southern state estimator network model is also inadequate for performing transient stability analysis, again primarily because of insufficient machine detail. A replacement external equivalent is generated from a reference planning base case that has a custom-built, dynamically reduced external model. The second step in creating a transient stability base case is to remove the external equivalent from the state estimate and to attach the dynamic equivalent. In order to balance the dynamic equivalent to the state estimate, first the boundary buses of the state estimate are temporarily treated as swing buses, with voltage magnitude and angle set to state-estimate values.

The transient base case is iteratively solved, distributing boundary swing mismatches to external buses using distribution factors based on short-circuit impedance from the swing to those external buses. This methodical distribution of the balancing injections helps to maintain a relative voltage profile in the equivalent external system that is similar to its preinjection state, which contributes to similarly robust performance during the power flow solution. Once convergence on each boundary swing is within some tolerance (i.e., 50 megawatts), the remaining mismatch is added as load to the boundary bus. The bus types of the boundary buses are then restored and the base case is finally solved using a standard Newton–Raphson iterative technique.

One unique aspect of this enhanced state estimator model is a highly reduced external system, which provides a virtually identical dynamic response as the much larger original from which it was derived. An empirical process was developed to build this special dynamics-ready external equivalent, which included use of a small-signal analysis program to identify the interarea modes of oscillation that should be retained. After extensive experimentation, a set of parameters relating to mode shape and unit participation was successfully developed to identify the critical generator buses that must be retained in order to successfully preserve the dynamic behavior of the much larger original model. For any interarea modes in which Southern Company units either contribute to or are impacted by the oscillations, all generators participating above certain thresholds are explicitly retained. A study area including the Southern control area and a few buses into the neighboring systems have also been preserved, along with the 345 kV and above network for the entire Eastern Interconnection. The EPRI DYNRED program was then used to identify coherent machines using the weak links method, and subsequently to aggregate any nonretained coherent generators into a single classical machine model.

6.3.1.3 Benchmarking the Results

A comparison of time-domain simulation results from the base case and the reduced case is shown in Figure 6-4. The solid line represents an observed power swing in the

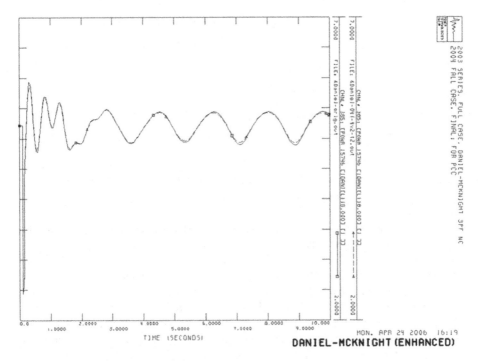

Figure 6-4. Power swing in base case and reduced case.

original base case, and the dashed line represents the same generator in the heavily reduced equivalent case. A severe disturbance consisting of a critical stabilizer out of service as the base condition, followed by a normally cleared three-phase fault on a 500 kV tie line at the edge of Southern Company's control area, was selected as one of the benchmark contingencies.

As illustrated in the previous figure, the dynamic behavior of the much larger original model is quite accurately captured in the custom-built, dynamically reduced model, and the results are obtained with only 15% of the CPU time required for the original case. The initial reduction yielded a CPU time of 30% of the original case. The reduction also eliminated some numerical instability, whereby the reduced model can now be simulated using a half-cycle, rather than a quarter-cycle, time step. A summary comparison of the two models is shown in Table 6-1. The CPU time represents a total number of seconds for the entire set of benchmark contingencies evaluated.

6.3.1.4 Preparations for Dynamic Simulation

Base-case preparation is performed in the SCADA/EMS environment on Unix servers. Once created, the transient stability base case is transferred to the DSA windows server for analysis. To allow machine dynamics data and contingency definitions to be defined on the DSA server, a static bus-numbering convention is used. The final step in creating the transient stability base case is to renumber the buses

Table 6-1. Size comparison of base and reduced case

	Original	Reduced	% of Original
Buses	43,138	7,977	18.5%
Branches	57,740	16,950	29.5%
Machines	7,019	2,548	36.3%
D-States	296,086	111,254	37.6%
CPU time (sec)	9,285	1,405	15.1%

and then to convert the case to PSS/E format. Once the case creation process is completed, the base case is transferred to the TSAT server, which is further described in Section 6.3.2.

When the transient case is initialized in TSAT, a number of checks are performed to ensure the quality of the simulations in the presence of spurious state estimation or extreme conditions. These checks include updating governor limits to reflect the measured unit output, netting out units with load when a steady-state negative field current is calculated, and adjusting the rating of devices to ensure a reasonable initialization. All these checks may be enabled or disabled while the system is online.

The contingency list is generated each cycle using generic criteria applied to the current power flow and dynamic data. At the Southern installation, these generic criteria specify three-phase faults on branches emanating from generator high voltage buses in the Southern area. The only input requirements are the area identifier for Southern, the minimum MVA rating of generators for a contingency, and the fault clearance times. Other contingency types are available as well, such as unit tripping or common tower events. By using generic criteria, the contingency list adapts to the current configuration of the system and ensures coverage throughout the Southern area.

6.3.2   System Architecture

Southern's online TSA system is based on two pieces of software from Powertech's DSATools™ package, TSAT and DSA Manager.

The TSAT (Transient Security Assessment Tool) is the computation engine that actually performs TSA. TSAT uses time-domain simulations as the core technology from which a number of TSA functions are built, including swing-based stability margin calculation, Prony analysis for damping determination, a versatile stability limit computation module. Special features for online applications are also included, such as early termination of simulations and automatic detection and correction of bad data in real-time cases.

The DSA Manager is the interfacing module for online operation. It provides three main functions: (1) data connection between SCADA/EMS and TSAT, (2) online system configurations, and (3) computation monitoring and result visualization. Shown in Figure 6-5 is the conceptual overview of Southern's SCADA/EMS system. Figure 6-6 shows the online TSA system architecture.

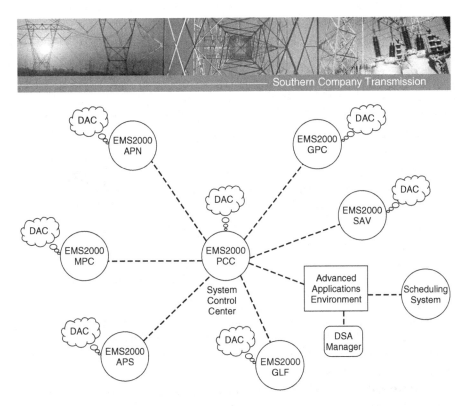

Figure 6-5. Conceptual overview of Southern Company Services' SCADA/EMS. The "EMS2000 PCC" bubble refers to the Power Coordination Center SCADA/EMS. The other bubbles, such as "EMS2000 APS," refer to the SCADA/EMS of the operating companies.

The TSA software (TSAT and DSA Manager) processes the cases primarily on the DSA Client (a workstation with dual 2.4 GHz Xeon CPU), with options to include multiple servers so as to distribute computation scenarios. Users can access TSA results on the DSA Client as well as on any workstations in the network on which the DSA monitoring software is installed.

The initial phase of the TSA project implemented at Southern in June 2005 has the following capabilities and features:

- The real-time system snapshots are created on an hourly basis.
- For each system snapshot, roughly 250 contingencies are processed to evaluate transient security of the system. This study cycle takes about 30 minutes without optional computation servers.
- Security assessment results, including insecure contingencies, stability margin information, swing curves, and so on, can be examined on any workstations properly configured by operators, operation support engineers, and planners.

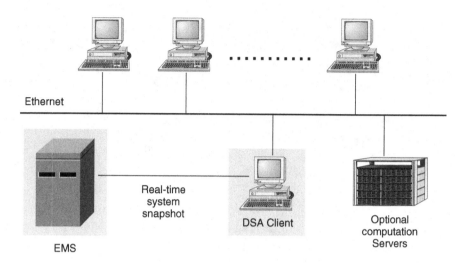

Figure 6-6. Southern's on-line TSA architecture.

- Real-time cases of up to one month are archived in a form that can be directly opened in TSAT. These cases can be readily used for various off-line scenario studies.

6.3.3 Examples and Operation Experiences

Southern Company has already derived significant benefit from incorporating key results of this real-time stability assessment into the reliability decision-making process. Following Hurricane Katrina in late August and early September of 2005, hourly real-time studies revealed problems with damping and transient voltage recovery in Mississippi and Alabama. Additional studies were quickly performed offline to determine safe generation limits in the Gulfport/Biloxi (Mississippi), Mobile (Alabama), and Pensacola (Florida) areas as the transmission system was being restored. On other occasions during periods of heavy transmission system maintenance, certain other generators were identified as contributors to stability limit violations, and appropriate system adjustments were then made to alleviate the reliability concern.

Although the initial online TSA implementation can presently be made available to Southern Company transmission system operators, results are presently monitored by an operations support engineer on an as-needed basis. Deployment to real-time floor operations is being deferred until the next phase of development (as described in the next section) is complete, and a suitable operator training program has been developed.

6.3.4 Future Development Plan

Recognizing the benefits that the initial implementation of the online TSA system bring, Southern and Powertech are working together on the next phase of the development which will include the following:

- Enhancing the selection of contingencies to be evaluated.
- Increasing the processing power; multiple servers will be used to distribute computation scenarios.
- Improving the system reliability; a dual client structure will be used to allow better failover protection. This is already a standard feature in DSA Manager.
- Including advanced analysis features; under consideration are stability limit computation and preventive control measure determination, which are also standard features of TSAT and DSA Manager.
- Including other security assessment options; voltage security assessment using Powertech's VSAT software will be added.
- Integrating summary results of the stability studies into Southern's SCADA/EMS alarms processing subsystem.

6.4 CONCLUSIONS

The implementation of online DSA systems is growing worldwide and the deployment of this advanced technology is expected to improve the real-time security and, hence, the reliability of power systems. Though not insignificant, the cost and effort required to install online DSA tools are minor compared to the benefits of reducing the volume of offline studies required and, more importantly, the benefits of identifying and avoiding potential security problems in the systems so as to reduce the risk of blackouts.

Based on practical experience, a process of DSA system integration is presented that can assist utilities and grid operators in addressing key issues during the specification, development, and installation of such tools. An online DSA project implemented at Southern Company Services is discussed to illustrate the viability and practicality of such applications, even for large complex power systems. The penetration of online DSA tools is expected to continue to grow as operators seek timely and cost-effective approaches to enhance system performance. In the meantime, work is continuing on new methods of online analysis, advanced preventive and corrective control tools, and improved hardware architectures.

6.5 REFERENCES

[1] IEEE/CIGRE Joint Task Force on Stability Terms and Definitions, "Definition and Classification of Power System Stability," *IEEE Transactions on Power Systems,* Vol. 19, No. 2, pp. 1387–1401, August 2004.

[2] CIGRE WG C4.6.01 Special Task Force publication on "Review of On-line Power System Security Assessment Tools and Techniques," January 2007.

[3] V. Vittal, P. W. Sauer, and A. P. Meliopoulos, "On-line Transient Stability Analysis Scoping Study," PSERC Final Report, 2005, http://www.pserc.org/cgi-pserc/getbig/publicatio/reports/2005report/ vittal_pserc_report_s-21_2005.pdf.

[4] K. Morison, L. Wang, and P. Kundur, "Power System Security Assessment," *IEEE PES Power and Energy Magazine,* September/October 2004.

[5]   J. Viikinsalo, A. Martin, K. Morison, L. Wang, and F. Howell, "Transient Security Assessment in Real-time at Southern Company," paper presented at the Panel Session of the IEEE PSCE Conference, Atlanta, Georgia (USA), October 2006.

[6]   Task Force on Blackout Experience, Mitigation, and Role of New Technologies, "Best Practices to Improve Power System Dynamic Performance and Reduce the Risk of Cascading Blackouts," IEEE-PES Power System Dynamic Performance Committee, 2006.

[7]   G. Andersson, P. Donalek, R. Farmer, N. Hatziargyriou, I. Kamwa, P. Kundur, N. Martins, J. Paserba, P. Pourbeik, J. Sanchez-Gasca, R. Schulz, A. Stankovic, C. Taylor, and V. Vittal, "Causes of the 2003 Major Grid Blackouts in North America and Europe, and Recommended Means to Improve System Dynamic Performance," *IEEE Transactions on Power Systems*, Vol. 20, No. 4, November, 2005.

ANNEX 6-1.   FURTHER DETAILS OF THE DSA SOFTWARE AND HARDWARE ARCHITECTURE

As noted in the introduction, online DSA systems have been implemented in many locations worldwide and the penetration of this technology is growing steadily as system operators recognize the potential benefits. The types of systems currently installed include tools for transient, voltage, and small-signal security, and the methods of analysis used range from measurement based systems to direct index methods to full simulation tools. Readers are referred to [2] as well as [3] for detailed descriptions of these various installations together with discussions on operating experience and ongoing research. To illustrate the practical applications of the online DSA technology, this section describes the implementation of one DSA software solution that is in use for online DSA in a number of major power grids.

Description of the Core DSA Software

The DSATools are leading-edge analysis tools designed for both online and off-line dynamic security assessment of power systems. The software is ideally suited for online use because of its modeling capabilities, computation features, speed, and high degree of automation for analysis tasks. All tools in the DSATools suite are able to assess base-case security (determination of the security for a given system condition under a specified range of contingencies) and are also able to perform transfer analysis in which the system is automatically stressed until security limits are located for all specified contingencies and transfers. The transparent use of distributed processing makes the software performance completely scalable, an essential feature for online use when large volumes of computations are required in relatively short cycle times. The main components of the DSATools suite include:

- *Voltage Security Assessment Tool (VSAT).* This is a power-flow-based tool designed for the steady-state analysis of voltage security. Its basic methods of analysis include P–V curves and modal analysis of the power-flow Jacobian. The software is able to assess system performance based on criteria related to voltage stability, voltage decline/rise, thermal overloads, and reactive power reserves.

- *Transient Security Assessment Tool (TSAT)*. This is a simulation tool designed for the comprehensive assessment of transient security. The fundamental methods of analysis include nonlinear time-domain simulation and transient security index computations. The software is able to assess system performance based on criteria related to transient stability, system damping, transient voltage and frequency excursions, and relay margins.

- *Small-Signal Analysis Tool (SSAT)*. This is a linear, dynamic simulation tool designed for the analysis of small-signal stability using eigenvalue analysis. SSAT has a myriad of analysis functions, primarily for off-line planning studies. However, some key features, such as the capability to compute small-signal stability indices for a given system condition and a set of contingencies are particularly useful for online application.

- *DSA Manager*. This software provides a single-point interface between the SCADA/EMS system and the DSATools online security-assessment applications. It includes an easy-to-use environment for configuring and controlling online DSA as well as full graphical capabilities for viewing results.

- *DSA Monitor*. This is the "view-only" version of the DSA Manager. It permits remote authorized users to view the DSA results over a network.

Online DSA Implementation Using DSATools

Figures 6-7 and 6-8 show typical hardware and software architectures for online DSA implementation using the DSATools VSAT and TSAT software.

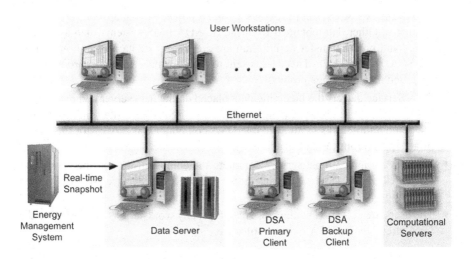

Figure 6-7. Online DSA hardware architecture.

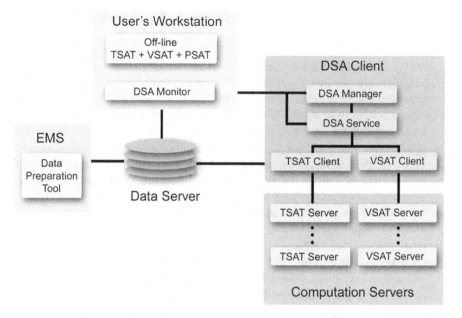

Figure 6-8. Online DSA software architecture.

The online system works as follows:

- The hardware and software systems are configured using DSA Manager, which resides on the DSA client machines. A separate module, DSA Service (part of DSA Manager), runs on the DSA client machine to manage all DSA activities.
- The SCADA/EMS system provides a solved power-flow snapshot via the state estimator. Other data obtained from the SCADA/EMS system (such as contingency and power transfer definitions) are created by a data preparation tool on the SCADA/EMS side. This data is sent to the DSA data server according to the specified data exchange protocol.
- DSA Service detects the incoming data placed on the data server and picks it up. DSA Service then creates the necessary VSAT/TSAT case files and distributes the computation tasks to all available DSA computational servers for the appropriate applications (VSAT and TSAT).
- The DSA computational servers complete the computations and the results are sent back to the DSA client and processed by DSA Service for display and archiving.
- The users are able to view the DSA results on any workstations networked to the DSA client via the DSA monitor software.

7

ONLINE SECURITY ASSESSMENT FOR THE BRAZILIAN SYSTEM—A DETAILED MODELING APPROACH

Jorge L. Jardim

7.1 INTRODUCTION

7.1.1 Problem Statement

Security assessment is one of the fundamental processes in the expansion and operational planning of power systems. Such assessments are based on simulation results that quantify power system security when subject to major disturbances. These computationally intensive simulations are traditionally performed offline and their results are subjected to expert engineering analysis. This process is performed for purposes such as upgrading the grid, planning outage schedules, validating economic dispatches, and imposing operational constraints. It is intrinsically an interactive process in which various load level and network configuration scenarios are simulated. Depending on the purpose and extent of the analysis, these studies can take anything from hours to months.

In the case of power system operational planning it is not possible to analyze all, or even most, of the operational states that the power system undergoes in real time, even when there is some degree of process automation. Because of uncertainties in equipment status, load levels, and generation dispatches, the number of scenarios resulting from combinations of different hypotheses can be huge, even for a modest-sized power system model. The problem can then become intractable due to time, computational, and human constraints. It is then necessary to rely on good engineering practice and system-related expertise to determine safe operational rules. A common approach to this is "worst-case scenario" analysis, which tends to give conservative results and noneconomic operation.

7.1.2 Need for Online Dynamic Security Assessment

An attractive way of approaching the "security versus economy" dilemma is to assess security online. In this case, the complexity of the problem is drastically reduced. When the system state can be measured and estimated with good accuracy, most of the uncertainties regarding load level and location, generation schedules, and network topology disappear. The effectiveness of this approach clearly depends upon good quality modeling and measurement of the power system, which can pose implementation problems. Nevertheless, as recommended for decades [5–8], it remains true that online security assessment offers the benefit of security without unnecessarily sacrificing economic operation. This approach has been adopted in Brazil in the form of the Steady-State and Dynamic Security Assessment (SDSA) system (also known by its commercial name, Organon) that is the subject of the present chapter. This SDSA is currently operational at the Brazilian National System Operator (ONS—Operador do Sistema Elétrico).

Power system security is estimated on the basis of steady-state and dynamic performance criteria. Real-time security must, of course, be assessed at the current or immediate-future operating point. But in addition, recognizing that excessive transfers can induce voltage collapse and, possibly, dynamic instability [18], the online assessment of secure transfer capacity is operationally important, in both the steady-state and dynamic senses.

In the case of steady-state security, voltages and power flows are required to remain within operational and equipment limits, even when subjected to credible contingencies (contingencies with significant probability of occurring). Typically these limits represent acceptable voltage level ranges, thermal capacities of transmission system components (transmission lines, transformers, and series capacitors), and maximum transfer capacities (loadability) of the transmission system. Thermal and voltage limit violations can damage equipment, and protection system actions can provoke cascading effects that eventually result in partial or total blackouts.

For many years, SCADA/EMS systems have performed online power-flow-based contingency analysis. This is the steady-state security assessment of the current or near-future system operating state for voltage and thermal limit violations. Using techniques such as linearized screening, this contingency analysis is typically very fast, and it can be distributed between CPUs when necessary. The analysis is also repeatedly performed within online security-constrained economic dispatch—a critical function in the implementation of power markets in North America and elsewhere [17]. Contingency analysis is performed even more repetitively within steady-state transfer capacity assessment, which, although not a standard SCADA/EMS function for reasons of both theory/technique and computing expense, is included in the present SDSA system.

Beyond steady-state security assessment, a further major technological step is dynamic security assessment. Secure dynamic behavior of a power system means that, when subject to a disturbance, the generators will remain in synchronism, power swings will be "quickly" damped, voltages will not violate transient limits, and the frequency will remain "near" the specified value. Monitoring all these factors requires extensive time-domain simulation. However, such time simulations are computationally much more expensive than power-flow calculations. In addition, recognizing model and parameter uncertainties, it is necessary to operate at a safe distance from instability limits. The calculation of such limits requires further simulations.

The computational complexity of dynamic security assessment is much higher than that of its steady-state counterpart. For many years, this prevented the implementation of dynamic assessment in real-time power system operation—sufficient computational power was not available at reasonable cost to meet online performance requirements. At the same time, these difficulties served to motivate the development of various methods (such as those outlined here) for fast power system instability detection and stability margin estimation. As a result, in combination with contingency screening techniques and high-performance computing, efficient online security dynamic assessment is today practical.

7.1.3 Brief Description of the Brazilian System

The installed capacity of the Brazilian power system in late 2007 was around 88,000 MW, of which 84% was hydro and the rest thermal (coal, oil, nuclear, and gas). The transmission system as shown in Figure 7-1 has 80,000 km of lines at 230 kV and above.

The peak load is around 57 GW. The system is owned by various utilities and is operated by the Operador Nacional do Sistema Elétrico (ONS). The system is connected to Paraguay and Argentina through HVDC links of approximately 6000 MW and 2000 MW, respectively. Many of the power plants, in particular the hydro ones, are located far from the load centers. Generation dispatch is mostly driven by optimization of the reservoirs, which can lead to significantly different power transfers across the network, depending on the water resources.

7.1.4 Operating Reliability Concerns at ONS

7.1.4.1 Transmission Security Issues
Within the past decades, the system has suffered partial and total blackouts. Recent transmission upgrades have been able to improve system security and keep pace with the increasing load levels, but security issues related to angle and voltage stability and voltage collapse are still a concern. In addition, operational planning has become more complex after the interconnection of the North–Northeast regions to the South–Southeast regions. Accordingly, online stability assessment and transmission-security enhancement at ONS became an important objective and led to the implementation of SDSA in system operations.

Although the primary role of the SDSA is in real time, it is also valuable in critical operational planning processes such as outage scheduling. The number of outage requests for the entire Brazilian bulk transmission network can be significant, and it is

Figure 7-1. Brazilian power system transmission backbone.

difficult to evaluate them rapidly enough using conventional simulation tools. Because of its very efficient simulation methods, process automation and parallel computation, the SDSA can speed up such a process by several orders of magnitude. That is, many more such operational planning cases can be run within the allotted time.

7.1.4.2 Underlying SCADA/EMS Architecture

The power system model for operational planning studies comprises approximately 4000 buses and 5600 branches. The data needed to represent, supervise, and control this extensive system are gathered in real time from the various Brazilian utilities. This information is concentrated in four regional centers and is then retransmitted to the main control center, as shown in Figure 7-2.

All centers perform their own topology processing, state estimation, and steady-state analysis. However, as a superset of the regional models, the model available at

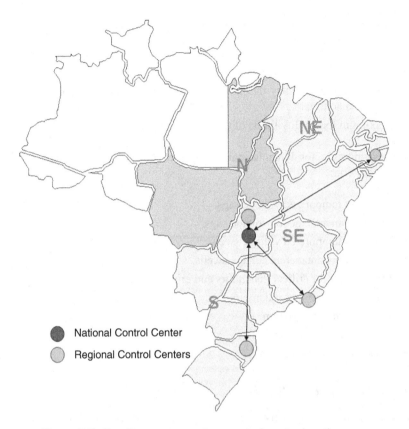

Figure 7-2. Brazilian power system control center locations.

the main control center is currently more suitable for dynamic security analysis. This includes a supervised network (mainly 230 kV and above) of approximately 1800 buses. Ongoing installations of new remote terminal units (RTUs) will increase this figure in the near future.

The North, Northeast, and South Regional control centers have AREVA SCADA/EMS systems. The Southeast control center runs the SOL technology developed by Furnas Centrais Elétricas (Furnas), a major power utility in Brazil. The main national control center runs SAGE™, a technology developed by the Brazilian electric energy research center CEPEL.

7.2 SECURITY CRITERIA AND FUNCTIONS

7.2.1 Security Criteria

The SDSA system's simulation results must be evaluated in accordance with established performance standards or criteria, and the violation or proximity to violation of any specific criterion must be reported. For steady-state analysis, the following performance criteria are used:

- Acceptable voltage range per bus.
- Normal and emergency thermal limits for transformers, transmission lines, and series capacitors.
- Maximum loadability limits of the transmission system.

Correspondingly, for dynamic analysis the user can elect to check some or all of the following criteria:

- Transient temporized voltage sags.
- Transient instantaneous voltage sags.
- Transient temporized voltage swells.
- Transient instantaneous voltage swells.
- MW margin for critical generators and critical cluster.
- Synchronous generator damping.
- Transient angle deviation.
- Steady-state angle deviation.
- Frequency deviation.

7.2.2 Security Functions

The automation of any security assessment process is dependent on the operation planning practices of the relevant utility or system operator entity. In the present SDSA case, three basic assessment functions apply:

1. Contingency analysis at the operating point (is the operating point secure?)
2. Maximum transfer (interchange) between two areas.
3. Maximum security region or transfer nomogram for three subsystems.

In addition, functions are provided for recommending preventive actions on the basis of such assessments. Preventive action is needed when one or more contingencies will bring the system state to a condition in which at least one security criterion is violated. These actions, such as generation redispatch, move the system state to a new secure operating point.

7.2.2.1 Contingency Analysis at an Operating Point
This is the most basic component of the SDSA system. Contingency analysis is normally performed for both the steady and dynamic states.

In steady-state contingency analysis, the critical aspect is the robustness of the power flow engine. For example, if a solvable contingency fails to converge, the interpretation might be that the operating point is beyond the maximum loadability limit. This can wrongly trigger an uneconomical preventive redispatch. To ensure maximum robustness, the SDSA employs multiple special algorithmic techniques, some of which are described in Section 7.4.

In dynamic contingency analysis, a stability diagnosis may not be produced at all if the numerical integration algorithm fails to converge. This kind of problem has been avoided by the careful choice of numerical integration methods based on variable–step–variable order algorithms, as described in Section 7.4.

If the number of contingencies is large, screening methods are highly desirable to improve performance. For steady-state analysis, we use DC and fast decoupled power flow methods as screening tools. For dynamic analysis, the key technique is to terminate as early as possible those simulations that are estimated to be stable. This estimation is based on heuristic methods such as maximum angle deviation or the single-machine equivalent model [4].

7.2.2.2 Import–Export Transfer Capacity

This SDSA function is useful for assessing the maximum secure transfer between two interconnected areas. It basically consists of performing contingency analysis at successively increasing/decreasing power-transfer levels. The precontingency operating point at which a security criterion becomes violated defines the maximum transfer capacity for that specific criterion. To identify this transfer level with relatively good accuracy, a binary search is used. To speed up this search, specific quantities (voltages, flows, etc.) and indices (MW stability margin, damping, etc.) are stored along the search and are used to estimate the violation point by interpolation and extrapolation. The search stops when two consecutive estimated points are sufficiently close to each other.

The changes in transfer level are effected by redispatching generation in the exporting and importing areas. All other generation and loads in the system remain constant. Then, neglecting changes in losses, the security region per criterion is defined by sets of points (line segments) belonging to the line $P_a + P_b = K$, where P_a and P_b are the respective generations in the areas and K is constant. The direction of search from the operating point is given by $\Delta P_a + \Delta P_b = 0$.

This function is obviously more complex than the single-contingency analysis at the operating point. On the other hand, the numbers of contingencies simulated for this kind of function are typically small, since only those that affect the transfer between the two areas are of interest. The function is useful for monitoring critical transmission corridors and it automatically provides the security margins for the current operating point, which is very desirable information in real time.

For steady-state transfer capacity assessment, the contingencies are computed by special Newton power flow versions. For dynamic transfer capacity assessment, contingency time simulations are performed.

7.2.2.3 Security Regions

Situations arise where transfer limits are highly dependent on the generation patterns in three areas. Therefore, redispatching generation in only two of them may provide inaccurate transfer limit estimates. At this point, the objective is to find secure regions in the two-dimensional surface defined by $P_a + P_b + P_c = K$, where P_a, P_b, and P_c are the respective generations in the areas and K is constant, if losses are neglected. This surface is embedded in three-dimensional space. Figure 7-3 illustrates a security boundary region for three network subsystems. It is also possible to replace generation P_c (say) by a load set. In this case, the security region is embedded in the surface defined by $P_a + P_b - L = K$, where L is the load.

A security region is much better visualized by its projection on a two-generation subspace. For example, for the security region of Figure 7-3, three projections are possible, as shown in Figure 7-4.

This figure shows three superimposed regions. The green region (displayed in medium gray) is secure. The yellow region (light gray) represents postcontingency thermal limit violation and the red region (dark gray) is unstable. The violation of any other monitored quantity can similarly be depicted by a contour.

The process of finding a point on the border of a particular region is the same as for the import–export transfer capacity, but now the direction of search is given by $\Delta P_a + \Delta P_b + \Delta P_c = 0$. These directions are radial from the operating point. The number of directions used for each contour plot determines the precision of the contour. But as this

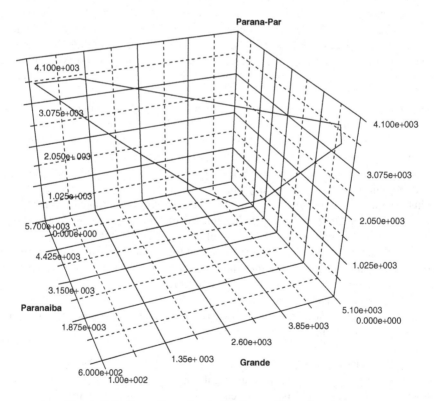

Figure 7-3. Security region in the three-dimensional space.

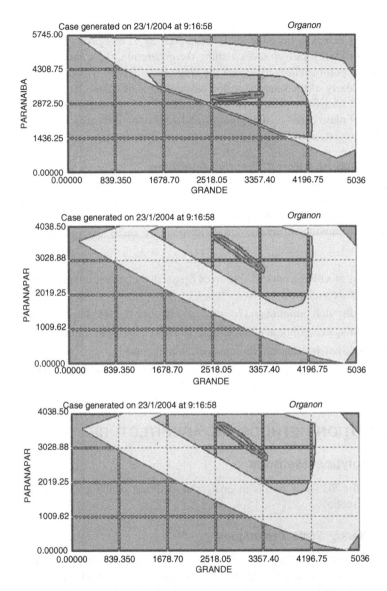

Figure 7-4. Security region projected on two generation subspaces.

number increases, performance decreases. Thus, a good compromise is required for online applications.

To achieve a good plotting contour, it is important to maintain a straight-line direction of search, keeping the ratios $\Delta P_a/\Delta P_b$ and $\Delta P_b/\Delta P_c$ constant.

If better accuracy in defining the security border is required, the generation units may be allowed to switch on and off along the redispatch direction to keep the spinning reserve within a realistic range (excessive spinning reserve would lead to optimistic security margins, since the total inertia and MVAr margin would be bigger than expected in practice). Before redispatching the generation, the spinning reserve at the new operating point is checked against the range and adjusted if possible.

The complexity of the security region calculation is obviously greater than that for import–export transfer capacity, but only a small set of contingencies is typically required. One of main benefits of computing and displaying a security region is the powerful and immediate insight that it provides to system operators.

7.2.2.3 Preventive and Corrective Functions
Preventive or corrective redispatch are by-products of the transfer capacity and security region calculations. For security regions, the required MW redispatch can be visualized from the nomogram, which is very useful for system operators. For contingency analysis, the following recommended actions can be easily derived from sensitivity analysis:

- Ranking of voltage control resources (generator or shunt compensation) to correct voltage violation,
- MW redispatch (amount and location) to alleviate thermal limit violation,
- MW redispatch (amount and location) to avoid angular instability,
- MW redispatch (location) to improve damping, and
- Suggested load shedding (amount and location) to move from alert/emergency states to secure state.

7.3 SOLUTION METHODS AND ARCHITECTURE

7.3.1 Analytical Methods

The core of the SDSA system is its analytical methods, which are described in some detail in this section.

7.3.1.1 Steady-State Techniques
NEWTON POWER FLOW. The Newton power flow is used for computing initial conditions for dynamic simulations, generation redispatch, and steady-state contingency analyses. The full Newton approach, in which all controls (generators, controlled shunts, taps, phase-shifters, and DC links) are simultaneously solved by the Newton method, was adopted because it is more reliable than other approaches. The DC power flow is used only for initialization and steady-state contingency screening. The following techniques are among those that improve the Newton method's convergence properties:

- The correction vector is scaled at each iteration to avoid unduly large state variations.

- Whenever a controlled voltage setting point is changed, the voltages at neighboring buses connected by low-impedance branches are reinitialized.
- The well-conditioning of the Jacobian matrix is reinforced by a special diagonal stabilizing approach.
- Constant power/current loads are converted into constant impedance loads for very low voltage levels (optional).
- Voltage control sharing or priority by multiple devices (generators, shunts, taps) is modeled, avoiding solution arbitrariness and convergence flip-flop.
- For contingencies, automatic rescue strategies are deployed to verify whether nonconvergence is due to lack of a solution or to a numerical problem. This includes techniques such as gradual pre- to postcontingency state transition and, when all else fails, submission of the case to the synthetic dynamic power-flow solver that is briefly explained below.

SYNTHETIC DYNAMIC POWER FLOW. This is an extremely robust power-flow approach [14], in which each main input variable is represented by a stable dynamic model whose equilibrium state coincides exactly with the power-flow solution (if this exists). For example, a generation bus can be represented by a MW and a voltage control as follows:

$$\dot{\delta} = P_{spec} - P$$

$$\dot{V} = V_{spec} - V$$

where:

$\dot{\delta}$ is the bus angle
P_{spec} is the specified power
P is the actual bus power
V is the bus voltage
V_{spec} is the specified voltage

All power-flow controls can be represented by similar synthetic dynamic models. The pseudodynamic system is solved by highly damped "time-domain" numerical integration. This method is computationally slower than the Newton method. Therefore, it is only used when absolutely necessary to confirm the existence of a power-flow solution.

CONTINUATION POWER FLOW. This method, based on the tangent vector method [9], is used for moving operating points during the search for security borders or for the computation of P–V curves. Continuation power flow methods can be slow near singularity points, but they are very efficient in normal situations. The cost of a predictor step is overcompensated by fewer corrector iterations. As the predictor estimates the new state and, consequently, approaches the solution, the chances of corrector convergence become higher.

Additionally, this method provides a mechanism for automatically adjusting the amount of MW to be redispatched, which makes it quite reliable for automated processes. The alternative would be to use a regular power flow with step-back and move-slow mechanisms in case of nonconvergence, but this was found to be less efficient. By allocating participation factors to the generators, it is possible to maintain the direction of search. It is necessary, however, to properly take into account the capacity of the generators and the losses. This is done by compensating mechanisms between the predictor and the corrector steps. For example, if a generator in a subsystem reaches its maximum MW capacity, the participation factors of the remaining machines in the same subsystem must be readjusted. If generator units are switched on and off to keep the spinning reserve within a specified range, the participation factors must also be corrected.

OPTIMAL POWER FLOW. An optimal power flow based on the primal–dual interior point method [2] is used to generate feasible base cases with no security criteria violations for near-real-time and days-ahead scenarios. The input data consists of network, generation, and load as in the regular power flow, but generation and other controls are adjusted to remove violations. Sensitivity analysis is used to rank voltage and flow controls and recommend preventive and corrective actions.

7.3.1.2 Numerical Integration

The most computationally expensive task in a detailed modeling approach for dynamic security assessment is that performed by the time domain simulation engine. Consequently, overall performance is very much affected by the numerical integration algorithms adopted. The time domain simulation in the SDSA is based on the ABM–BDF numerical integration method [1,3] associated with the variable-step–variable-order (VSVO) approach and the simultaneous solution of the algebraic and differential equations.

These techniques yield high numerical stability and improved performance (more than 10 times faster) compared with more traditional nonsimultaneous fixed-time-step approaches. The shortcomings of the latter are explained as follows. To avoid numerical instability [3], such approaches have to use very small integration time steps. Typically, their step size should not be greater than the smallest time constant in the dynamic models. But under stiff numerical conditions, even smaller time steps are required to avoid numerical instability. Obviously, the impact on the performance is severe, particularly when fast-acting control devices such as static VAr compensators or DC links need to be represented.

By contrast, in a simultaneous solution approach, the size of the time step is bounded by the accuracy of the simulation rather than its numerical stability. In practical terms, the desired accuracy can be met with small time steps during fast transients and larger time steps on smoother trajectories. The time step size is optimized by a dynamic adjustment mechanism as described below.

The differential and algebraic equations describing a power system model are represented by the following equations:

$$\dot{y} = f(y, x, t) \tag{1}$$

$$0 = g(y, x, t) \tag{2}$$

where:

$\dot{y} \in R^n$ is the vector of the state variables (or phase variables) that represent the dynamic models of control components such as synchronous machines, voltage regulators, DC links, etc.

$x \in R^m$ is the vector of algebraic variables, which are basically network voltages, current injections, and certain variables of the control components.

The ABM and BDF can be represented as

$$\sum_{i=0}^{j} \alpha_i y_{n+1} = h \sum_{i=0}^{j} \beta_i f_{n+1}$$

where:

α_i and β_i are parameters dependent on the specific integration method
j is the number of steps of the method and h is the time step

Using the appropriate parameters, the first-order ABM and BDF methods correspond to the Euler method:

$$y_{n+1} = y_n + h f_n' \qquad \text{(predictor)} \tag{3}$$

$$y_{n+1} = y_n + h f_{n+1} \qquad \text{(corrector)} \tag{4}$$

The second-order ABM is usually known as the trapezoidal method:

$$y_{n+1} = y_n + 0.5h(3f_n - f_{n-1}) \qquad \text{(predictor)} \tag{5}$$

$$y_{n+1} = y_n + 0.5h(f_{n+1} + f_n) \qquad \text{(corrector)} \tag{6}$$

The second-order BDF is given by

$$y_{n+1} = 3y_n - 3y_{n-1} + y_{n-2} \qquad \text{(predictor)} \tag{7}$$

$$y_{n+1} = \tfrac{4}{3}y_n - \tfrac{1}{3}y_{n-1} + \tfrac{2}{3}f_{n+1} \qquad \text{(corrector)} \tag{8}$$

The ABM method is used for differential equations, whereas the BDF method is used for algebraic equations and dynamic equations with very small time constants (< 10 ms). For improved efficiency the current and past information is stored in Nordsieck [3] vector form.

By applying the numerical integration formulae (3–8) to the model equations (1–2), the following set of algebraic equations is obtained:

$$0 = y_{n+1} - \beta_{n+1}hf(y_{n+1}, x_{n+1}) - C \tag{9}$$

$$0 = g(y_{n+1}, x_{n+1}) \tag{10}$$

where:

C is the weight sum of y and $\dot{y}$ terms

β_{n+1} is the constant that multiplies f_{n+1} in the integration formulae (3–8)

The solution of this set of equations is obtained by a dishonest Newton method, in which the Jacobian matrix is updated only when there is a time-step change, when the method does not converge, or when a large hard discontinuity occurs. Typically, two to three iterations are needed to converge at a time step.

The mechanism to change the time step is based on the estimation of the local truncation error (LTE) at the end of each time step. If the LTE is smaller than the required tolerance, the current step is accepted and the possibility of increasing its size is evaluated. If the LTE is above the tolerance, the current step is rejected and the step size is reduced sufficiently to bring the error to half of the tolerance.

The LTE estimation is based on the first neglected term of the Taylor series:

$$E_k = \frac{h^{k+1}y^{(k+1)}}{(k+1)!}$$

where k is the current integration order (1 or 2). The maximum time step $\bar{h}$ is calculated by considering the truncation error equal to the tolerance:

$$\bar{h} \approx h(\tau/E_k)^{1/(k+1)}$$

where τ is the tolerance. Considering that there is no error margin in this estimation, a conservative approach is adopted for the next step, say half of the estimated value.

Whenever the time step is changed, the best order is also evaluated. The criterion to choose the best order is the decreasing pattern of the truncated Taylor terms. Thus, the Taylor series expansion behaves as expected for the second order if the magnitude of the third-order terms form a decreasing sequence. Otherwise, first-order integration is used.

A difficult problem in time domain simulation is the treatment of discontinuities. The one-step (self-starting) methods can handle this better, but are generally less efficient as they require smaller time steps when compared to higher-order methods. The multistep methods need step size and order changes to deal effectively with such situations. Reinitialization with first-order integration is one of the possible approaches.

This implementation deals with discontinuities in different ways, depending on their types. The main sources of discontinuities are network and control switching and state variable nonlinearity. Switching operations can be specified by the users (e.g., in the contingency definition) or are automatically activated by controls such as excitation limiters or protection systems such as line tripping. Depending on the severity of the switching operation the program reinitializes the numerical integration process by zeroing the time step. The severity is measured by the norm of the first order derivative of the state variables. The reinitialization of the integration process nullifies the past (previous steps) information and the integration order is set to one. If necessary, the time step is reduced so that a switching operation occurs at the precise specified time.

The severity of the effect of nonlinearity on state variables, such as integrator saturation, is taken into account at the end of the step. If the local truncation error is greater than the tolerance, the step is rejected and decreased.

7.3.1.3 Energy Functions

The program uses numerical energy functions and a modified version of the SIME method (see Appendix B) for energy/power margin computation, instability detection and identification of oscillatory machines.

System stability is detected via the following dot product [11]:

$$f_{ip} = -P_{ac}^T \Delta\theta \qquad (11)$$

where:

$$P_{ac}^T = (pac_1, pac_2, \ldots, pac_{ng})$$
$$\Delta\theta = (\Delta\theta_1, \Delta\theta_2, \ldots, \Delta\theta_{ng})$$
$$\theta_i = \delta_i - \theta_{coi}$$
$$\theta_{coi} = \frac{1}{M_t} \sum_{i=1}^{ng} M_i \delta_i$$
$$pac_i = pm_i - pe_i - \frac{M_i}{M_t} pcoi, \qquad i = 1, 2, \ldots, ng$$
$$pcoi = \sum_{i=1}^{ng} pm_i - pe_i$$
$$M_t \sum_{i=1}^{ng} M_i$$

M_i, δ_i, pe_i, pm_i, and pac_i are, respectively, the inertia constant, rotor angle, electrical output, mechanical power, and accelerating power of machine i

ng is the number of synchronous generators

θ_i is rotor angle of machine i, referred to the center of inertia θ_{coi}

P_{ac} and $\Delta\theta$ are the vectors of generator accelerating power and angle deviation, respectively. Both quantities are referred to the center of inertia.

For classical synchronous machine models, system instability is detected when f_{ip} < 0. For higher-order synchronous machine models, a lower level is used: $f_{ip} < \tau$, $\tau <$ 0.

Individual energy functions are also computed to determine machines with low damping. The potential energy function is given by

$$Vpe_i = \int Pac_i \, d\theta$$

the kinetic energy function is

$$Vke_i = \tfrac{1}{2} M \omega_i^2$$

and the total energy is

$$Vt_i = Vpe_i + Vke_i$$

The rate of decay of Vt_i indicates those machines with lower damping. These are selected for Prony decomposition analysis.

7.3.1.4 Single-Machine Equivalent Methods

The computation of energy and power margin is performed by a SIME-like approach every time the system total kinetic energy,

$$V_{ke} = \sum_{i=1}^{ng} Vke_i$$

is at a minimum or if system instability is detected. This requires storage of the machine angles and power. The margins are computed for individual machines, a specified group of machines, and the estimated critical cluster. The margins of individual machines are useful for corrective redispatch. The specified groups are those previously known as consisting of coherent machines for the considered contingencies. The estimated critical cluster is the set of generators estimated by the method as the one that will first loose synchronism with the rest of the system.

The margins are estimated for forward- and backward-swing modes. If the critical cluster is in a net import area, it is likely that the instability will be due to a backward swing. This information is very important because the corrective action in this case is to increase generation in this group. On the other hand, if the instability is due to a forward swing, the corrective action requires decreasing generation in the group.

The margins computed by this method are used only as estimates. More accurate margins are computed by the binary search.

7.3.1.5 *Prony Analysis*

The Prony method [15] is used for spectral analysis (damping assessment) of synchronous machine angle trajectories. The objective is to compute the following spectral decomposition for a given signal, say rotor angle,

$$\hat{\delta}(t) = \sum_{i=1}^{n} R_i e^{\lambda_i t} \tag{12}$$

or in discrete time

$$\hat{\delta}_k = \sum_{i=1}^{n} R_i z_i^k \tag{13}$$

where $R_i \in C$ is the residue for pole $\lambda_i \in C$. The objective is to identify residues, poles, and the order n of the model to minimize the least square of $\delta(t)$.

A summary of the method is:

1. Given a sampled signal δ_i, $i = 0, 1, \ldots, m$, determine the coefficients a_i of the characteristic polynomial by fitting a linear prediction model.
2. Using the a_i's, compute the discrete-time eigenvalues by solving for the roots of the characteristic polynomial z_i's.
3. Compute the R_i's by curve-fitting (13) to the signal.
4. Calculate the continuous-time eigenvalues λ_i's using $z_i = e^{\lambda_i t}$.

Some signal preprocessing, such as windowing and offset removal, helps to improve the modal estimate.

7.3.2 Solution Architecture

7.3.2.1 *Parallelization*

To meet the performance requirements, the SDSA adopts a distributed processing approach in a manager/worker (master/slave) configuration, as shown in Figure 7-5. The

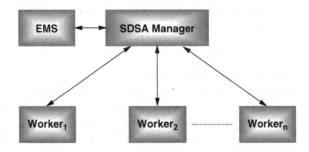

Figure 7-5. Manager/worker distributed processing environment.

manager process contains the high-level instructions needed to perform the security assessment functions. The calculation engines (power flow, time domain simulation, etc.) are part of the worker processes. The manager is responsible for generating base cases, distributing tasks among servers, collecting the respective reports, communicating with the SCADA/EMS, managing distributed resources, and storing/displaying results and plots. Workers receive tasks from the manager on a first-to-ask, first-to-get basis, process them using the task-specified power-system simulation tool, send the respective results back to the manager, and ask for another task. The workload per processor tends to be well balanced as the number of tasks increase. This is certainly the case for security region calculation and contingency analysis with a large number of contingencies. The idle time per processor is relatively small and occurs only at the end of an assessment cycle when some have finished their tasks and there is no additional work to be done. The system scaled almost linearly within the tested range (1–15) of processors. One critical factor impacting the scalability is the initial (base case) data broadcasting from the manager to the workers. Therefore, performance improvements can be achieved by broadcasting only changed data from previous assessments, when possible. During the assessment, the data exchange (task assignment and report) between workers and the manager is very small.

Worker processes can run in silent mode or can be attached to a console. The manager can run in silent mode or be attached to a console or the graphical user interface.

This level of parallelization is implemented via the MPI (Message Passing Interface) [10]. The initial data is broadcast to the workers. Task assignments and diagnosis reports are implemented via send/receive communication. The tasks are assigned in a first-to-ask, first-to-get order. In addition to MPI parallelization, the software can also be compiled with OpenMP [16] directives for lower granularity parallelization, allowing improved performance in multiple-core hardware architectures. This is especially useful for time-domain simulations in which significant parts of the numerical integration process can be done in parallel.

7.3.2.1   Software Layers

Internally, the software is organized in layers, as shown in Figure 7-6, where the arrows indicate the data-dependence relationship. The graphical user interface can be removed if not desired. The shell consists of input/output interfaces. The API communicates with the security assessment functions as well as allowing direct access to their analytical engines, such as power flow and contingency analysis. This design facilitates different levels of both SCADA/EMS integration and the offline use of the software.

7.3.2.3   Integration to SCADA/EMS

The general design for SDSA hardware integration with the SCADA/EMS systems from ONS is shown in Figure 7-7. This is valid for all five control centers.

A loose-integration approach was initially adopted. The base cases generated by the SCADA/EMS real-time network analysis subsystem were saved in the form of flat files and were retrieved by the SDSA periodically but asynchronously. The security assessment results were displayed in the SDSA GUI. They could also be viewed either

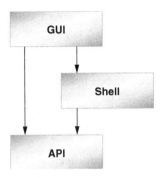

Figure 7-6. Software layers.

on dispatcher consoles or projected on the control room displays. This kind of integration is relatively simple to implement. The SDSA can also periodically retrieve base cases through FTP. In this case, the SDSA can be located in a remote place and be used in study mode.

In tight software-integration mode, the SCADA/EMS servers and SDSA servers are in the same local network, as shown in Figure 7-7. Any of the nodes dedicated to

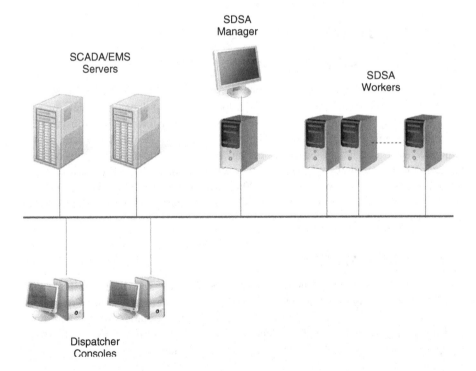

Figure 7-7. SCADA/EMS network for SDSA.

SDSA can be the manager process. There is no communication between the SCA-DA/EMS and SDSA workers. The SDSA is integrated in this mode in the Brazilian main control center. Its tightness of integration in the other control centers is currently at varying levels.

A failover procedure still needs to be implemented. The general idea is the following. Failure of a worker process can be detected by the manager process, which reassigns the task to another process. A monitor process in the SCADA/EMS detects failures of the manager process and restarts the SDSA, reallocating the manager process to another node if necessary.

The sequence of events for security assessment is as follows.

- The SCADA/EMS generates a bus-branch model based on the state estimation output.
- The SCADA/EMS sends a message to the SDSA to start a new assessment cycle.
- The SCADA/EMS sends the data and calculation parameters to the SDSA.
- The SDSA performs the assessment and sends the results back to the SCADA/EMS to be stored in its database.
- Results are displayed on the dispatcher's consoles and control room projection board.

7.4   PRACTICAL IMPLEMENTATION ASPECTS

7.4.1   Bus Numbering

The regular bus-branch network model used in planning studies needs to be adapted for online dynamic security assessment. The locations of the dynamic models are usually specified by bus numbers, but these numbers are not fixed in real-time models. Therefore, mapping is necessary. To help with this, a bus identification scheme was implemented. It consists of extending the bus number by adding the section number. For example, bus section 1 of bus 100 can be represented by 100.1. If there is only one section in this bus, it is represented by 100 as usual without section information. This also helps to model bus split/merge events in contingencies without loosing the original bus number identity.

7.4.2   Dynamic Models

Most of the models used in planning studies are currently represented in the online Brazilian model. These include synchronous generators, automatic voltage regulators, power system stabilizers, overexcitation limiters, governors, static VAr compensators, controlled series capacitors, voltage dependent loads, and out-of-step relays. The main exceptions are the DC links because their states are not presently estimated in the SCADA/EMS.

7.4.3 Network Size

Network size obviously impacts performance. Therefore, network equivalents can be quite useful. If that is the case, the recommended approach is the following:

- Reduce the network using the Ward method [21], but retain the buses of generators that are above a certain capacity.
- Similar generators at a power plant can be aggregated with negligible effect on the simulation results.
- Generators electrically too far away from the internal area can be represented by low-order dynamic models. These generators can be determined by steady-state sensitivity analysis. The parameters of their low-order models (internal impedance and damping coefficient) can automatically be found by least-square fitting of their frequency responses to the reference model.

7.4.4 Contingency Set

For transfer capacity and security region studies, the set of contingencies to be simulated is relatively small, since the contingencies are restricted to the transmission paths of interest. These contingencies are predefined by planning engineers based on their experience.

Security assessment by contingency analysis for the entire network demands a relatively large set of contingencies. In this case, a screening process can significantly improve the performance. However, the methods that have been proposed for screening of dynamic contingencies are typically based on short-time simulation which, although faster than complete simulation, is still expensive. Further development is needed in this area. Artificial intelligence algorithms are potential candidates [12].

7.4.5 Quality of Real-Time Data

Good quality data is essential for online security assessment. Among the well-known prerequisites are sufficient observability of the internal system, robust state estimation, and a carefully maintained database. When used to initialize security assessment, a state estimator's requirements are much higher than merely providing correct values for voltages and flows. The estimator needs to produce a solved model that is valid for power flow solutions, including all controls.

For example, it is not uncommon to see real-time cases with the following problems:

- Unrealistic MVAR injections in neighboring buses, sometimes canceling each other
- Generators with MW outputs significantly above their specified maximum capacities
- Generators with unbounded (very high) capacity
- Generators, shunts, or on-load tap changing transformers controlling extremely remote buses

- Parallel transformers controlling buses at opposite sides or controlling different buses on the same side
- A tap-changing transformer controlling a bus on the high-voltage side, but with no MVAr-controllable source in the system on the low-voltage side

Some of these problems are caused by inadequate state estimation, particularly in the presence of topology errors. These can be resolved by better estimation algorithms and topology error detection [20]. Other problems are related to errors in the database, for example, wrong controlled-bus information, and need to be fixed by database maintenance. At the cost of some extra computation, various conflicting control problems can be intercepted and resolved by extensive input-data-checking functions of the security assessment software itself. Problems of poor observability (generally, an inadequate measurement set) are more difficult to solve and, depending on the required accuracy, may be the major barrier for an online security assessment implementation.

7.4.6 Impact on Processes

The implementation of online security assessment functionality affects (for the better) existing planning and operating processes, which may need to be reengineered. Operating orders must be adapted to avoid possible conflicts with online assessments.

7.5 USER INTERFACE AND PERFORMANCE

7.5.1 User Interface

Clear visualization of the security assessment results by the system operators is paramount. In this sense, the security region display (as illustrated in Figure 7-4) is very powerful. A dispatcher can immediately perceive if the operating point, in generation coordinates, lies in the secure (green) or alert (yellow or red) region. If the operating point (OP) is in the yellow region, at least one of the credible contingencies will cause one or more thermal limit violations. If the operating point is in the red region, at least one of the contingencies will cause instability. Under the GUI, mouse-positioning tips and report tables provide detail information per violation. Preventive and corrective MW redispatches can be visually determined by the operators.

Transfer capacity between two subsystems can also be nicely displayed in an import/export scale. The graphical reporting of contingency analysis is more difficult. The approach adopted so far is to list the cases in report tables with optional sorting by severity. Another possible approach is to use a rotary-gauge-like interface to display global or local security indices.

Detailed information about the security assessment is better displayed in report tables, like the one illustrated in Figure 7-8.

The SDSA also displays single-line diagrams as shown in Figure 7-9. These are useful to quickly inspect out-of-service components and to edit data in study mode. They can be also a powerful visualization tool for violations.

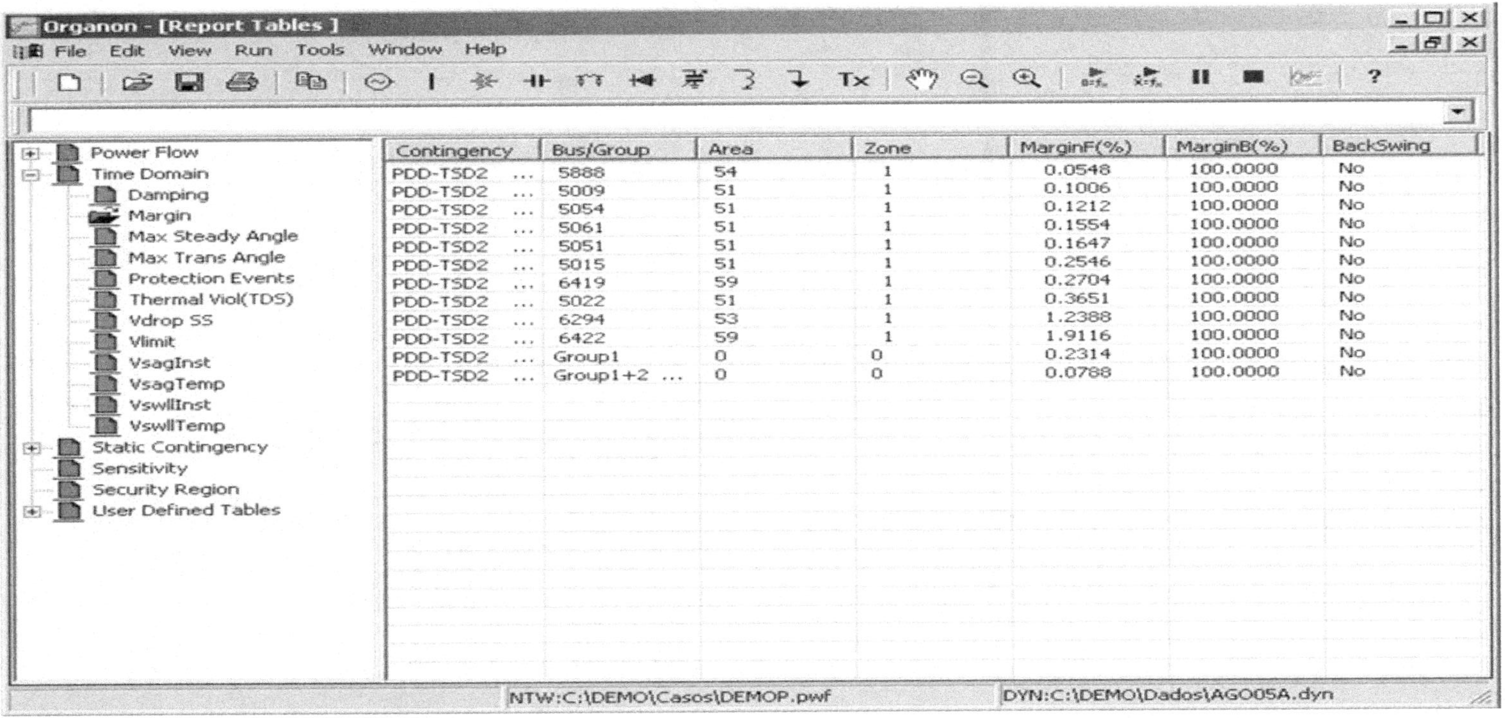

Organon – [Report Tables]

File Edit View Run Tools Window Help

Contingency	Bus/Group	Area	Zone	MarginF(%)	MarginB(%)	BackSwing
PDD-TSD2 …	5888	54	1	0.0548	100.0000	No
PDD-TSD2 …	5009	51	1	0.1006	100.0000	No
PDD-TSD2 …	5054	51	1	0.1212	100.0000	No
PDD-TSD2 …	5061	51	1	0.1554	100.0000	No
PDD-TSD2 …	5051	51	1	0.1647	100.0000	No
PDD-TSD2 …	5015	51	1	0.2546	100.0000	No
PDD-TSD2 …	6419	59	1	0.2704	100.0000	No
PDD-TSD2 …	5022	51	1	0.3651	100.0000	No
PDD-TSD2 …	6294	53	1	1.2388	100.0000	No
PDD-TSD2 …	6422	59	1	1.9116	100.0000	No
PDD-TSD2 …	Group1	0	0	0.2314	100.0000	No
PDD-TSD2 …	Group1+2 …	0	0	0.0788	100.0000	No

Power Flow
Time Domain
Damping
Margin
Max Steady Angle
Max Trans Angle
Protection Events
Thermal Viol(TDS)
Vdrop SS
Vlimit
VsagInst
VsagTemp
VswllInst
VswllTemp
Static Contingency
Sensitivity
Security Region
User Defined Tables

NTW:C:\DEMO\Casos\DEMOP.pwf   DYN:C:\DEMO\Dados\AGO05A.dyn

Figure 7-8. Report table: generator security margins.

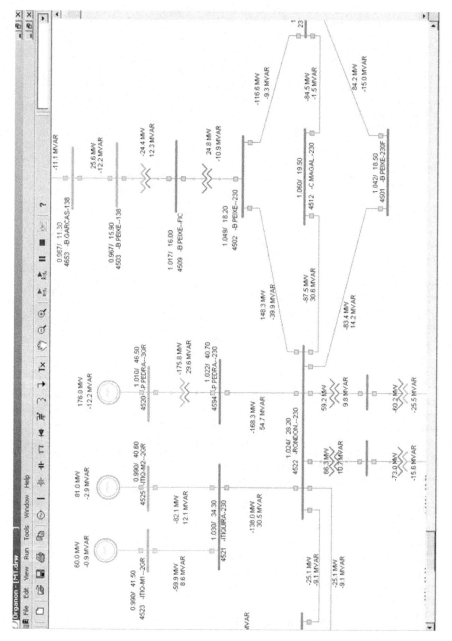

Figure 7-9. Single-line diagram.

For example, different color bands can aid quick visualization of critical components such as transmission lines, loads, and generators. A transmission line with red color indicates that its outage, as a result of a contingency, will cause system instability. If it is orange, the stability margin will be small, and so on. The respective critical cluster of synchronous machines will also show the same color.

7.5.2 Performance

7.5.2.1 Factors Affecting Performance
The response times of calculations performed by the SDSA system are affected by the following factors:

- Choice of methods and algorithms, as mentioned in the previous paragraphs,
- Size of the network model,
- Number of contingencies to be simulated, and
- Computational resources available.

Ideally, security assessment analysis should be completed a few seconds after the system state has been estimated. In practice, a delay of some minutes is tolerated, on the assumption that the system state does not change significantly during this period. This can be dangerous; events causing large discontinuities, such as the tripping of a major transmission line, can invalidate the assessment results and leave the dispatcher temporarily without reliable information. As a counterargument, the existence of real-time security assessment implies that major events (disturbances) have already been considered in previous analyses and should at most move the system to alert state.

7.5.2.2 Performance Requirements
Current dynamic security assessment performance requirements are typically in the range of 2–15 minutes. As an example, the target performance per function for the implementation in Reference [19] is shown in Table 7-1. These requirements were met in the main control center with twelve 3 GHz Pentium IV processors.

The actual response time varies with system operating conditions, that is, the computational cost of power-flow and time-domain simulations is generally more expen-

Table 7-1. Performance requirements

Functionality	Contingencies	Performance
Static contingency analysis	100	≤ 10 s
Dynamic security analysis	100	≤ 1 min
Static transfer limit	10	≤ 5 s
Dynamic transfer limit	10	≤ 30 s
Static security region	10	≤ 1 min
Dynamic security region	10	≤ 2 min

sive for stressed cases. For the implementation in [19], the state estimation is processed every 2 minutes, within which interval the most time-consuming function (dynamic security region calculation) should run. As previously stated, the analysis is highly suited to parallel computing so that, when the system size increases and/or more functions or contingencies are added, the desired response can be maintained by adding more processors.

7.6 CONCLUDING REMARKS

This chapter has described an online security assessment system whose development was dominated, particularly on the dynamics side, by the need for consistent and solid analytical methods, process automation, and high-performance computing. Future improvements will also focus on these three aspects. For instance, a faster dynamic contingency screening approach is still desirable. Inexpensive multi-CPU hardware will continue to facilitate large-scale implementations. A natural evolution should be the integration of dynamic security assessment with security-constrained optimal power flow.

In summary, until recently it was very difficult, if not impossible, to implement online dynamic security assessment for a nonsmall power system; the analytical methods for security margin estimation were not mature enough and sufficiently economical computing power was not available. These barriers no longer exist. Today, the state of the technology allows us to perform huge numbers of detailed simulations in a few minutes. This is of use not only for online security assessment, but also for power system planning. It is a change in paradigm, bringing benefits to all areas of power system analysis.

7.7 ACKNOWLEDGMENTS

The author is grateful to Dr. Brian Stott and Dr. Ongun Alsaç for their valuable comments. Special thanks are given to the Directors of the National System Operator (ONS) in Brazil, who have enthusiastically supported this project, and to the real-time and planning technical teams whose commitment and efforts were fundamental to its success.

7.8 REFERENCES

[1] Astic, J. Y., Bihain, and A., M. Jerosolimski, "The Mixed Adams—BDF Variable Step Size Algorithm to Simulate Transient and Long Term Phenomena In Power Systems," *IEEE Transactions on Power Systems,* Vol. 9, No. 2, May 1994.

[2] Granville, S., "Optimal Reactive Dispatch Through Interior Point Methods," *IEEE Transactions on Power Systems,* Vol. 9, No. 1, Feb 1994.

[3] Lambert, J. D., *Numerical Methods for Ordinary Differential Systems: The Initial Value Problem,* Wiley, New York, 1991.

[4] Pavella, M., Ernst, D., and Ruiz-Vega, D., *Transient Stability of Power Systems: A Unified Approach to Assessment and Control,* Kluwer, Norwel, MA, 2000.

[5] DyLiacco, T. E., "Control of Power Systems via the Multi-level Concept," Case Western Reserve University System Research Center, Report no. SRC-68-19, June 1968.

[6] Limmer, H. D., "Security Applications of On-line Digital Computers," presented at Second Power Systems Computation Conference, Stockholm, June 27, 1966.

[7] Hayashi, S., "Power System Security Assessing by Digital Computer Simulation—Basis Control," in *Proceedings of PICA Conference,* Denver, Colorado, May 18–21, 1969.

[8] Debs, A. S., and Benson, A. R., "Security Assessment of Power Systems," in *Proceedings of System Engineering for Power: Status and Prospects,* Henniker, NH, 1975.

[9] Ajjarapu, V., and Christy, C., "The Continuation Power Flow: A Tool for Steady State Voltage Stability Analysis," in *Proceedings of IEEE PICA,* May 91, pp 304–311.

[10] Gropp, W., et al., *MPI: The Complete Reference,* MIT Press, Cambridge, MA, 2000.

[11] Pai, M. A., *Power System Stability,* North Holland, New York, 1981.

[12] Jardim, J. L., "Online Dynamic Security Assessment: Implementation Problems and Potential Use of Artificial Intelligence," in *Proceedings of IEEE Power Engineering Society Summer Meeting,* vol. 1, July 16–20, 2000.

[13] Jardim, J. L., Neto, C. S., and Kwasnicki, W. T., "Design Features of a Dynamic Security Assessment System," in *Proceedings of IEEE Power System Conference and Exhibition,* New York, Oct 13–16, 2004.

[14] Jardim, J. L., and Stott, B., "Synthetic Dynamics Power Flow," in *Proceedings of IEEE General Meeting,* San Francisco, June 12–16, 2005.

[15] Hauer, J. F., "Application of Prony Analysis to the Determination of Modal Content and Equivalent Models for Measured Power System Response," in *Proceedings of IEEE Winter Meeting,* 215-4 PWRS, 1991.

[16] Chandra, R., *Parallel Programming in OpenMP,* Academic Press, San Diego, 2001.

[17] Alsaç, O., Bright, J. M., Brignone, S., Prais, M., Silva, C., Stott, B., and Vempati, N., "The Rights to Fight Price Volatility," IEEE Power and Energy Magazine, Vol. 2, No. 4, pp. 47–57, July–Aug. 2004.

[18] Taylor, C. W., *Power System Voltage Stability,* McGraw-Hill, New York, December 1993.

[19] Jardim, J. L., Neto, C., and Santos, M. G., "Brazilian System Operator Online Security Assessment System," in *Proceedings of IEEE Power System Conference and Exhibition,* Atlanta, GA, Oct 30–Nov 2, 2006.

[20] Vempati, N., Silva, C., Alsaç, O., and Stott, B., "Topology Estimation," in *Proceedings of IEEE Power Engineering Society General Meeting,* June 12–16, 2005, pp. 806–810.

[21] Monticelli, A., Deckmann, S., Garcia, A., and Stott, B., "Real-Time External Equivalents for Static Security Analysis," *IEEE Transactions on Power Application Systems,* Vol. PAS-98, No. 2, March/April 1979.

8

DYNAMIC NETWORK SECURITY ANALYSIS IN A LOAD DISPATCH CENTER

Guenter Beissler, Olaf Ruhle, and Roland Eichler

8.1 INTRODUCTION

Electrical systems worldwide are growing or getting interconnected to allow new economic objectives for operation. In the process, the power transfers under open transmission access in deregulated markets are pushing the grids to their limits. As a result, unexpected cascading events, weak interconnections, high loading of lines and transmission corridors, and hidden protection failures may cause the systems to loose stability, possibly leading to catastrophic failures or blackouts. In recent years, the numbers of blackouts as well as their negative consequences have increased. Analyzing these catastrophes shows that the operating guidelines that have been used for a long time were based on offline stability studies, which tend to be conservative for normal conditions and inaccurate for unexpected unusual events.

Oscillations and dynamics can compromise grid reliability and poorly understood dynamic constraints can unnecessarily narrow system limits. The complexity of large electrical systems with different primary and secondary control mechanisms, emphasis on economic objectives in operations, use of extremely fast acting FACTS devices, rapidly changing power flows, and, last but not least, the complex protection philosophies of today cannot be handled just by static security assessment. Accordingly, dynamic security assessment (DSA) tools become increasingly important in modern power system control centers.

Real-Time Stability Assessment in Modern Power System Control Centers. Edited by Savu C. Savulescu   **183**
Copyright © 2009 The Institute of Electrical and Electronics Engineers, Inc.

With the advent of electricity markets and competitive deregulated structures, the electric power industry is now driven by economics and requires highly sophisticated systems to assess, maintain, and enhance their operating reliability (Figure 8-1). In many cases, static security assessment cannot achieve the necessary security under changing grid and generation conditions.

The need for fast and versatile DSA was emphasized by the major blackouts that affected the grids in 2003 in United States and Europe. For example, the Italian blackout started with 6545 MW imported to Italy when a series of cascading events isolated the Italian system from Europe, and the loss of generation in Italy, coupled with insufficient load shedding, shut down the entire power system. Once triggered, the phenomena leading to blackout took less than 3 minutes, but they had been preceded by approximately 15 minutes when the problem started to evolve from a normal situation to an alert and then to an emergency state.

Subsequent analysis indicated that, among the factors that affect the operating reliability, improved power system monitoring and preventive actions are the most important items. Other factors that affect system security are significant changes in genera-

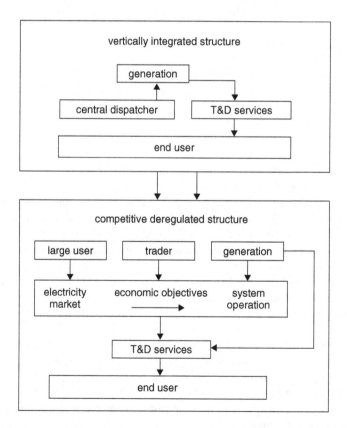

Figure 8-1. Change of system structure in deregulated electric markets.

tion in deregulated markets; combined cycle power plants and distributed generation reduce the controllability, and wind farms in locations not related to load centers affect the system security through the changing wind availability.

At the planning stage, systems are typically designed to meet or exceed previously specified voltage, frequency, stability or other adequate criteria. A key planning objective is to guarantee a secure transmission capacity subject to established criteria by adopting a suitable system design. In this context, the objective of security assessment is to design and operate networks whose operating integrity would not be jeopardized by unforeseen events, and to provide for preventing "cascading outages when the bulk power supply is subjected to severe disturbances." According to CIGRE, the "power system security is the ability of the system to cope with incidents without the operator being compelled to suffer uncontrolled loss of load."

The system operating procedures must ensure that operations are consistent with sound network design philosophy and criteria. In operations, such criteria must be applied to complete as well as to degraded topologies because the network topology may change due both to unforeseen events and to scheduled maintenance outages. This is fundamentally different from long-range system planning, in which the degraded topologies are known at any time and the transmission capacity is the only system variable.

The security limits provided to the system operator are power-flow values that guarantee that a given topology is secure for every one of a list of contingencies. The most restrictive are the dynamic security limits, whether from steady-state, voltage, or transient-stability considerations. There are various methodologies to define dynamic security limits such as system stability calculations, sensitivity studies, and security margin calculations that optimize security limits in terms of various network parameters.

Dynamic security has to be guaranteed in order to maintain electric service reliability and quality. Dangerous events such as short circuits, loss of transmission equipment, and loss of generation produce electromechanical transients and must be represented. The response of protection and automatic control devices must also be considered. In order to model all these phenomena, fast-time domain simulation of the most severe contingencies is of paramount importance. Figure 8-2 shows the different operational states of a system. The DSA task is to verify that none of he plausible contingency scenarios would cause the system to enter an emergency state.

For any given system, the DSA requirements depend upon the particular topology, generation mix and interconnection to other systems, among other factors. For practical purposes, however, DSA requirements can be broadly classified as:

- Margins to thermal limits,
- Margins to loading limits,
- Stability,
- Margins to instability, and
- Damping.

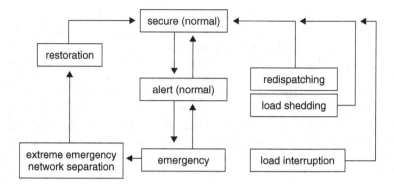

<u>Figure 8-2.</u> Power system operating states and means to restore normal secure operation.

In order to be applicable, or quantifiable, these requirements must be expressed in practical terms, such as:

- Critical fault clearing time (generator stability),
- Oscillation time (damping),
- Generators out of step and machine load angle (instability),
- Line loading,
- Critical under/over voltages,
- Critical under/over frequencies, and
- Angle differences between system parts.

8.2 SIEMENS APPROACH TO DYNAMIC SECURITY ASSESSMENT

8.2.1 DSA Overview

The structure of a DSA system provides for selecting different load-flow scenarios and building individual contingencies to be evaluated automatically. The contingencies are checked using user-defined criteria. The process is depicted schematically in Figure 8-3.

On the simulation level, several user-selectable load-flow scenarios are available. The contingency builder allows the user to select and calculate the most severe contingencies. The program checks the security criteria (stability, under/over voltage, damping, etc.) that have been defined with the help of the security criteria builder. The security criteria can be combined to define sets of criteria that describe the system limits and are suitable for the individual user's needs. The DSA reports and documents the contingencies that cause system limit violations, for example, generator instability, voltages below 80%, angles between node i and node j larger than 40°, and so on.

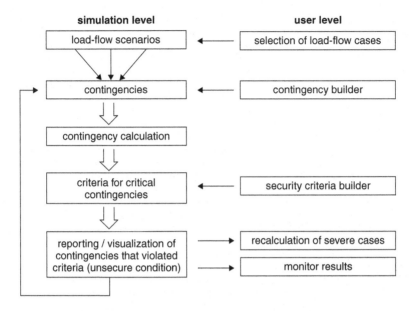

Figure 8-3. Functional structure of the DSA system.

These cases can be recalculated very easily and all typical characteristics can be visualized to provide the analyst with a deeper view. In parallel, the critical contingencies can also be monitored for the operator. Key design requirements for a DSA system should include the ability to:

- Simulate the important components of the passive grid equipment, such as lines, cables, transformers, and so on, as well as the active switching or control elements, including capacitor banks and FACTS devices, together with their control schemes.
- Represent the action of the important protection when simulating cascading faults.
- Use a simple contingency building process when simulating contingencies.

In addition, the decision criteria must be flexible, user-definable, and capable of addressing the critical system aspects. As an example, Figure 8-4 illustrates an online DSA implementation based on the PSS™ NETOMAC simulation package [10,11].

The time-domain simulation allows the most accurate description of the system from transient stability to voltage collapse. The DSA can handle dozens of contingencies per minute, based on the actual state of the system and potential system failures. In a typical study, approximately 10 main scenarios, each of which can consist of up to 20 major contingencies, are checked and reported in 10 minutes, as shown in the example described in Section 8.3.2. The system size of this study was 420 buses, 2700 branches, 52 generators, and 400 controllers. Eigenvalue analysis is also available to assess system interarea oscillation and damping [12,13]. As shown in Figure 8-4, the simula-

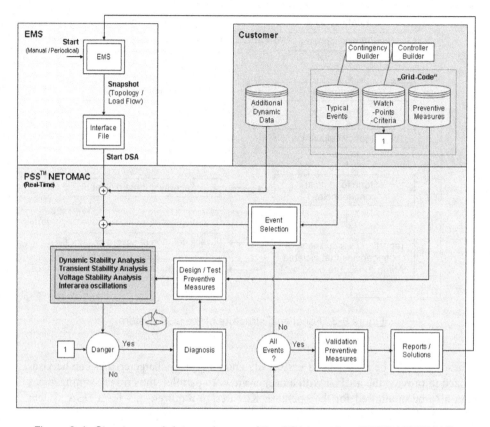

Figure 8-4. Structure and data exchange of the DSA based on PSS™ NETOMAC.

tion tool is structured to be deployed with data imported from the SCADA/EMS. The user can define contingencies and control the building of security criteria. Preventive measures can also be designed and tested to support the operator when looking for countermeasures in case of critical system situations.

8.2.2 DSA Tools Available on Siemens' SCADA/EMS Platform

The DSA tools available on Siemens' SCADA/EMS platform are basically the well-proven power system planning tools for system dynamics. At the core is the time domain functionality, or transient stability analysis (TSA). Figure 8-5 shows the capabilities of the software package in simulating electrical systems. There are two alternative options in the time domain: instantaneous value mode and stability mode.

The instantaneous value mode (EMT mode) allows electrical systems to be represented phase-wise. Symmetrical systems are entered as single-phase models and completed to three-phase systems internally. Asymmetrical systems can be represented by means of elements in the individual phases. This is also possible for any kind of DC

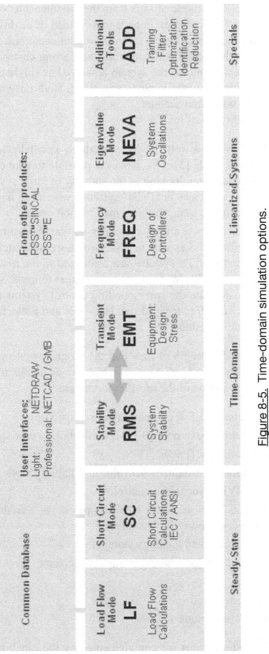

Figure 8-5. Time-domain simulation options.

system. Therefore, the instantaneous value mode provides for the total solution of any electromagnetic or electromechanical problem.

The stability mode (RMS mode) differs from the instantaneous value mode in that it simulates the network with complex impedances instead of as differential equations. Controllers and machines are modeled with differential equations. Machines are used with reduced order in the differential equations (neglecting changes in flow in the d- and q-axes). In the stability mode, the system is viewed as single-pole. Typically, the stability of multimachine systems is being examined. Both unsymmetrical and symmetrical faults can be taken into account.

In addition, the ability exists to:

- Supplement calculations in the stability mode by parallel calculations in the instantaneous value mode [3]. This makes it possible to take complex short-time events into account when looking at the stability of large systems, for example, the commutation processes in HVDC systems, which can have an effect on the stability of the entire system, should a malfunction occur.
- "Sap" sequentially between the two modes so that short-time events that arise during the stability studies can be assessed more accurately. In the stability mode, HVDC and FACTS systems are connected to the network through variable admittances, variable loads or variable sources (current, voltage, power). So it is possible to simulate the corresponding control systems in detail. Different kinds of voltage- and frequency-dependent loads and protection systems can be modeled in a similar fashion.

During the considerable length of time that the program system has been in use, a large number of models have been created, which are available as macros or can be called up from a library. Some of the most important are:

- Excitation systems (IEEE specification or user-specific),
- Turbines and turbine governors (IEEE specification or user-specific),
- Power system stabilizer (PSS),
- HVDC models for the instantaneous value and stability mode, including control,
- Multiterminal HVDC, including control,
- FACTS elements such as static VAr compensators, variable series compensators, and universal power-flow controllers,
- Models for superconducting energy storage,
- Models for circuit breakers, taking arcing into account,
- Load models,
- Transformer models, and
- Generic wind turbine models.

The time domain simulations are complemented by the capability to perform frequency domain simulations as well as eigenvalue and modal analysis. These additional functionalities are briefly addressed in Annex 8-1.

8.3 CASE STUDIES: CHALLENGES, IMPLEMENTATION APPROACH, AND SOLUTION FEATURES

The integration of transient-stability-analysis-based DSA into a SCADA/EMS system is not a trivial exercise. It poses many challenges that need to be addressed on a case-by-case basis. In the following, we describe the implementation approach and solution features of two successfully conducted projects in which DSA was integrated in the power system control-center environment. The key findings of a planning study conducted by the Union for the Coordination of Transmission of Electricity (UCTE) are also summarized to further illustrate the use of DSA time domain simulation tools.

8.3.1 National Control Center in Kuwait

8.3.1.1 Kuwait Power System

The power system of Kuwait forms an electrical island. Links to neighboring countries are planned but not yet in operation. The power system comprises 275 kV and 132 kV levels for transmission, and 33 kV and 11 kV levels for distribution. The system load varies by a factor of more than two between about 4000 MW in wintertime to about 9000 MW in summertime (Figure 8-6).

8.3.1.2 Kuwait Control Center Architecture and Tasks

The National Control Center (NCC) is responsible for the dispatch and control of the generation system as well as for the operation of the 275 kV, 132 kV, and 33 kV voltage levels. The NCC as well as one of the four District Control Centers (called Jabriya

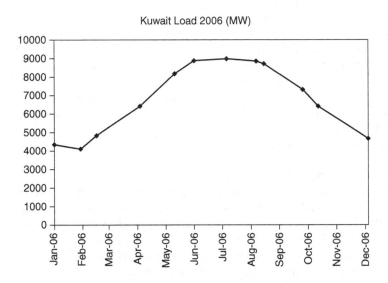

Figure 8-6. System MW load in Kuwait in 2006.

DCC) at the same location are equipped with a common SCADA/EMS; three more DCCs are connected via ICCP links. The first release of the SCADA/EMS at the NCC covering 130 substations became operational in April 1994. Meanwhile, after continuous extension and maintenance of the SCADA/EMS, the NCC together with the DCC covers about 180 substations. From the beginning, the SCADA/EMS comprised an integrated package for transmission network security analysis.

8.3.1.3   The SCADA/EMS

The SCADA/EMS at NCC is based on Siemens SINAUT Spectrum. SINAUT Spectrum at NCC consists of self-contained subsystems that intercommunicate via defined interfaces. One of these subsystems is the Network Analysis (NA), which is an integral part of the SCADA/EMS. NA includes features such as engineering all application data with the same tool as the SCADA data and displaying application calculation results in SCADA one-line diagrams. The SCADA/EMS architecture is predicated on the principles of modularity as well as functionality distributed across several servers and workstations connected by a LAN (Figure 8-7). All the servers and workstations run under a Unix operating system.

The NA subsystem comprises commonly used modules such as state estimation, online load-flow calculation, online short-circuit calculation, and contingency analysis. Additional modules include bus (load) scheduling, transmission loss penalty factor calculation, online network reduction, and a set of different optimal power flow applications. The NA functional modules can be seen on the real-time execution control display, except for the load-flow module, which is available only in study mode (please refer to Figure 8-10 in Section 8.3.1.7).

The functional modules of the NA subsystem use several common services such as sequence control, user interface, study-case management, and data export. The sequence control and data export services play a major role in providing the real-time stability calculation functionality as described below.

The NA functions can be activated in real-time mode and study mode, as well as in the security checked switching mode (SCS). In real-time mode, the current state of the power system is analyzed; in study mode the operator can freely set up study cases based on network switching situations and injection/load patterns as needed. In SCS mode, the power system state that will occur after an intended switching action has been performed is automatically checked for steady-state load-flow feasibility beforehand. Optionally, the operator can request contingency analysis and online short-circuit calculation to be included in SCS as well [15].

8.3.1.4   Operational Problem

The electrical power system of Kuwait is mainly a cable network. Therefore, in all voltage levels about 80 shunt reactors are installed with different rated powers. There are smaller ones of 30 to 50 MVAr located primarily at 33 kV bus bars, at the tertiary windings of the 275/132 kV transformers (45 MVAr), and at the 132 kV level (50 MVAr). But there are also large 250 MVAr shunt reactors that are located in selected line switching bays (Figure 8-8) and on 275 kV bus bars.

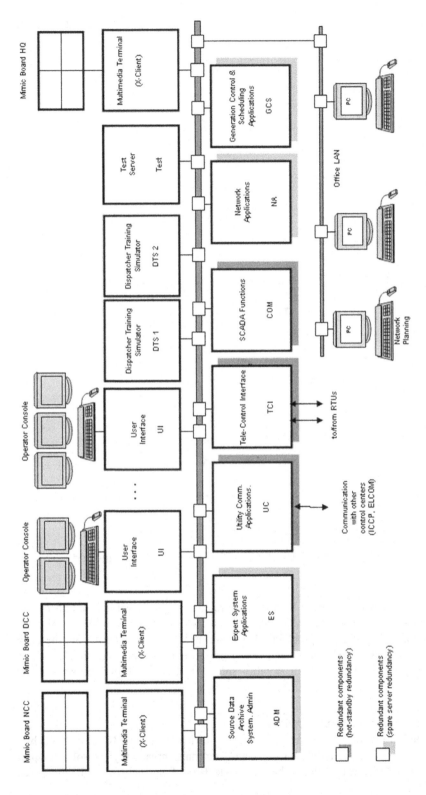

Figure 8-7. Conceptual system configuration of MEW Kuwait National Control Center.

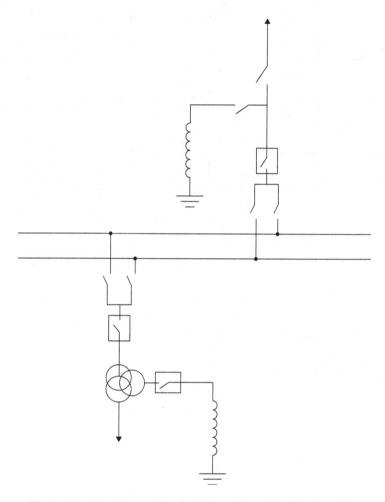

<u>Figure 8-8.</u> Principles of connecting shunt reactors to the grid.

Since the power system of Kuwait is operated as an electrical island, transient stability is an issue of special importance. The switching of large 250 MVAr shunt reactors may cause stability problems. Therefore, it was foreseen right from the beginning of the NCC SCADA/EMS upgrading project that the operators should have the capability to perform transient stability studies just before switching such a large shunt.

8.3.1.5 *Solution Architecture*
At the time when the upgraded SCADA/EMS was installed at the NCC, its NA subsystem did not contain a module for the evaluating transient stability. To solve the problem, the stability calculation submodule STABSI of the network planning system

PSS™ SINCAL was chosen to provide the missing functionality. In today's Siemens PSS™ product suite, the STABSI functionality is a subset of PSS™ NETOMAC, as described in Section 8.2.2 and Annex 8-1, and PSS™ NETOMAC is integrated in PSS™ SINCAL for a broad range of stability calculations.

PSS™ SINCAL is a product used for planning purposes beyond the operational problem mentioned above and comes with its own database, user interface, and engineering subsystems, so it is entirely independent from the SCADA/EMS. In Kuwait, PSS™ SINCAL was installed on a separate server (identified as "Network Planning" server in Figure 8-7) and a file transfer from the SCADA/EMS system to SINCAL was established. The simulation algorithms available in STABSI are outlined in Section 8.2.1. For the purpose of providing online stability calculation functionality in the SCADA/EMS environment, however, only a subset of the SINCAL features was needed: the calculation capabilities per se and the dataset import feature. Neither the network planning system database nor the network planning system user interface were used. All effort was done on the SCADA/EMS side so that the operator can use the stability calculation module STABSI at a level of integration with the SCADA/EMS that is fully sufficient for the operational needs. In particular, this relates to the ability to use the SCADA/EMS user interface for all stability-related operator actions. For that purpose, STABSI was installed on the NA server of the SCADA/EMS (shown as NA server in Figure 8-7), entirely integrated in the software environment of the SCADA/EMS.

Application integration mainly on the user interface level as opposed to a full integration of the stability calculation module was considered a good trade-off between implementation risk, implementation time schedule, and cost on one hand, and additional operator benefits gained from full integration on the other hand. It should be noted that the key information the operator needs from the stability calculation module is just a YES/NO response concerning system stability after an intended switching action of a 250-MVAr shunt reactor.

8.3.1.6 Implementation

The real-time stability-check function is entirely processed in the Unix environment of the SCADA/EMS. The function is initiated by the operator by means of a button in the real-time execution control display of the NA subsystem. The further process steps are handled by a script called "check_stab." The script opens a Unix shell window at the operator desk for the subsequent user dialog.

The core stability calculation program STABSI does not have access to the operational database of the SCADA/EMS. Rather, the necessary data are read from the operational database by means of a utility program and written to an ASCII file using BETINA file format. These steps are also controlled by the script "check_stab." The results of the real-time stability calculation are displayed in the Unix shell window mentioned above. Results printed on screen are reduced to an absolute minimum. They indicate as clearly as possible if the proposed switching will cause instability of generators or not. Besides the yes/no indication concerning system stability, STABSI also produces an ASCII file with detailed calculation results for all generators. This file could additionally be evaluated if desired.

The time needed to execute the complete real-time stability check procedure (excluding state estimation) is approximately 5 seconds.

8.3.1.7 Data Export for Stability Calculations

The data transfer between the SCADA/EMS and the stability calculation modules is carried out via an ASCII file. The data flow is shown in Figure 8-9.

The data transfer must enable STABSI to perform the real-time stability check as well as the network planning package SINCAL to perform load-flow, short-circuit, and stability computations based on the current network state. For that purpose, the following data are transferred from the SCADA/EMS to the network planning package:

- The node–branch topology representing the actual switching state
- The network structure is transferred in form of the node–branch topology. The node–branch topology is the basic description of the network topology for network analysis functions. It does, however, not include information to display a single line diagram of the transferred network by SINCAL.
- The parameter data of network elements such as lines, transformers, and loads.
- The network equivalents, that is, the equivalent lines and equivalent injections calculated by the online network reduction and state estimator applications of the SCADA/EMS. The equivalent lines are defined in the network planning package at an individual network level. For planning purposes this level can be replaced by networks stored in the network planning package database.
- The bus voltages and the active and reactive power injections of the "own" network (i.e., the network controlled by the NCC), and the "subordinate own network" (i.e., the network controlled by DCCs).

The results of the most recent state estimate are used. If the time elapsed since the last run of the state estimator is too long, the operator can manually request a run of the state estimator and then activate the data transfer function.

In order to enable stability studies also for planned or past network situations, there is as well the capability to transfer the output of the online load-flow application as input to SINCAL on operator request. The transfer file has the same contents and format as outlined above.

Data describing the dynamic behavior of the generators and their controllers are not part of the SCADA/EMS database and, therefore, are not included in the ASCII file described above. These data are instead maintained in the database of the network planning package SINCAL. In case these data have been changed, an ASCII file with the data subset needed by STABSI can be exported from the SINCAL database on request of the SINCAL user. The file is stored in a dedicated directory of the NA server of the SCADA/EMS. This file will automatically be included in each subsequent run of STABSI until a more recent file version is stored on that directory. In order to properly merge the data from the two ASCII files, one from the SCADA/EMS and the oth-

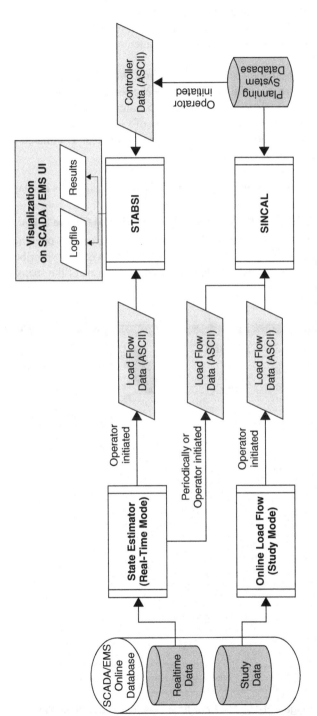

Figure 8-9. Real-time stability-check data flow.

er one containing the generator and controller data, the data engineers must ensure that the same naming convention for the node–branch topology is used both in the SCA-DA/EMS and in the network planning environments.

8.3.1.8   User Interface for Real-Time Stability Checks

As mentioned above, the real-time stability checks are initiated by the operator by means of a button in the real-time execution control display of the NA subsystem (CheckStab button in the upper middle section of Figure 8-10).

By pressing the button, the following sequence of steps is initiated (controlled by the script "check_stab"):

• A window opens in which a list of all switchable shunts is displayed (Figure 8-11).

• The user selects the ID of the shunts to be switched (Figure 8-12).

• A utility program is activated that exports data from the operational SCADA/EMS database and creates the input file for STABSI as described above. The name of the created ASCII file includes date and time of the snapshot.

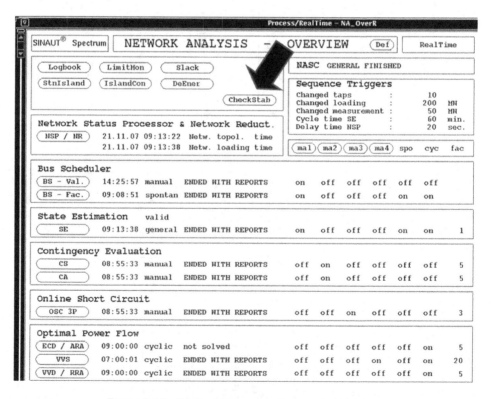

Figure 8-10. NA Real-time execution control display.

```
cmdtool - /home/s/bin/SPUX/NA_CHECKSTAB
-------------------------------------------------------------------------------
|                          NA - Transient Stability Check                      |
---- --------------------------- ----------- ----- ---- --------------------------- ----------- -----
| Nr | Station  VL  Line       Shunt  | Stat |  | Nr | Station  VL  Line       Shunt  | Stat |
---- --------------------------- ----------- ----- ---- --------------------------- ----------- -----
|  1 | FINT W   300             250 SR1 | off |  | 26 | JABR W   300  SALM W F3         |  on |
|  2 | FINT W   300             250 SR2 | off |  | 27 | JABR W   300  SALM W F4         |  on |
|  3 | FRRD W   300             250 SR  |  on |  | 28 | DWPS     300  OMAR W F1         |  on |
|  4 | JABR W   300  TOWN W F1  250 SR1 |  on |  | 29 | DWPS     300  OMAR W F2         |  on |
|  5 | JABR W   300  TOWN W F2  250 SR2 |  on |  | 30 | DWPS     300  OMAR W F3         |  on |
|  6 | SHKH W   300  TOWN W     250 SR1 | off |  | 31 | DWPS     300  OMAR W F4         |  on |
|  7 | SHUB W   300             250 SR  | off |  | 32 | ZSPS     300  JABR W F1         | off |
|  8 | SRRD W   300  SSSM W F   250 SR  | off |  | 33 | ZSPS     300  JABR W F2         | off |
|  9 | JABR W   300  TOWN W F1          |  on |  | 34 | ZSPS     300  SRRD W F1         |  on |
| 10 | JABR W   300  TOWN W F2          |  on |  | 35 | ZSPS     300  SRRD W F2         |  on |
| 11 | SHKH W   300  TOWN W F            |  on |  | 36 | ZSPS     300  SHUB W F1         |  on |
| 12 | OMAR W   300  SHKH W F1          |  on |  | 37 | ZSPS     300  SHUB W F2         |  on |
| 13 | OMAR W   300  SHKH W F2          |  on |  | 38 | FINT W   300  SRRD W F1         | off |
| 14 | JABR W   300  OMAR W F1          |  on |  | 39 | FINT W   30(  SRRD W F2         | off |
| 15 | JABR W   300  OMAR W F2          |  on |  | 40 | FINT W   300  SHUB W F1         | off |
| 16 | OMAR W   300  FRRD W F1          |  on |  | 41 | FIN  W   300  SHUB W F2         | off |
| 17 | OMAR W   300  FRRD W F2          |  on |  | 42 | DWPS     300  SLBY W F1         |  on |
| 18 | FRRD W   300  SSUR W F            |  on |  | 43 | DWPS     300  SLBY W F2         |  on |
| 19 | SRRD W   300  SSUR W F1          |  on |  | 44 | SLBY W   300  JAHR W F1         |  on |
| 20 | SRRD W   300  SSUR W F2          |  on |  | 45 | SLBY W   300  JAHR W F2         |  on |
| 21 | SRRD W   300  SSSM W F            |  on |  | 46 | SLBY W   300  FINT W F1         |  on |
| 22 | FINT W   300  SSSM W F1          |  on |  | 47 | SLBY W   300  FINT W F2         |  on |
| 23 | FINT W   300  SSSM W F2          | off |  | 48 | SRRD W   300  JABR W F1         |  on |
| 24 | JABR W   300  SALM W F1          |  on |  | 49 | SRRD W   300  JABR W F2         |  on |
| 25 | JABR W   300  SALM W F2          |  on |  | 50 | SLBY W   300  SBPS    F1        |  on |
|    |                                  |     |  | 51 | SLBY W   300  SBPS    F2        |  on |
|    |                                  |     |  | 52 | JAHR W   300  SBPS    F1        |  on |
|    |                                  |     |  | 53 | JAHR W   300  SBPS    F2        |  on |
---- --------------------------- ----------- ------ ---- --------------------------- ----------- ------

please select a feeder [1..53 q=Quit] : 4
```

Figure 8-11. SCADA/EMS window showing the list of all switchable shunts on 275 kV line.

- STABSI is executed by a command that specifies the path name of the recently created input data file, and the ID of the shunts to be switched.
- STABSI writes the results in a file that is displayed in the above-mentioned window.

In the view of the operational usability of the real-time stability check in the control room environment, besides sufficiently fast execution time, an adequately compressed display of results is decisive. Since curve displays of the calculated generator rotor an-

```
please select a feeder [1..53 q=Quit] : 4

"JABR W 300 TOWN W F1 250 SR1"   on --> off

begin data export
export successful

checking transient stability - please wait
```

Figure 8-12. User dialog for real-time stability check.

gles would be too complex to comprehend for the control room operator, the critical information conveyed by the maximum distance between a generator rotor angle and the system center of angle is derived. The system center of angle is calculated as the arithmetic average of all generator rotor angles. Thus, the results displayed are reduced to the minimum:

- In case the network is instable, list all the instable generators, for example, Generator_abc in PowerStation_xyz, and the distance to the center of angle is 96.70 degrees.
- In case the network is stable, list the generator with largest rotor angle, for example, the rotor angle of Generator_abc in PowerSation_xyz is 10.20 degrees, and the distance to the center of angle is 49.66 degrees.

The real-time stability-check function does not write any data back to the SCADA/EMS operational database. The results are just displayed in the shell window. Due to this handling concept, the quite complex function of real-time stability checking became extremely clear and easy to use.

8.3.1.9 User Interface for Planning Calculations
As mentioned above, besides the ability to perform stability checks in real time there is also the capability to do comprehensive planning calculations using the full features of the SINCAL package offline. The data transfer from the SCADA/EMS to SINCAL is performed via the same ASCII file mentioned above. The file is produced cyclically (hourly) or on operator request in real time as well as in study-mode. For this purpose, buttons for the activation of the automatic (cyclic) or manual data export are provided to the operator. In real-time mode, the transferred data are based on the results of the last SE run, whereas in study mode these data are based on the results of the last online load-flow execution. The data transfer can also be activated upon user request from the separate SINCAL server.

8.3.2 National Dispatching System of ONE in Morocco

8.3.2.1 Morocco Power System
The electric transmission and distribution grids and the National Dispatching System in Morocco are planned and operated by the Office National de l'Electricitè (ONE). The electric power system of Morocco is linked to the European UCTE system via Spain and to neighboring North African countries (Figure 8-13).

Figure 8-13 suggests how important is Morocco's transmission grid for extending the UCTE system with the so-called Mediterranean Ring, which interconnects the power systems in North Africa, the Middle East, and Turkey with the UCTE network. Several planning studies about this future interconnection show that transient stability becomes more and more important to ensure secure and reliable operation of the interconnected system.

The transmission grid, schematically depicted in Figure 8-14, comprises 400 kV, 225 kV, 150 kV, and 60 kV lines with a total length of 17.186 km (2004). The total in-

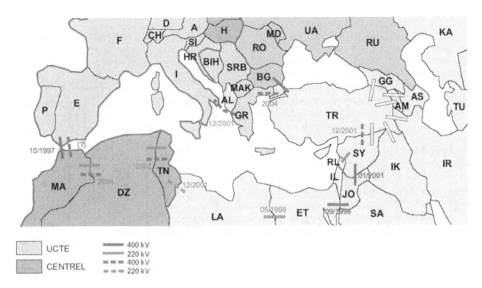

Figure 8-13. Morocco in the Mediterranean context.

stalled generation capacity at the end of 2004 was 4621 MW, including 3069 MW in thermal power plants and 1265 in hydro stations. In 2004, the total energy consumption was 17,945 GWh and the energy exchange with Spain was 1554 GWh.

The electrical energy demand in Morocco shows high evolution rates. The system load curves of two comparable days in 2005 and 2007 show the rapid growth of electricity demand (Figure 8-15).

8.3.2.2 The SCADA/EMS

In 2004, ONE undertook the renewal of its national control center. The project aimed at providing ONE with a modern, reliable, and high-performance control tool to:

- Provide remote control of 85 facilities as opposed to the 54 current ones.
- Process 13,000 items of remote information, compared with the 6000 remote data items processed previously.
- Generalize the remote control of the substations in the very high and high-voltage grids.
- Implement a hierarchical control strategy whereby the control of the national network (400 kV, 250 kV network and interconnections) is performed by the National Dispatching system, whereas the control operations on injections and the 60 kV network are performed by the regional divisions.
- Control and optimize the utilization of energy generation, transmission, and interconnection facilities.
- Use advanced network analysis applications to monitor and optimize the transmission grid in real-time and/or study-mode operation, including state estima-

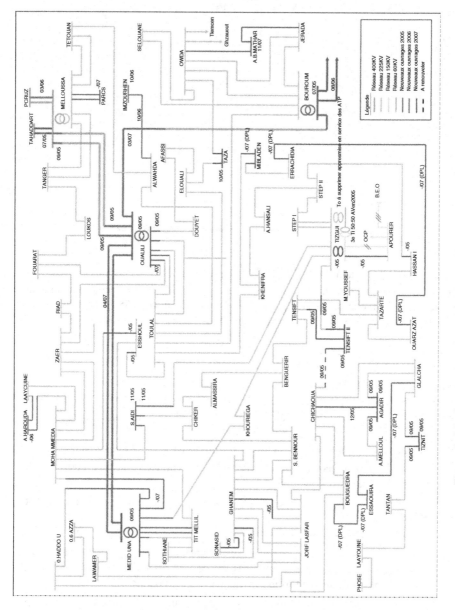

Figure 8-14. One-line diagram of the transmission grid of Morocco.

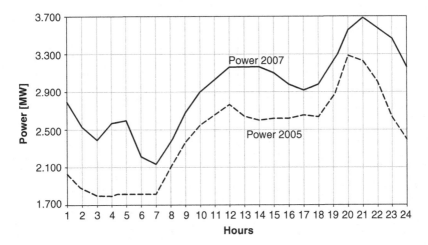

<u>Figure 8-15.</u> Comparison of load consumption of 9/19/2005 with 9/17/2007.

tion, short-circuit calculation, dispatcher power flow, optimal power flow, and dynamic security assessment.

8.3.2.3 Solution Architecture

The SINAUT Spectrum SCADA/EMS, including the NA subsystem, delivered to the National Dispatching System of ONE did not contain a module for transient stability assessment. In order to provide the missing functionality, the stability tool PSS™ NETOMAC described in the Section 8.2.2 of this book was loosely integrated with the SINAUT Spectrum system. Accordingly, the SINAUT Spectrum native network analysis applications, such as state estimator, dispatcher power flow, and so on, are executed on the NA server; DSA and Transient Stability Analysis (TSA) are performed by PSS™ NETOMAC, which runs on its own Windows PC. This is shown in Figure 8-16.

8.3.2.4 Implementation

The integration of PSS™ NETOMAC-based DSA with the NA subsystem that runs on SINAUT Spectrum was performed as shown earlier in this chapter (Figure 8-4). At the present time, the solution implementation for ONE does not return DSA reports to the SCADA/EMS system, but this might be a future option.

The data flow between NA and DSA is illustrated in Figure 8-17. From the DSA PC, the user can request a data snapshot from state estimator (SE) or dispatcher power flow (DPF). Those data are automatically exported by using the standard NA data eXport (NAX) tool of SINAUT Spectrum and copied to the local disk drive of the DSA PC. PSS™ NETOMAC automatically imports those data files and also reads additional dynamic data from its own database since, as discussed earlier, dynamic data are not available in the SINAUT Spectrum data model.

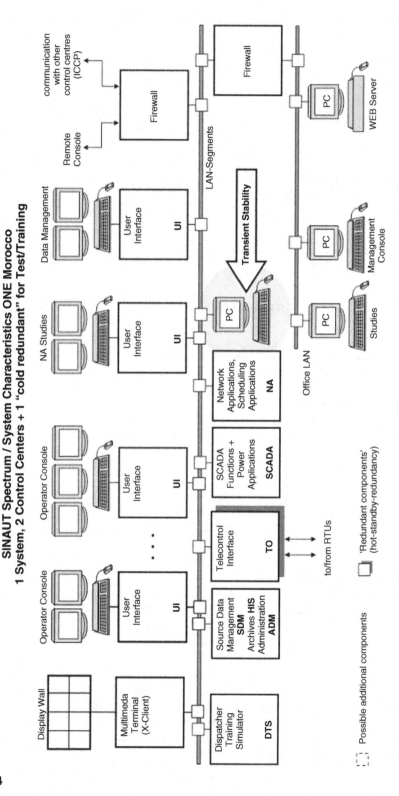

Figure 8-16. Conceptual system configuration of ONE at the National Control Center.

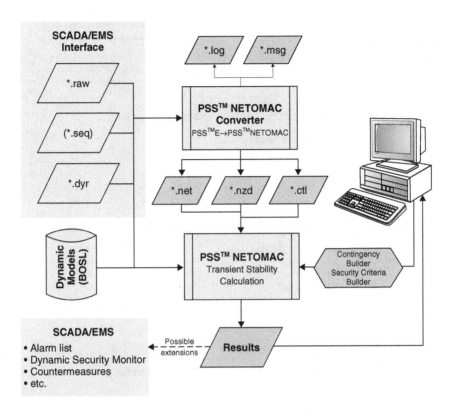

Figure 8-17. DSA–NA data flow.

Illustrated in Figure 8-18 is the user interface that allows building the contingencies to be investigated by DSA. The contingency builder provides a user-friendly interface to prepare a set of events or contingency lists. Those generated files are used by PSS™ NETOMAC in order to simulate the degraded states of the main network during contingencies. After the user has defined a set of contingencies on the contingency builder template, the file containing all the contingencies is saved on the project folder in Extensible Markup Language (XML) format. After the creation of the contingencies, the contingency builder can quickly determine if the power system working point is harmless or not. It gives the operator a general overview of the system, classifying the state of the system as "normal" (green), "warning" (yellow), or "harmful" (red). This is a first approach to the contingency evaluation.

The security criteria builder shown in Figure 8-19 is a program which provides a user-friendly interface to prepare a set of controllers for the evaluation of criteria. Basic input to this program are the static and dynamic network data and the XML file created before by the contingency builder. These macros created by the criteria builder check that typical network parameters remain in a range or do not overcome a predefined value. Typical parameters are, for example, angle of generators, voltage at generator nodes,

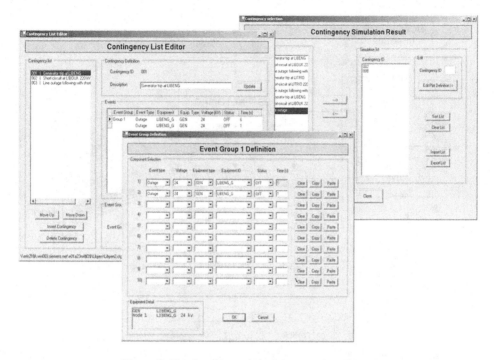

Figure 8-18. Contingency builder user interface.

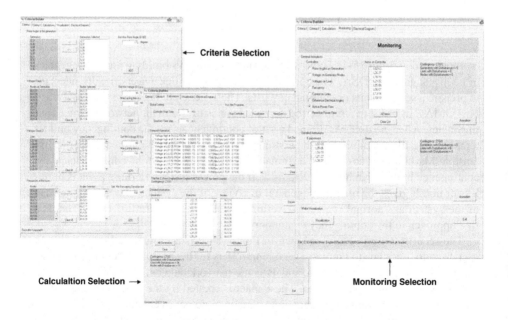

Figure 8-19. Criteria builder user interface.

current on lines, voltages between nodes, active and reactive power flow, electrical angle differences, and frequency deviations. The generated criteria files not only check the defined criteria during the calculation but also cause PSS™ NETOMAC to write an "Alarm" to the report file in the case of unsuccessful contingency calculation.

In order to get more information about the system under evaluation, recalculations have to be done based on the "result protocol." The re-evaluation of the contingencies is user-defined. In order to save time during the calculation, the recalculation procedure has been divided into "general information" and "detailed information." The results are taken for each contingency and shown in checklist boxes. Here, the user can choose to reevaluate the case he considers more urgent.

The experience has shown that human beings are able to process and to analyze visual information faster than numerical information (text information). With the help of graphical representations for voltages, currents, and power-flow profiles it is possible for the operator to recognize in a short period of time the weakest points of the network. It allows the operator to make the right decision at the right time in order to avoid further consequences for the network. For example, the red color of increasing or decreasing brightness means that a transmission line is overloaded, the brightness suggesting how dangerous this situation might be. The same information would take several minutes to review and digest if it was to be gathered in text format. As fast as these weak points are found, the operator can take preventive measures to avoid undesirable situations that could lead to the further development of the failure. Figure 8-20 illustrates several examples of this visualization approach.

8.3.3 UCTE Study—The European Interconnected System

The UCTE system was used to demonstrate the performance of the DSA. In the year 2004, the system had an installed capacity of approximately 530,000 MW with a peak

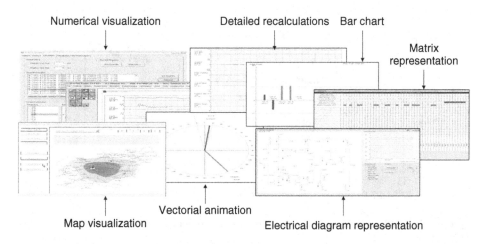

Figure 8-20. Visualization of DSA and TSA results.

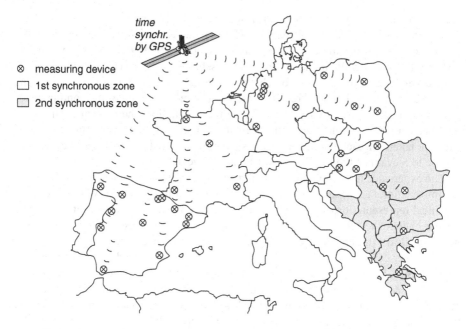

<u>Figure 8-21.</u> UCTE Wide Area Measurement System (2004 configuration).

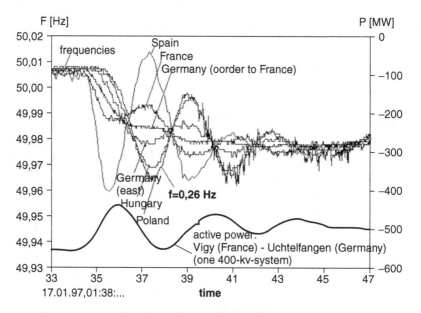

<u>Figure 8-22.</u> Actual WAMS readings of interarea oscillations after tripping a 300 MW unit.

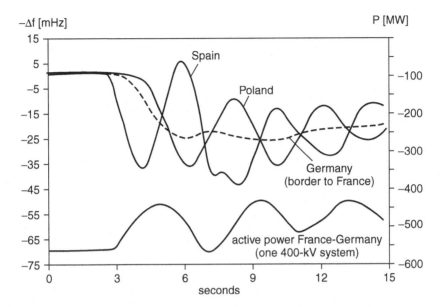

Figure 8-23. Computer simulation of interarea oscillations after tripping the 300 MW unit.

load demand of about 386,000 MW. A system model was built with 610 generators, 4400 buses (electrical nodes), 12,000 grid branches and 1050 controllers. This model was tested using measurements collected from the Wide Area Measurement System (WAMS) shown in Figure 8-21 and described in [14].

The measured interarea oscillation after tripping a 300 MW power station is illustrated in Figure 8-22 and the simulation of the event for 15 seconds is illustrated in Figure 8-23. The results demonstrate that the model represents faithfully the overall electromechanical system behavior.

Shown in Figure 8-24 is the time domain simulation of the UCTE system extended with the Mediterranean Ring mentioned in Section 8.3.2. The limit to run the system slower than real-time conditions was set to time steps of 10 ms. As illustrated in Figure 8-25, the accuracy of simulating the electromechanical behavior under real-time conditions was found suitable with time steps of 20 to 50 ms.

Using the eigenvalue mode of the program system, the interarea oscillations of the power system can be easily monitored and the program shows how and which generators are involved in the oscillation (Figure 8-26).

Because of the flexible change from time-domain to frequency-domain calculation, remedial actions and preventive measures can be determined very quickly. Figure 8-27 depicts countermeasures at different generators to increase damping in the system, here shown in the time domain, but analyzed in the frequency domain by system eigenvectors and residues.

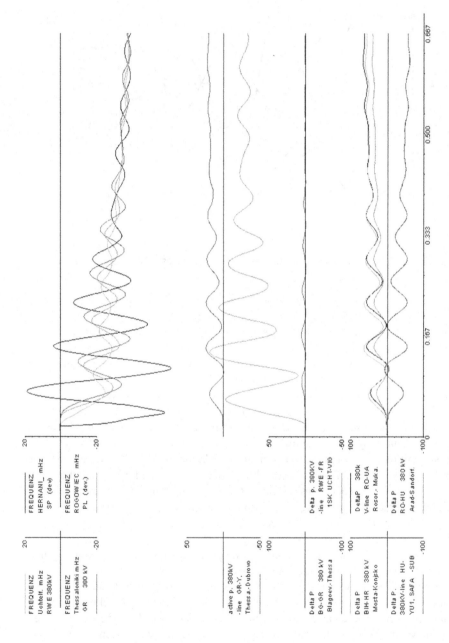

Figure 8-24. Time domain simulation of 300 MW trip in the UCTE extended system.

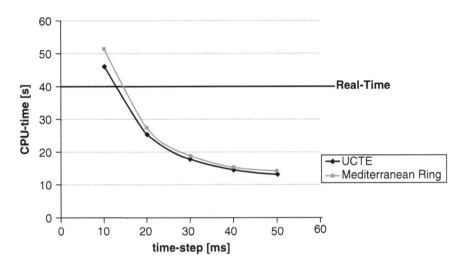

Figure 8-25. Computation time corresponding to selected time steps.

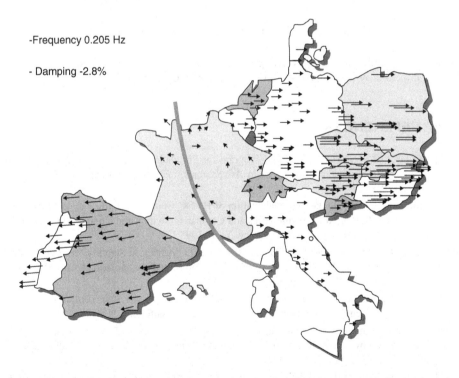

Figure 8-26. Monitoring the geographical mode shape of an interarea oscillation in the UCTE system. Spain oscillates against Central Europe and the CENTREL countries.

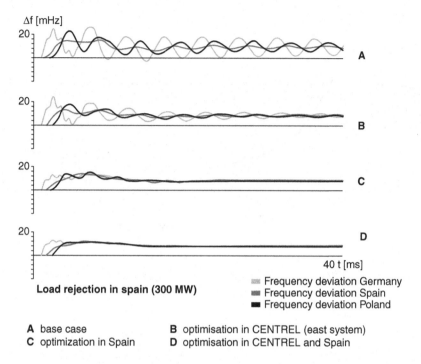

Load rejection in spain (300 MW)

▨ Frequency deviation Germany
▬ Frequency deviation Spain
■ Frequency deviation Poland

A base case   **B** optimisation in CENTREL (east system)
C optimization in Spain   **D** optimisation in CENTREL and Spain

<u>Figure 8-27.</u> Countermeasures to improve system stability and reduce interarea oscillation, checked by frequency-deviation monitoring in different countries.

8.4 REFERENCES

[1] Kulicke, B., "NETOMAC Digital Program for Simulating Electromechanical and Electromagnetic Transient Phenomena in AC Systems," *Elektrizitätswirtschaft,* Vol. 1, pp. 18–23, 1979.

[2] Rittiger, J., and Kulicke, B., "Calculation of HVDC Converter Harmonics in Frequency Domain with Regard to Asymmetries and Comparison with Time Domain Simulations," presented at IEEE Winter Meeting, New York, 1995.

[3] Povh, D., Rittiger, J., Kulicke, B., and Clodius, D., "Advanced Stability Program for the Simulation of HVDC in Large Power Systems," presented at International Conference on Power System Transients, Lisbon, Sept. 1995.

[4] Povh, D., Lei, X., Lerch, E., and Kulicke, B., "Optimization—A New Tool in Simulation Program System," presented at IEEE Winter Meeting, Baltimore, Jan. 1996.

[5] Lei, X., Povh, D., and Lerch, E., "Global Parameter Settings of FACTS-Controllers for Improving Power System Stability," presented at International Conference on Power System Transients, Lisbon, 1995.

[6] Lehn, P., Rittiger, J., and Kulicke, B., "Comparison of the ATP Version of the EMTP and the NETOMAC Program for Simulation of HVDC Systems," presented at IEEE Winter-Meeting, New York, Feb. 1995.

[7]   Winter, W., and Kulicke, B., "Automated Closed Loop Testing Of Microprocessor Protective Relays with the Real Time Digital Network Model," presented at ICDS '97, Montreal, Canada, 1997.

[8]   Kulicke, B., Pannhorst, D., Winter, W., and Eickmeyer, D., "Hardware-in-the-Loop Test Using the Real Time Simulator NETOMAC," presented at ICDS '97, Montreal, Canada, 1997.

[9]   De Wilde, V., Duschl-Graw. G., Pannhorst, D., and Ruhle, O., "Sophisticated Real-Time Tests on Protection Relays and Turbine Controllers with DINEMO-II and PSS™ NETOMAC," presented at EUROSIM 2007, Ljubljana, Slovenia, 2007.

[10] Lei, X., Lerch, E., Povh, D., and Ruhle, O., A Large Integrated Power System Software Package—NETOMAC, presented at Powercon, Beijing, China, August 1998.

[11] Lerch, E., Kulicke, B., Ruhle, O., and Winter, W., "NETOMAC—Calculating, Analyzing and Optimizing the Dynamic of Electrical Systems in Time and Frequency Domain," 3rd IPST '99, Hungary, June 1999.

[12] Wang, X., *Modal Analysis of Large Interconnected Power Systems,* VDI Verlag, Düsseldorf, 1997.

[13] Witzmann, R., Rittiger, R., and Winter, J., "Inter-Area Oscillations during Development of Large Interconnected Power Systems" presented at CIGRE Symposium, Working Plant and Systems Harder, London, 7–9 June 1999.

[14] Breulmann, H., Grebe, E., Loesing, M., Winter, W., Witzmann, W., Dupuis, P., Houry, M. P., Margotin, T., Zerenyi, J., Dudzik, J., Machowski, J., Martin, L., Rodriguez, J. M, and Urretavizcaya, E., "Analysis and Damping of Inter-Area Oscillations in the UCTE/CENTREL Power System," presented at CIGRE 2000, Paris, pp. 38–113, 2000.

[15] Beißler, G., and Schellstede, G., "Advanced Network Analysis Functions for Supporting Power System Operation," in *Proceedings of IFAC Symposium on Control of Power Plants and Power Systems,* March 9–11, Munich, Germany, 1992.

ANNEX 8-1.   FURTHER DYNAMIC SIMULATION CAPABILITIES

Time Frame for Dynamic Simulations

The bandwidth in which DSA can be carried out at present ranges from extremely fast traveling-wave phenomena on overhead power lines to the slow control phenomena of steam turbines (Figure 8-28). Real-time simulation of electromechanical transients of large systems is also possible and is useful to interactively test actual equipment, for example, protective relaying [7,8,9] or control equipment by using the test hardware DINEMO-II (Digital Network Model).

Simulation in the Frequency Domain

In addition to the time domain simulation functionality described in Section 8.2.2, the DSA software incorporates frequency domain analysis as well. For this, starting from a load-flow case, an automatic linearization of the entire system, including network, machines, control systems, machine shafts, and so on, is performed around the system operating point.

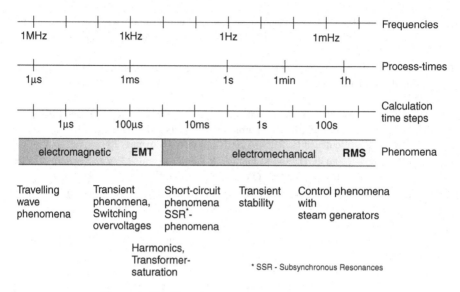

Figure 8-28. Time frame for dynamic simulations.

This provides access to the small-signal behavior of the whole system. Network elements, machines, and control systems can be represented via transfer functions (Bode diagram and Nyquist diagram) so that ordinary conventional methods can be used for the design of control hardware; for example, Figure 8-29 shows a typical calculation in the frequency domain.

Eigenvalue and Modal Analysis

In the present context of subcontinental and continental grid interconnections, small-signal stability analysis and, in particular, the analysis of interarea oscillations become more and more important. Many electric systems worldwide are experiencing increased loading on portions of their transmission systems that can, and sometimes do, lead to poorly damped, low-frequency (0.2–0.8 Hz) interarea oscillations. This topic has been extensively addressed for a long time in conjunction with power systems for which the extension of the grid and the high level of power transfers led to stability problems. Interarea oscillations can severely restrict system operations by requiring the curtailment of electric power transfers as an operational measure. These oscillations can also lead to widespread system disturbances if cascading outages of transmission lines occur due to oscillatory power swings.

Eigenvalue or modal analysis describes the small-signal behavior of the system—the behavior linearized around one operating point—and does not take into account the nonlinear behavior of system components, for example, controllers, during large system perturbations. Therefore, time-domain simulation and modal analysis in the frequency domain complement each other in analyzing power systems. This is schematically illustrated in Figure 8-30.

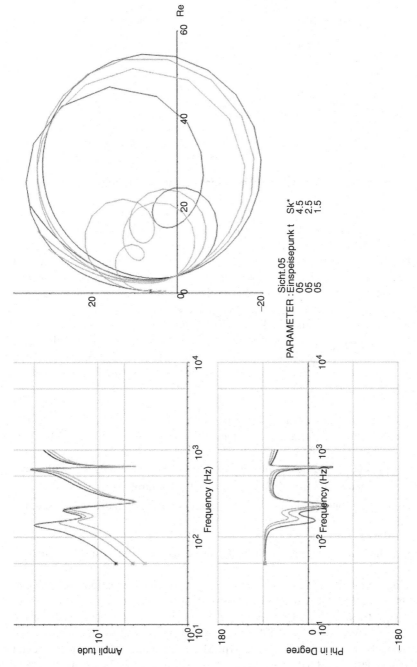

Figure 8-29. Presentation of calculation results in frequency-domain coordinates.

215

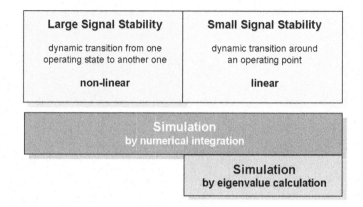

<u>Figure 8-30.</u> Large signal versus small signal stability calculations.

The *eigenvalue analysis* investigates the dynamic behavior of a power system under different characteristic frequencies ("modes"). In a power system, it is required that all modes be stable. Moreover, it is desired that all electromechanical oscillations be damped out as quickly as possible. The results of an eigenvalue analysis are given as frequency and relative damping for each oscillatory mode to make them easier to understand. A damping ratio of 5% means that in three oscillation periods the amplitude is damped to about 32% of its initial value. The minimum acceptable level of damping is not clearly known. A damping ratio less than 3% must be accepted with caution. Damping is considered adequate if all electromechanical modes have a predicted damping ratio of at least 5%. Figure 8-31 shows how the damping of a system can be easily analyzed.

In addition, the *modal analysis* allows a much deeper view of a system by not only interpreting the eigenvalues but by analyzing the eigenvectors of a system, which are automatically calculated during the modal analysis:

- The right eigenvector gives information about the observability of oscillation.
- The left eigenvector gives information about the controllability.
- The combination of right and left eigenvectors (residues) indicates the setting of controllers.

For example, the eigenvectors of a 0.3 Hz interarea oscillation are shown in Figure 8-32.

Damping the interarea oscillations is very important. The oscillation can be damped when extra energy is injected into the system, which is instantaneously decelerated, and/or when extra energy is consumed in the system, which is instantaneously accelerated. In actual power systems, the damping energy is obtained by the modulation of load or generation for a period of time, typically in the range of five to ten seconds. The damping energy must have the correct phase shift relative to the accelerated/decel-

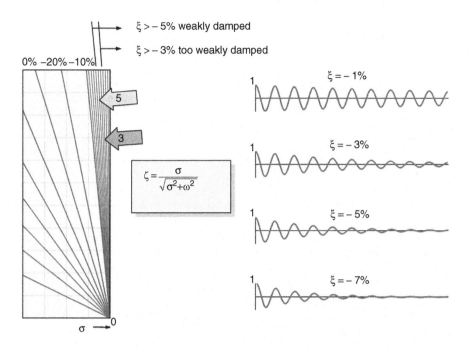

Figure 8-31. Criteria for weak and well-damped systems.

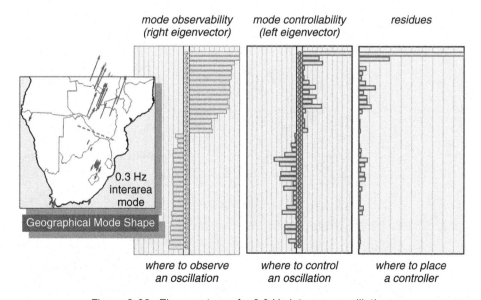

Figure 8-32. Eigenvectors of a 0.3 Hz interarea oscillation.

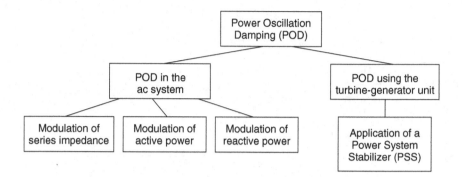

<u>Figure 8-33.</u> Strategies to dump power oscillations.

erated systems. Wrong phase angles can even excite power oscillations. Figure 8-33 shows different strategies to damp power oscillations.

The best damping location can be found by using the system eigenvectors, as shown in Figure 8-32. Depending upon the chosen damping strategy (Figure 8-33), the residues chart will show the location(s) for power system stabilizers (PSS) (generator bar chart), SVCs (bus bar chart), TCSCs (line bar chart), and so on.

9

REAL-TIME TRANSIENT SECURITY ASSESSMENT IN AUSTRALIA AT NEMMCO

Stephen J. Boroczky

9.1 INTRODUCTION

9.1.1 Overview of the Power System under NEMMCO's Oversight

Australia's largest interconnected power system encompasses the states of Queensland, New South Wales, Victoria, South Australia, and Tasmania, with transmission voltage levels ranging from 500 kV down to 110 kV. This interconnected system is illustrated in Figure 9-1 and represents a power system that stretches 5000 km and has a maximum demand of 33 GW. NEMMCO is both the Independent System Operator and the Market System Operator and is responsible for the secure operation of this power system under the provisions of the National Electricity Rules.

The National Electricity Market (NEM) is a regional-based market managed by NEMMCO. Generation is optimized and dispatched on a five minute market cycle that is binding. Constraints are applied in the dispatch solution to ensure that the power system will remain in a secure condition.

These constraints, which represent both the static and the dynamic security requirements on the system, can be applied as and when required and are usually formulated using offline system analysis. NEMMCO has the responsibility to invoke appropriate constraints in order to ensure secure operation. Control of the power system is thus

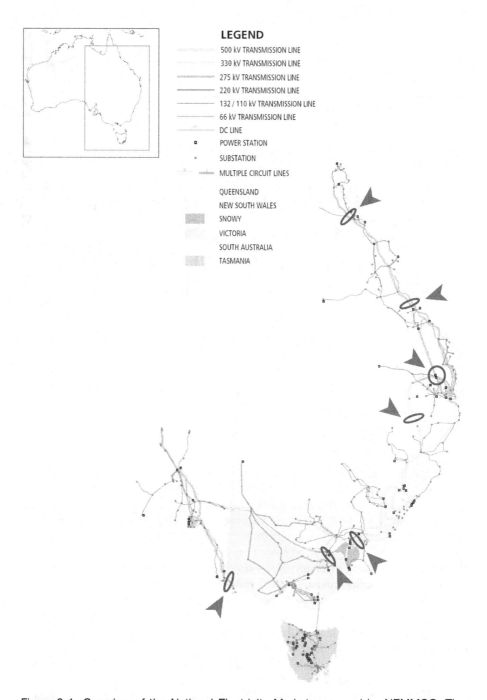

LEGEND
- 500 kV TRANSMISSION LINE
- 330 kV TRANSMISSION LINE
- 275 kV TRANSMISSION LINE
- 220 kV TRANSMISSION LINE
- 132 / 110 kV TRANSMISSION LINE
- 66 kV TRANSMISSION LINE
- DC LINE
- POWER STATION
- SUBSTATION
- MULTIPLE CIRCUIT LINES

QUEENSLAND
NEW SOUTH WALES
SNOWY
VICTORIA
SOUTH AUSTRALIA
TASMANIA

Figure 9-1. Overview of the National Electricity Market managed by NEMMCO. The market regions roughly correspond to state political boundaries. Interfaces prone to dynamic security issues are indicated.

achieved by the application of constraints in the dispatch process. In addition, NEMM-CO uses real-time security assessment tools to monitor the security of the power system and achieves closed-loop control via operator intervention through the invocation of discretionary constraints to limit flow.

9.1.2 Key Operational Reliability Requirements

NEMMCO has the responsibility to operate the National Electricity Market under the National Electricity Rules that require the power system to be returned to a secure operating state within 30 minutes of having entered an insecure postcontingency operating condition. Credible contingencies to be examined are defined as the loss of a single item of plant (e.g., generator or line). Credible faults, for the purposes of transient security, are generally considered to be two-phase earth faults of single lines. Double-line faults are not normally considered as credible contingencies, unless there are unusual conditions, such as adverse weather conditions or bushfires that can increase the chance of such contingencies occurring. Some double-line faults are considered where there are a number of lines that share a right of way and have been declared as credible.

Thus, during the normal course of operating the system, NEMMCO can declare certain contingencies to be credible or noncredible. In these situations, it is important to be able to add or remove contingency definitions for analysis. Similarly, in the dispatch process, different constraints can be invoked, depending on what contingencies are considered to be credible at the time.

Adjustments to the power system are usually achieved by the invocation of constraints, which can only be applied at each five-minute dispatch cycle. Thus, if the security assessment tool returns a result within 10–15 minutes, then there are only two or three dispatch cycles left for an operator to interpret the results and take the appropriate action to restore the power system to a secure operating state. Hence, the desired performance requirement would be for a security assessment tool to be able to return a result within no more than a 10–15 minute time frame.

9.1.3 General Description of Security Analysis at NEMMCO

NEMMCO uses a variety of tools to assess the real-time security of this power system. The network model comprises approximately 2100 buses, 1400 lines, and almost 350 generating units, representing a system with a maximum demand of about 33 GW. As the Independent System Operator, NEMMCO is rather fortunate in that it has oversight of the entire physically interconnected power system and does not have to deal with issues of neighboring interconnected systems.

9.1.3.1 Static Security

NEMMCO uses the SCADA/EMS Contingency Analysis program to analyze static security. $N - 1$ contingency analysis is performed for the loss of every single item of plant within NEMMCO's direct oversight. There are also a number of $N - 2$ contingencies defined which are enabled when they are declared to be *credible contingen-*

cies. The effect of special protection systems are also evaluated during the analysis. The assessment involves the monitoring of

- Line and transformer thermal limits,
- Bus over- and under-voltage limits,
- Voltage angle limits between pairs of load buses as provided by state estimation rather than by telemetered quantities, and
- Interface flow limits, which represent limits derived from offline studies of other stability criteria.

In addition to contingency analysis, NEMMCO also runs online short-circuit analysis to ensure that there are no violations of terminal equipment short-circuit ratings. There are a number of areas in which the short-circuit levels have grown to critical levels over the years of operation.

From the market perspective, constraints to limit postcontingency branch flows are formulated offline for a large variety of system conditions. In real time, the appropriate constraints are applied in the market dispatch engine to enforce limits on these branch flows. NEMMCO is currently implementing a constraint-automation tool that will bring the thermal constraint-formulation process to the near real-time frame. When the contingency analysis indicates that there is potential for a postcontingency branch violation, a constraint can be created and posted on the market system, thereby limiting the postcontingency flow on the branch.

9.1.3.2 Voltage Security

Real-time voltage security assessment is rather limited at present to the assumption that an unsolved contingency analysis solution, that is, a diverging postcontingency load flow, is possibly a result of the proximity to the state of voltage collapse. A number of voltage security limits are determined in the offline environment in order to maintain the required reactive margin as specified under the provisions on the National Electricity Rules. This is predominantly based on Q–V analysis, though some P–V and modal analysis is also performed. A limited number of transient security limits have also been formulated with voltage security considerations as part of the offline time-domain simulations. However, work needs to be done to improve real-time assessment of voltage security, and several options are being investigated to address this issue.

9.1.3.3 Transient Security

A power system that is long and thin, stretching for some 5000 km hugging the Australian coast, is likely to be subject to transient stability problems. The critical points are along interregional boundaries where the transmission corridors are narrow, and in some cases only a double circuit line spans these regions. Accordingly, transient security is one of the most studied forms of dynamic stability in the Australian context.

The Australian power system is particularly sensitive to the control system modeling details and a lot of effort has also gone into developing large-signal models for transient stability analysis, so much so that there is now a disproportionately large number of user-defined models required for the Australian power system. Offline studies are performed to assess the transient- tability characteristics of the power system and to derive interconnection operating limits that encompass the many scenarios examined in these studies. These operating limits, which are specified in terms of system variables such as demand, generation pattern, and inertia are then applied by NEMM-CO as constraints in the electricity market dispatch process.

To determine the transient stability of the power system, transmission network service providers and NEMMCO are largely dependent on time domain simulations, predominantly using PSS/E. Direct methods are usually treated with suspicion, largely because of the importance of control system limit action and other nonlinearities in the simulation of the Australian power system. Although these limits are applied to the dispatch process to ensure that the power system remains stable, NEMMCO has also developed a real-time security assessment tool that monitors the real-time transient security of the power system, based on state- stimator snapshots of the system.

9.1.3.4 Small-Signal Security

Small-signal stability is of major concern to NEMMCO and network service providers in Australia. At present there is no real-time postcontingency assessment of small-signal security and NEMMCO is reliant on offline studies to produce operating limits. In addition, NEMMCO has two tools that physically measure the system damping modes present in the power system: the Psymetrix Monitor (Psymetrix) and the Oscillatory Stability Monitor (OSM). These tools are integral to the operation of the power system. When sufficiently low damping is detected, additional constraints are enabled in the dispatch engine to limit power transfers across various interfaces.

Psymetrix extracts modal damping information from the constant perturbations present in interregional line flows and analyzes a three-minute window of sampled tie-line flow data at various points in the system. It monitors a number of telemetered buses spread across the mainland, each sampling at a rate of 50 samples/second. At each of these buses, the different modes can be detected to varying degrees. Because of its short sample window, Psymetrix can detect instances of poor damping rather quickly, in which case NEMMCO can take corrective action.

The OSM is a wide-area damping measurement tool that has been installed in Tasmania and on the mainland. It measures the damping modes within the two islands. Every five minutes, it uses a three-hour rolling window of phasor and voltage samples from a number of data acquisition points in the system to estimate the system damping modes. It has been used for the calibration of the offline small-signal models and for stabilizer performance measurement.

9.1.3.5 Frequency Security

Frequency control ancillary services (FCAS) ensure that the power system frequency is maintained within the frequency standards for credible contingency events such as loss of generation, loss of load, and network separation. They are dispatched in terms of fast (6 seconds), slow (60 seconds), and delayed (5 minutes) time frames, and both raise and lower ancillary services, based on bids. A "raise" service is one that raises system frequency. The time frames fit together so that a rapid and sustained response can provide each of the three services sequentially. Governor control usually provides most fast and slow FCAS, but delayed FCAS is usually provided by manual control of generating units. Load shedding or other load control can also provide these services.

In addition to the energy market, NEMMCO manages six ancillary service markets to dispatch the required raise and lower FCAS for the three time frames. Two more markets are used to dispatch raise and lower regulating services. The amount of each service needs to be sufficient to maintain frequency within the required bands for the contingency with the largest FCAS requirement. The allocated FCAS is simply the amount of generation or load at risk, less the expected amount of load relief, as a result of load frequency dependence at the relevant frequency standard limit. This method, however, suffers from a number of assumptions. It ignores the power system inertia and assumes that the size of the disturbance is sufficiently small and that the rate of change of frequency is zero at 6, 60, and 300 seconds after the event.

With the completion of the HVDC link between the mainland and Tasmania, where the size of the link represents a substantial portion of the Tasmanian load and the island can have a relatively low inertia, a better method of determining the requirement was necessary. In addition, a special protection system (SPS) would trip a number of generators or loads for the contingent loss of the HVDC link, resulting in significant inertia or load changes on the island. A new method is now being used to calculate the FCAS requirements in Tasmania. This method models inertia, load relief, and the various ancillary services, and requires the following inputs:

- Total regional demand and generation,
- Total regional (generation) inertia,
- Amount of generation or load armed for SPS operation,
- Inertias of generating units that are armed for SPS operation, and
- Modeled load relief factor (% change in load/% change in frequency).

This information is readily available in the SCADA/EMS and, as the calculation of the FCAS requirements is an iterative solution of a number of nonlinear equations, it is most practical that this be performed on the SCADA/EMS. The solution is passed to market systems to dispatch the various FCAS markets.

Benchmarking against time domain simulation has shown the accuracy of this method to be superior. It is intended that it will later be used for dispatching the FCAS markets of the mainland.

9.2 TRANSIENT SECURITY ASSESSMENT AT NEMMCO

9.2.1 SCADA/EMS Environment Overview

NEMMCO maintains two coprimary control centers, each fully equipped with redundant AREVA SCADA/EMS systems.

Figure 9-2 illustrates the computer configuration of the SCADA/EMS systems and the DSA servers. Note that the two sites operate totally independently and collect field

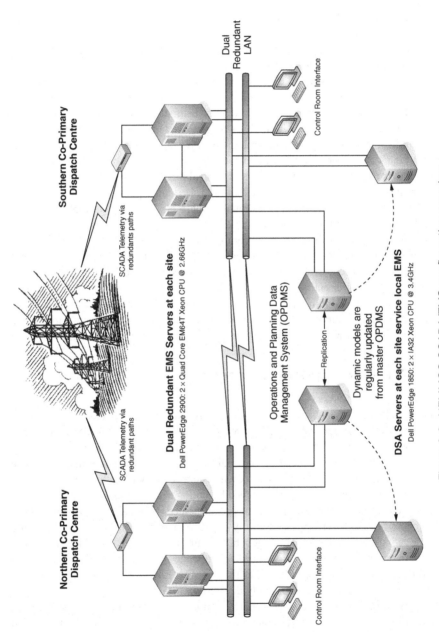

Figure 9-2. NEMMCO SCADA/EMS configuration overview.

Southern Co-Primary Dispatch Centre

Northern Co-Primary Dispatch Centre

Dual Redundant LAN

Control Room Interface

Control Room Interface

SCADA Telemetry via redundant paths

SCADA Telemetry via redundants paths

Dual Redundant EMS Servers at each site
Dell PowerEdge 2900: 2 × Quad Core EM64T Xeon CPU @ 2.66GHz

Operations and Planning Data Management System (OPDMS)

Replication

Dynamic models are regularly updated from master OPDMS

DSA Servers at each site service local EMS
Dell PowerEdge 1850: 2 × IA32 Xeon CPU @ 3.4GHz

225

measurements via redundant telecommunication paths. All communication paths and network adapters are duplicated for redundancy and all critical plant is powered by dual redundant UPS supplies.

Each control room has access to both SCADA/EMS systems. The two sites are operated as coprimary control centers, each site sharing the responsibility to operate the power system and the energy market, but can, nevertheless, take over the operation of the other center should the need arise.

The SCADA/EMS servers operate as primary/hot standby pairs and provide the primary function of supervisory control and data acquisition as well as the energy management functions. These functions include state estimation every minute and contingency analysis and short-circuit analysis every five minutes. In addition, AGC provides power system frequency control and ensures that scheduled generations reach their targets as dispatched by the market-clearing engine. AGC can only be active at one site and is the main form of telemetered control that NEMMCO has in the National Electricity Market. Other plant is usually controlled and operated by the asset owners under NEMMCO's direction. Additionally, there is a plethora of applications for demand estimation and for monitoring market and power system performance.

Once the estimation of the power system state is complete, the SCADA/EMS passes the resultant state estimator snapshot together with a list of contingencies to the DSA servers for further processing. Communication between the SCADA/EMS and DSA servers is by file transfer via file sharing. The DSA combines this with the dynamic plant models, which are maintained by the Operations and Planning Data Management System, to analyze the transient behavior of the power system. The results are passed back to the SCADA/EMS servers for display on the standard SCADA/EMS interface.

9.2.2   Online DSA System Architecture

The Dynamic Security Assessment (DSA), as the application is known, is the main online facility NEMMCO has for the estimation of dynamic security. As shown in Figure 9-2, there is one DSA server at each site servicing the site SCADA/EMS. It was considered that redundant DSA servers were not necessary at each site because the alternate site server could always perform the analysis for both sites. The primary purpose of the DSA is to assess the postcontingency transient security of the power system under NEMMCO's oversight. This is achieved by taking state-estimator snapshots of the power system and simulating a number of predefined contingencies in a power system simulator on the DSA server. The stability of the system is ascertained by observing the rotor swing characteristics of the generators during the simulation. The greater the rotor angle separation between any two machines, the more unstable the case is considered to be. The results are fed back to the SCADA/EMS for display and alarming. The overall DSA process is conceptually depicted in Figure 9-3.

The DSA process is essentially initiated and controlled by an application on the SCADA/EMS. All operator interaction with the DSA is performed on the SCADA/EMS. Only planning and support personnel have access to the DSA server directly. Thus, the SCADA/EMS is the source, not only for the steady-state network model, but also for the dynamic contingency models. This is necessary because of the

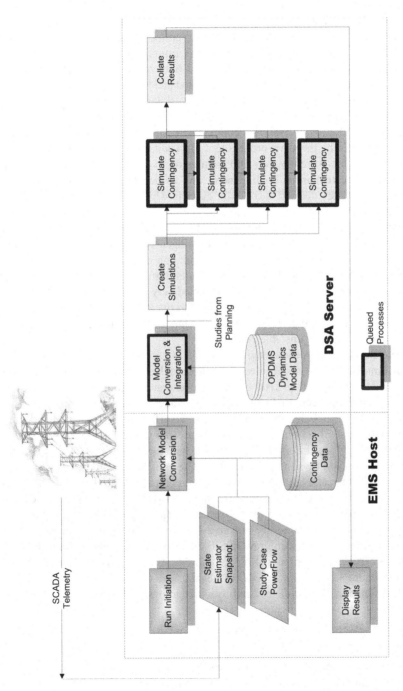

Figure 9-3. NEMMCO DSA Architecture. Upon run initiation, either the currently estimated power system state or a study-case power flow can be combined with the contingency data to create a case for analysis within the DSA. This is then processed on the DSA server, which passes the results back to the EMS for alarming and display.

need to be able to select and change the contingencies to be analyzed from the SCA-DA/EMS. A real-time DSA run is only initiated by the SCADA/EMS application when a valid state estimation of the system is available. The estimator solution is written to a flat file in a form that reflects the data structures of the SCADA/EMS. At this point, it is necessary to convert the network model from the SCADA/EMS representation, in the form of nodes and breakers, to a bus–branch form that is suitable for the PSS/E simulation engine. Thus, nodes are collapsed and breakers are removed.

Once it is passed on to the DSA server, the network model is combined with the dynamic models and input files are created that are suitable for the dynamic simulation engine, PSS/E. In addition, the selected contingencies are also dumped to a flat file and the data is converted by the same process to a form suitable for the simulation engine to interpret. There are currently over 100 contingencies defined, of which 60 are normally selected. The remaining contingencies are defined for unusual system conditions, most of which are defined to cover for cases in which there is a prior outage of a double circuit line that should be considered. Other contingencies also cater for multiple line faults for those times they are declared to be credible. The contingencies are grouped for rapid selection and, if necessary, contingencies can also be defined online. A control program on the server initiates a simulation of each contingency, starting with the contingency that produced the worst results from the previous run. In this way it is likely that an unstable condition can be detected very early in the DSA run. The results of each simulation are collated and returned to the SCADA/EMS as soon as they are available. All results are displayed in the SCADA/EMS in a form that ranks the results starting with the most unstable contingency.

The DSA servers, currently a single Dell 1850 at each site with dual 3.2 GHz hyperthreaded Intel IA32 CPUs, can adequately process the 60 contingencies for a model that consists of approximately 2100 buses, 350 generators, and almost 800 dynamic models. Even with the recent inclusion of HVDC models and the reduction of integration time step down to 1 ms, the configuration is sufficient to perform five second PSS/E simulations of all contingencies in less than ten minutes.

The handshaking and information exchange between the SCADA/EMS and the DSA server is achieved by simple file sharing. The applications at both ends are triggered by the appearance of key files in a file share. Figure 9-3 also shows that the DSA can take input from a power-flow study case in the SCADA/EMS. This gives the operator the ability to assess in advance the dynamic security of possible future power system configurations, for example, for outage assessment. The planning input provides planners the opportunity to submit raw PSS/E cases to the DSA and perform the standard analysis. These cases do not get reported to the SCADA/EMS, but the results are deposited in an output directory for later retrieval.

The DSA functional components can be divided into the following:

- SCADA/EMS application interface,
- DSA server master control program,
- Node-to-bus network model conversion module, and
- PSS/E simulation engine and associated scripts.

9.2.3 SCADA/EMS Application Interface

The SCADA/EMS has primary control over the DSA, where runs are initiated and the results displayed. It is the only interface available to the control room operator. The DSA displays are tightly integrated into the SCADA/EMS environment and can be called up from any SCADA/EMS console. From here, overall control of the process is maintained. As soon as a state estimate becomes available, DSA runs are automatically initiated. A user can abort and restart a new cycle at any time. One can also submit study cases from the SCADA/EMS for analysis at this point.

Contingencies are defined on the SCADA/EMS and, using prior understanding of the contingencies, they are associated with individual cutsets. Some may actually belong to a number of cutsets. This will make it easier, when results are displayed, to identify which cutset needs attention should an unstable contingency be detected. Contingencies can be rapidly selected on a group basis or individually as well.

As shown in Figure 9-4, the results are displayed in the SCADA/EMS native interface in tabular form and are ranked in terms of their maximum swing angle. The greater the maximum swing, the more unstable the contingency is considered to be. Both the current run and previous run results are displayed, so that, though the current run remains incomplete, there will always be at least one complete set of results available. It has been found that the maximum swing angle is a crude measure of the stability of the system. For the NEMMCO system, the angle could increase up to about 140 degrees before the system becomes unstable. With experience, the operator can get a "feel" for how close the system is to instability.

Although the maximum angle spread does give an experienced operator a feel for the stability of the power system, there is still no true quantitative assessment of the transfer limit because the DSA only performs base-case analysis. The only other indication is given by the offline transfer limit constraints. Furthermore, NEMMCO would need to gain a better understanding of what might be the best way to adjust interface transfers in order to find the transfer limit in real time.

Figure 9-5 shows a typical onset of instability on the system. The maximum rotor swing angles slowly creep up from about 80 degrees until they reach about 120 degrees, after which the cases become unstable. It can be seen that there is little movement between a stable case and the onset of instability.

Figure 9-6 shows the detailed information about an individual contingency result, which includes the two machines that contributed most to the maximum rotor angle swing. This information, together with the cutset association, gives the operator a crude indication of where the problem could be. The operator can then take remedial action by applying discretionary constraints along the cutset that is yielding the unstable contingency.

The DSA also has an interface to the SCADA/EMS alarms and will raise alarms for the following conditions:

- Unstable condition is detected.
- There was a simulation run failure and results could not be returned.
- The initial mismatch in PSS/E was too large.

EXECUTING Ctg 4 of 55	RUNNING	ABORT	Realtime DSA	Current Transient Ctg Results

UNSTABLE

Activation:	Groups	5 of 6	Cutsets	14 of 15	Trans Ctgs	55 of 90	Must Run	0	Ineligible	0	Invalid	0
Latest Results:	14-Jul-2006 17:51:37		Cutsets		Trans Ctgs		Worst Ctg	TRNG_CLVL_8810			Max Angle	180
Previous Results:	14-Jul-2006 17:43:37		Cutsets		Trans Ctgs		Worst Ctg	TRNG_CLVL_8810			Max Angle	85

Id	Description	Latest Run			Previous Run			
		Result	Max Ang	Time	Result	Max Ang	Time	
TRNG_CLVL_8810	TARONG - CALVALE 8810 FAULT AT TARONG	UNSTABLE	180	1.55	Stable	85	1.60	Details
CAL_C_UN	LOSS OF LARGEST CALLIDE C	Stable	71	1.78	Stable	69	1.77	Details
BOYNE_LD	LARGEST BOYNE ISLAND LOAD TRIP	Stable	63	1.96	Stable	60	1.92	Details
FA_SH1	FARRELL - SHEFIELD 1 LINE FAULT AT FA	Stable	55	0.46	Stable	54	0.46	Details
PARA_BUNGAMA	BUNGAMA - PARA FLT AT PARA	Queued			Stable	55	0.49	Details
GLAD_GNGN_813	GLADSTONE - GIN GIN 813 FLT AT GLADSTONE	Queued			Stable	46	0.76	Details
NPS_UN	TRIP LARGEST NORTHERN UNIT	Queued			Stable	52	1.34	Details

Figure 9-4. Screen shot of DSA results display showing contingencies ranked by the severity of the result. Unstable results are immediately alarmed and colored red to draw attention.

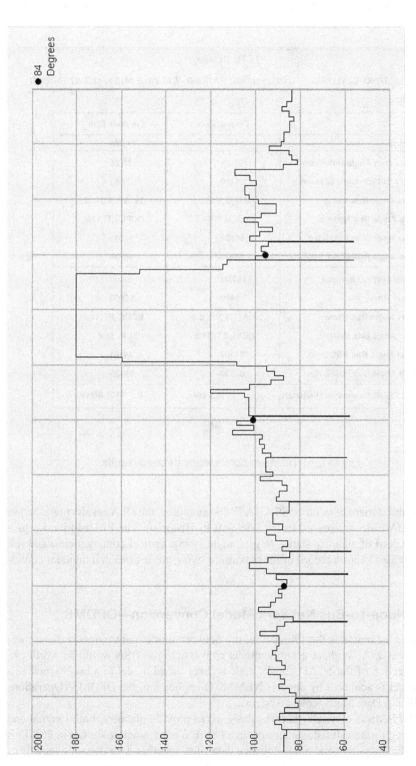

Figure 9-5. Plot of maximum swing angles showing a typical onset of instability at around 120–130 degrees. Note how quickly the system can become unstable. There are only a few runs on either side of the instability that indicate that the system is approaching instability.

231

TCTG Details

TCTG ID: **TRNG_CLVL_8810**   Description: **TARONG - CALVALE 8810 FAULT AT TARONG**

Results:

	Current Run	Previous Run
Result	UNSTABLE	Stable
Maximum Angle Difference	180.09	85.25
Time of Max Angle Difference	1.5490	1.5980
Max Angle Bus Name	CAL_B_PS20.0	TVLLE_PS11.8
Min Angle Bus Name	OCPL-LTF11.5	OCPL-LTF11.5
Max Angle Bus PSS/E No.	44302	44841
Min Angle Bus PSS/E No.	85930	85930
Final Angle Difference	180.09	26.03
Final Time	1.5490	5.0008
Max Angle Bus Name	CAL_B_PS20.0	BCDE_PS-11.0
Min Angle Bus Name	OCPL-LTF11.5	PELP 15.8
Max Angle Bus PSS/E No.	44302	44701
Min Angle Bus PSS/E No.	85930	85920
SE-PSS/E Summated Mismatch	1115.1 MVA	1101.3MVA

OK

Figure 9-6. Transient contingency detailed results.

In a similar manner to other SCADA/EMS quantities, the DSA results presented in the SCADA/EMS are passed to the OSI-Soft PI Historian. The PI Historian keeps a running record of stability status, largest angle swing, critical contingencies, and run failures. It also keeps a record of the maximum swing angle of each individual contingency.

9.2.4 Node-to-Bus Network Model Conversion—OPDMS

If the state estimator is the foundation, the network model conversion is the cornerstone of the DSA. Without accurate model conversion, the DSA would be worthless. The conversion of the SCADA/EMS node-breaker-based model to a bus–branch network model is achieved by another NEMMCO application, the OPDMS (Operations and Planning Data Management System).

The OPDMS is a system that was designed to provide planners, both internal and external, with state-estimator snapshots in a PSS/E format, suitable for use in PSS/E. It provides both positive and zero-sequence data files, together with dynamic models of generators, excitation systems, stabilizers, governors, SVCs, and any other items of plant that require dynamic models for the purposes of time-domain simulation. The OPDMS is the master repository for all the dynamic model data and provides the interface for the management of this data. It is also capable of creating data files that can be used for small-signal analysis. One of the key features of the OPDMS is its ability to create snapshots with out-of-service plant components included and ready to be connected to their in-service buses. This allows planners to easily manipulate the snapshot, taking plant components in and out of service.

Bus numbers and dynamic models are mapped to the network model on an SCADA/EMS node basis. As all the data files, including the dynamics data file, are created by the OPDMS when a snapshot is requested; the bus numbers will be synchronized across all data files. This is particularly important when various items of plant can end up on different bus numbers depending on the breaker topology at the station at the time of the snapshot. Furthermore, the dynamic model parameters themselves can also be a function of bus numbers. This is useful when the model parameters specify sources of input signals from remote buses or even lines.

The DSA uses the OPDMS model conversion module, but is independent of the OPDMS. This is designed to improve the availability of the DSA. It has its own cache of dynamics models and bus number mappings, which are regularly refreshed from the live OPDMS. The OPDMS conversion also has a number of other features to cater for PSS/E idiosyncrasies, including:

- Conversion to equivalent loads for those generators that do not have dynamic models associated with them, and
- Conversion of switched shunts to fixed bus shunts as a work-around when PSS/E is not able to handle operational switched shunt combinations.

The files created by the OPDMS should be able to run directly in PSS/E without further manipulation. The level of dynamic modeling includes:

- Full machine modeling, both salient pole and round rotor with saturation,
- Excitation systems,
- Stabilizers,
- Governors in some regions,
- Static VAr compensators,
- HVDC controls, and
- Static nonlinear voltage-dependent load models.

In addition, an attempt is made to estimate the amount of distribution capacitor loading as a function of the high-voltage bus (measurable) load and power factor.

Basically, the level of modeling is equivalent to that used in the offline studies to determine the limits equations in the first place. The one exception is the level of modeling used for loads, where in one region the network service provider uses composite load modeling in determining limits. In composite load modeling, an attempt is made to estimate the level of induction-motor load, distribution capacitance, a certain amount of distribution transformer saturation, and a percentage of load shake-off as a result of the fault. Estimates of these loads are based on load characteristics obtained from the analysis of fault recordings. For the purposes of real-time dynamics security assessment, it was considered that composite load modeling was computationally onerous for the time being and would gravely impair the performance of the DSA to warrant the improvement in accuracy that could be achieved. One also needs to consider the uncertainties in the forecasting of the load composition at each state-estimator snapshot.

The OPDMS also performs the conversion of the contingency definitions. The contingencies are defined on the SCADA/EMS in terms of "EMS identifiers." From these definitions, the OPDMS creates PSS/E scripts that are synchronized with the generated PSS/E load-flow file to apply faults and open lines to clear them. Contingency definition is capable of catering to a variety of events including:

- Loss of single item of plant, including generators, loads, lines, transformers, reactive plant, and HVDC plant;
- Loss of largest generation at a station or in a region;
- Loss of largest load in a region;
- Application of balanced and unbalanced faults at any location along a line. Calculation of unbalanced fault impedance is done during the contingency operation; and
- Independent clearing of either end of the faulted line or transformer.

The contingencies that have been chosen are well understood in the planning environment, where the mechanisms to alleviate unstable outcomes are well known. Other contingencies could possibly yield unstable outcomes also, but without the prior experience in dealing with the contingency, it is difficult to interpret how to alleviate them and justify their inclusion.

9.2.5 DSA Server Control Program

The control program on the DSA server orchestrates the various tasks on the DSA server once the job has been handed over to the server. This includes

- Initiating the OPDMS process to create PSS/E data files for the given snapshot;
- Converting the contingency definitions to PSS/E scripts;
- Setting up the base-case power flows;
- Queuing the various PSS/E simulations and running them when resources are available. This has two advantages, in that it is actually more efficient in terms of overall CPU performance and this allows the more critical contingencies to be processed first, thereby providing results sooner and allowing some results to be displayed within two minutes of the state-estimator snapshot. This also makes

the application *multiple-processor ready* for when multiple server architectures are adopted.

- Collation of results from the PSS/E simulations and the packaging of these results onto the DSA client application on the EMS. These results are passed back to the EMS application by simple file transfer where they are displayed immediately.
- All the general housekeeping on the DSA server;
- Recovering from errors in the simulation when they occur;
- Archiving of result files in the event that simulations need to be reproduced. There is enough information archived to reproduce the results of the simulation if necessary. Also, for cases that are unstable or where there is some form of simulation error, all working files are archived. This makes it easier to analyze the case further or to diagnose any problems.

9.2.6 PSS/E Simulation Engine

At the heart of the DSA is the simulation engine. The stability of the system is determined by time-domain simulation of contingencies on the power system. PSS/E uses a fixed-time-step modified Euler algorithm for the integration of differential equations to perform time-domain simulations. A step size of 1 ms is required because of the time constants involved in the control systems. This fixed time step size has meant that the simulation speed is not the most efficient, and considerable numerical capability is required from the processors to achieve the performance requirements.

The choice of PSS/E as the simulation engine was largely dictated by the fact that PSS/E was widely accepted as the tool of choice among the network asset owners in Australia. This was to provide a level of confidence in NEMMCO's DSA among the various planning bodies that have an interest.

Individual simulations are queued for each contingency to be analyzed, with the worst contingency from the previous run at the head of the queue. This would provide the best possible chance of early detection of unstable conditions. Thus, a stability issue could be reported to the operators within minutes of the analysis starting, even before all the analysis is complete.

The PSS/E simulations are controlled by scripts, which perform initialization and then apply the contingency during the simulation. For unbalanced faults, the equivalent zero-sequence impedance is also calculated using the supplied sequence data. The simulation time is currently set to five seconds. Although a ten second simulation time would have been preferable, it was considered that limiting the simulation to five seconds would not introduce too much error in the interface limit and was worth the improved performance benefit. This too can be easily adjusted as a configuration option.

A special dynamic model in the simulation monitors the rotor angle swings of each machine in the simulation and assesses the stability of the simulation. If the deviation in angle between any two machines reaches a predefined threshold, then the simulation is stopped and declared unstable.

9.2.7 Measure of Stability

The stability measure is a swing criterion; an attempt is made to measure the amount of swing any two generators can undergo during a contingency. Figure 9-7 shows the plot of various rotor angles in the system for a stable simulation. The maximum difference between any two rotor angles (ignoring their initial offsets) is shown in Figure 9-8.

The rotor angle swing monitor is implemented as a user-defined dynamic model in PSS/E and has a number of distinguishing features:

- It ignores the initial angle spread. The swing monitor is trying to measure the amount of swing between any two machines, not the steady-state angle separation. This should avoid any bias toward different generators, particularly at the extreme points in the system where the generators have a large starting angle. The swing monitor measures the deviation of each rotor angle from its original starting point. At the starting point, the angle deviation of each machine is zero. At each time step, the monitor calculates the maximum difference between the angle deviations of any two machines. If this angle reaches a predetermined threshold, the simulation is declared to be unstable. Once the system reaches a state of instability, the rotor angles quickly separate, so the actual value of the threshold is not particularly important, just as long as it is greater than any angle that could be considered stable. This threshold is currently set at 180 degrees.

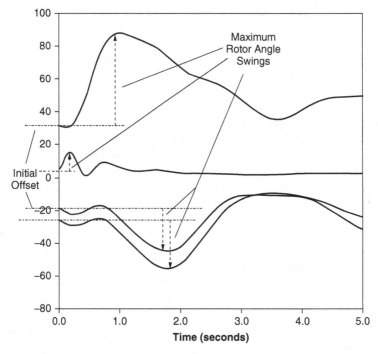

Figure 9-7. Individual rotor angle swing for a single stable simulation. Each curve corresponds to a machine rotor angle response for the contingency.

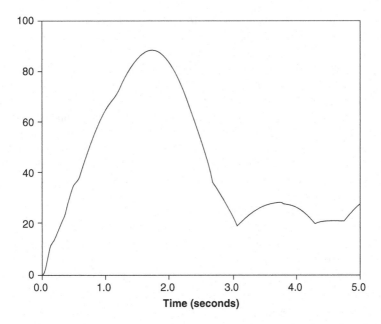

<u>Figure 9-8.</u> Maximum difference between any two rotor angles (ignoring their initial offsets).

The maximum value of this maximum angle difference is then reported for each simulation, which has become a rather crude indicator of the degree of stability. The larger the indicator, the closer the system is to instability. This simple measure has proved to be surprisingly reliable.

- It has the ability to look at swings within each individual island, even when islands are created as a result of the contingency during simulation.
- It has the ability to exclude specific generators from being monitored. This is to avoid problems with equivalent generators that are used to model voltage source converters. It can also be used to exclude certain small generators at the extremities of the power system, which can be considered as insignificant if they were to lose synchronism.

In addition, the monitor also has an early termination criterion, so if the rotor angle swings do not appear to be growing significantly then the simulation is terminated early to improve performance.

9.2.8   Other Features

Confidence in the state estimator solution is measured by the amount of mismatch the state-estimator solution has at each bus. If the mismatch is too high, then the solution is flagged as such and none of the network applications, including DSA, will proceed until a reasonable solution is obtained.

Since the network model conversion is such a significant step in the DSA, it would be beneficial to monitor the accuracy of the conversion process itself. Ideally, the DSA would end up with an accurate representation of the current state of the power system, but errors and approximations introduced by the conversion process serve to blur this. In an attempt to monitor confidence in the conversion process, the initial mismatch is measured as soon as the load-flow is read into PSS/E, prior to any attempt at a load-flow solution. An alarm is raised if the initial mismatch becomes too high to indicate that there will be some uncertainty in the DSA solution and that it should be treated with a bit more caution. There are a number of other tests that could be performed to try to measure confidence in the converted solution, such as the mismatch of solved power flow variables (e.g., voltages, angles, and taps) with the estimated equivalents.

Other features of the DSA include:

- The DSA can accept study cases from the EMS in the same manner as it does for real-time cases.
- Facility to accept study requests from planners who submit PSS/E files directly for analysis.
- Ability to choose contingencies on a group basis.
- Ability to exclude governor models on a regional basis. This feature is required because it is sometimes difficult to initialize governor models and the feature allows a solution to be obtained, albeit slightly less accurately. As governors are relatively slow acting, the removal of governors has limited impact on the accuracy of the solution for the purposes of transient-security assessment.
- Contingency definitions are defined on the EMS and are translated by the same process that generates the power-flow files. This ensures that the contingency definitions always map to the correct buses in the power flow.

9.3 PERFORMANCE AND RELIABILITY

9.3.1 DSA Performance Issues

A lot of work has gone into improving the performance and reliability aspects of the process. A number of trends were degrading the performance of the DSA. These included increasing number of contingencies, growth in the power system dimensions, and increasing complexity of the dynamic models. Initiatives taken in 2002–2005, aimed at improving performance, included:

- Installation of faster hardware,
- Development of early termination that would detect stable contingencies early and terminate the simulation,
- Improvements in the OPDMS network model conversion, which was originally taking a significant portion of the available processing time.

The recent implementation of HVDC modeling for Basslink and the need to halve the integration time step have further degraded the performance to the point where the complete cycle takes almost ten minutes and is just sufficient for base-case analysis.

The existing hardware is currently nearing end of its life and now (2008) NEMMCO is looking at replacing the existing technology with dual quad-core IA32 CPUs, such as the Dell PowerEdge 1950. The performance boost is expected to be on the order of 10- to 20-fold and it will allow NEMMCO to keep up with the demands of dynamic models that are continually increasing in both number and complexity. This performance boost will also provide the opportunity to include some additional single- and multiple-contingency definitions. It should also allow for the improvement of governor modeling and maybe even the introduction of composite load modeling where appropriate.

This improvement, however, still falls far short of the requirements for the eventual development, when iteration will be required to estimate the transfer capability across a number of different interconnectors. Much more efficient integration algorithms will be required to make this development viable.

9.3.2 DSA Reliability Issues

NEMMCO's DSA has been crucial over the period of time that it has been in service. NEMMCO has become reliant on the DSA, to the point where 24x7 support is necessary in the event of failure. For this reason, a lot of effort has gone into making the DSA more reliable. Reliability should be considered in two contexts:

1. Solution accuracy, or the false indication of stability or instability. This will be discussed in Section 9.4.1.
2. Reliability in obtaining a valid result and not experiencing solution failure.

Much work has gone into the DSA to make it more reliable. The DSA is a critical tool for NEMMCO and is expected to run successfully at each run. Figure 9-9 shows the daily failure rates the DSA has experienced over a twelve month period. This represents an overall availability of 98.2% over this period.

It is evident that the dominant issue is initialization failure. Modeling changes often correlate with periods of higher failure rates. The problems exhibited around the November–January time frame, for example, were largely a result of the introduction of HVDC models for Basslink, the new Tasmanian–mainland interconnection. Initialization issues around June 2006 were as a result of tolerances being retightened as NEMMCO became more confident with the HVDC models. Power-flow divergence is another significant issue and this can usually be traced back to poor state-estimator solutions and load models in the SCADA/EMS.

DSA failures can be basically classified into several categories: dynamic model robustness, state-estimator solution, network model conversion, telemetry, and SCADA/EMS model issues. These aspects are briefly reviewed in the following sections.

DSA Daily Failure Rate

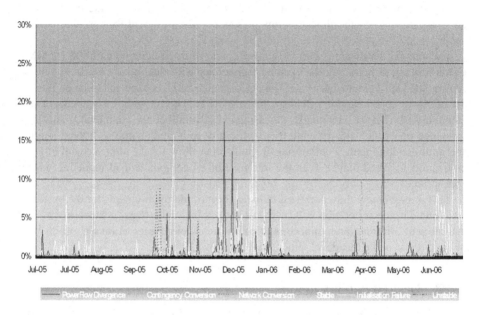

Figure 9-9. DSA daily failure rate highlighting the dominant issues that contribute to DSA run failure.

9.3.2.1 Dynamic Model Robustness

This is the primary cause of initialization failure. This is a particular issue for NEMM-CO as a significant number of models are user-defined models, provided by network service providers. Most of the dynamic models come from a planning environment that works on well-behaved planning studies but often cannot cope with the unusual operating conditions that can be experienced in the real-time environment.

Teething problems can often be experienced for a period when a new model is implemented until it is tuned for the real-time environment. Typical examples are governor models that rarely allow for overload capability of the generator, which is now a regular occurrence in the market-driven dispatch. Basically, any model that has limits is prone to some form of initialization issues. It is important that the model be valid for the full operating range, not just the rated range of the plant. With the introduction of the market, it has been found that plant components are often operated well above their nameplate rating. If the dynamic models do not take this into consideration, this usually results in some form of initialization failure.

Other examples include models that rely on a particular topology. The summated flow on parallel lines, for example, could be used as input to a stabilizer. It would be foolhardy, however, to assume that parallel lines would always be connected to the same bus in real-world examples. It is also important that models be up to date. An outdated excitation model can be the primary cause for a series if initialization failures;

for example, when generator terminal voltages wander beyond the operating range of an inappropriately defined excitation model. Another important aspect to consider here is that it requires the failure of only one dynamic model to cause complete solution failure. Often, it is the same model that causes initialization problems, and these are the models that need attention. It would also be useful if there were features to remove specific models online when a model happens to become particularly troublesome.

9.3.2.2 State Estimator Solution
The state estimator solution itself can drive models into unexpected operating conditions. This often is as a result of poor telemetry coverage. It is important to remember that an issue in only one small part of the state-estimator solution is sufficient to render total failure in the DSA. Even though only valid state estimates are passed to the DSA and NEMMCO has achieved a reliability of over 98.5% for its state estimator, the state estimator solution can still be unsatisfactory for the DSA. The state estimator sometimes has bad estimates of loads, whereby loads end up with very poor power factors, or even end up generating. This can often lead to power-flow solution failure in PSS/E. Similarly a poor state estimator solution can lead to initialization issues, particularly if it results in violation of control system limits. It is important that machine reactive limits be maintained. Sometimes, the telemetry for a generator's reactive output can go full scale, resulting in reactive loading that is far beyond the capability of the unit. Placing reasonability limits on such measurements gives the state estimator an indication that there is something wrong with the measurement and allows it to make better estimates of the unit output.

9.3.2.3 Telemetry
Often, there is insufficient telemetry of important statuses, such as HVDC control modes and AVR statuses, so an assumed state may work much of the time, but can cause initialization failures when the physical operating conditions change. The AVR status of an SVC, for example, may cause an optimistic outcome if the assumed SVC state is the voltage-control mode, whereas it is actually in the MVAr control mode. Other models also require analogues that are not available in the SCADA/EMS. Some governor models require lake levels to initialize correctly, so there is no single set of default model parameters to suit all conditions. The only viable solution has been for telemetered lake levels to be fed to the DSA. Until suitable lake levels can be telemetered directly, these governor models will have to be ignored to prevent initialization issues. It is considered that any error introduced as a result would not be too significant for the nature of the simulations in question. Although the obvious solution would be to improve telemetry, this is not always practical, and a suitable compromise can sometimes be reached by representing such inputs as manual inputs that can be hand dressed in SCADA. This is only feasible for points that do not change regularly, and can in itself be a source of error if they are not dressed accurately.

9.3.2.4 Network Model Conversion
The PSS/E format does not always lend itself to the detail that is contained in the SCADA/EMS and often approximations or assumptions need to be made in the con-

version process from the SCADA/EMS node-breaker representation to the simulation engine's bus–branch representation. When these assumptions break down, we are often left with errors in the PSS/E solution. This can either lead to PSS/E solution failure or, worse still, to an inaccurate solution. Though not ideal, the only way to address these situations is to implement work-arounds either in the SCADA/EMS or in the PSS/E model itself. Topology-dependent models are particularly sensitive to the conversion of the SCADA/EMS network model to a bus–branch representation. These are typically dynamic models that refer to buses or lines as part of their parameter data. These parameters need to be mapped correctly to EMS nodes in OPDMS so that correct bus numbers will be generated during the conversion.

9.3.2.5   SCADA/EMS Model Issues

Usually, the SCADA/EMS model is more detailed than the PSS/E model, but in some instances this is not the case. One example is voltage source converter HVDC links, which are modeled as equivalent generator pairs in the state estimator. There is no linkage between the two generators and sometimes, but surprisingly rarely, the state estimator can come up with solutions that are infeasible for the DC link and cause initialization problems in PSS/E. In one instance, the state estimator even estimated contraflow indications on either end of a HVDC link as a result of some telemetry failure at one of the ends. This lead to some initialization issues. To overcome this, some SCADA/EMS calculations were substituted to calculate flow from one end from the other, until the data quality issues were resolved.

9.4   EXPERIENCE, BENEFITS, AND OUTLOOK

9.4.1   Operational Experience

NEMMCO's DSA has been operational since 2001, and in that time many improvements have been implemented as a result of operational experience. Originally, it was planned that the simulation load would be shared across the two sites at NEMMCO, with half the contingencies being processed at one site and the other half at the other site. However, operators were more comfortable if the full set of contingencies were enabled at both sites. Usually, instabilities are quite marginal and small differences in the state-estimator solution can mean the difference between a stable and an unstable result. Since the estimator solutions being fed to the DSA can be slightly different at the two sites, the DSA can yield different results at each site. Also note that the timing of the results from the two DSAs are not necessarily synchronized. An unstable situation first needs to be confirmed at the alternate site before NEMMCO takes any remedial action.

The calendar in Figure 9-10 shows the days in a period of 12 months on which unstable operation had been detected. It was a surprise to note the large number of unstable indications, particularly when one considers that the constraints in the dispatch process should be preventing insecure operation. During this period, there was a concerted effort to investigate any unstable indications and there were a number of illustrative examples identified. The volume of cases involved means that it is far easier to

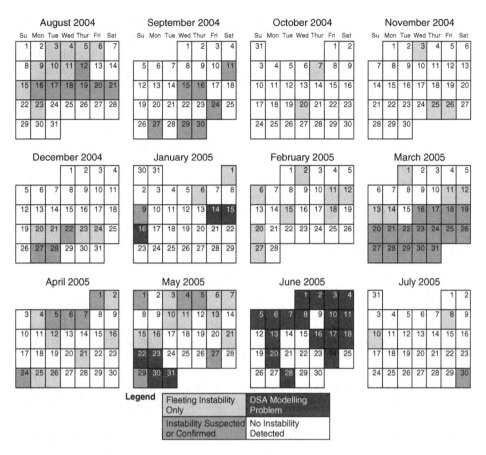

Figure 9-10. Calendar showing days when DSA detected unstable contingencies.

check for false indications of instability than false stable indications. Approximately 40% of the unstable conditions detected were fleeting conditions, appearing for a single run only. These typically occur during switching operations or for postcontingency conditions prior to rearrangement of the power system back to a secure operating condition. However the bulk of the remainder consists of bona fide unstable situations.

As shown in Figure 9-10, March was highlighted by an instability for which the mechanism of instability was puzzling and difficult to alleviate, as there were no interconnections operating near their stability limit. Investigations later showed that the instability was actually for a hitherto neglected transmission corridor and no constraint was available to control its flow. Constraints have now been created that can be enabled in the dispatch to cater for this issue.

On the other hand, May and June highlighted an instability that was erroneously detected because of inadequate distribution capacitor modeling. This significantly contributed to a false alarm rate of almost 20%. The subsequent inclusion of distribution capacitor modeling yielded as much as 50 MW improvement in interchange capability. It would be worth noting that in over six years of operation, no more than six different contingencies have ever indicated insecure conditions. However, this does not preclude the other contingencies from being insecure during more abnormal operating conditions.

From the early days of operation, investigations showed that there were a number of underlying issues that yielded false unstable indications, such as:

- SCADA and Estimator Issues. Some parts of the network have little or no telemetry, which can cause issues. A good example is for estimated generator terminal voltages. At some generator locations, there is no SCADA telemetry of terminal voltage or transformer tap position. The state estimator is thus free to solve at any valid voltage and tap position. Unfortunately, when the state estimator solves for low terminal voltages, the dynamic excitation models run into underexcitation. This usually results in local unstable modes for those generators. This was addressed by adding pseudomeasurements that would bias the terminal voltage near 1 p.u. The pseudo measurement was a calculation based on the in-service status of the unit, that is, the terminal voltage is set to the typical value when the generator breakers indicate that the generator is in service.

  A similar issue can occur when there is a local loss of SCADA at a power station. In these instances it is quite normal that the generator output would be hand dressed where possible. However, for correct DSA operation, it is now also important to hand dress the transformer tap position and the terminal voltage as well.

  Other unstable cases can be attributed to poor state-estimator results, such as generators with high MVAr output or SVCs solving near the end of their dynamic range. Improved telemetry and state estimator load models can help rectify most of these issues.

- Dynamic Models. Sometimes, there would be a problem with dynamic models that end up in unexpected operating modes, which can cause unstable indication. This could be attributed to dynamic model robustness.

- Network Model Conversion. Some problems with the network model conversion process may lead to unstable indication, though total solution failure is more likely.

A number of unstable conditions have also occurred during unusual operating conditions in which there are multiple prior outages in the system. In these conditions, the limit equations provided by the transmission network service providers no longer hold true.

In other situations, the DSA has detected unstable local oscillatory modes rather than a system-wide transient problem in which a single machine or station would have an undamped oscillation against the rest of the power system. However, the major issue is to understand how to deal with an unstable situation when it arises. Usually, a particular contingency may be affected by the flow in a particular cutset, but this is not always the case. Hints are provided by grouping contingencies into cutsets and by reporting on the two machines that contributed most to the instability, but this is often insufficient. The two machines reported are often on the extremities of the system and it is difficult to see where the separation occurs. There is no mechanism to view how groups of machines are separating. The incident in March is one such incident where this was highlighted.

9.4.2 Relays and Special Protection System Modeling

The importance of modeling relays and special protection systems came to the fore when there was prior outage in a narrow portion of the system. The prior outage left only one transmission line interconnecting two parts of the system in the main transmission network. However, for the contingent loss of this remaining line, an unstable indication was immediately indicated. Figure 9-11 shows that the contingent loss of the remaining line would have islanded-off the power system into two separate networks save for an additional weak subtransmission network that connected the two islands.

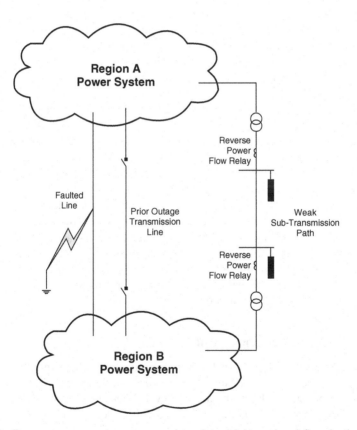

Figure 9-11. Reverse power relays are used to detect interregional flow in the sub-transmission network and separate the two regions in the event that there is a complete loss of the main interconnector. It is important to model such schemes in order to avoid false indication.

Reverse power relays in the subtransmission network would immediately separate the two islands for such an event, and an unstable situation would thus be averted. The modeling of these reverse power relays would have avoided any false unstable indication. This example highlights the importance of modeling such relays and other special protection systems that can have an affect, particularly on the topology of the system. Such modeling would need to be defined as part of the dynamic modeling.

9.4.3   Proven Benefits

The most exemplary benefit came to NEMMCO when a new transmission corridor prone to stability problems was detected. This previously unknown issue shows how the DSA is being used to avoid exposure to optimistic or unknown technical limits. With the addition of further contingencies, it is hoped that any new problem areas can also be identified. The DSA has again proved itself to be useful following a system separation incident, whereby the DSA indicated insecure operation during the restoration process. The operators were able to take immediate action to ensure that the system remained dynamically secure during the restoration [11].

The DSA provides confidence in the offline limits that drive the dispatch process and acts as a monitoring tool that ensures that the constraints are working as they should. It enables NEMMCO to run the power system close to the envelope with confidence. In addition, the DSA offers the ability to assess and maintain the power system security while under abnormal operating conditions in which there are no previously derived offline constraints to match the system condition. This seems to become a more common occurrence under the market-driven environment.

At present, there is no financial benefit to the electricity market while the offline constraints are being used in the dispatch engine. The DSA only operates as a monitoring tool. However, with confidence there may come a time when the traditional dynamic constraints to maintain security are replaced by real-time constraints derived by the DSA. This will avoid the use of overly conservative constraints that are derived offline to encompass a range of system conditions. At this point, the market would benefit immensely by virtue of the increased transfer capabilities that can be achieved as a result. Studies have shown that even small improvements in transfer capabilities can yield annual benefits of many million dollars.

9.4.4   Planned Enhancements

The current DSA at NEMMCO is far from the ultimate development. It was set up as a proof of concept and was much more successful than originally anticipated. After over six years of operation, the next stage of development is long overdue and involves some major enhancements. These enhancements include:

- Continuous effort in improving the robustness of user-defined models.
- Implement a much larger set of contingencies. This will likely include the need for contingency screening.

- Determine secure operating margin in real-time. This will iteratively determine the limiting transfer level along transmission corridors. It will provide a better measure of the security of the power system and indicate the proximity to the security envelope. A number of transmission corridors would need to be examined. Furthermore, if an insecure state is entered, it will indicate how much the transfer level needs to be reduced to bring the system back to a secure condition. This feature will go a long way toward providing recommendations about what an operator needs to do in such situations.

- Perform oscillatory and voltage-security analysis in addition to the transient-security analysis. This will add an extra dimension to dynamic security analysis.

- Provide useful recommendations for alleviating insecure conditions.

- Implement faster integration algorithms. With the orders of magnitude improvement in computation required to support additional contingencies and iterate to a limit, it is becoming impractical to use the existing fixed-time-step modified Euler integration algorithm of PSS/E. Major efficiency improvements are required to support this and alternatives are being examined that could have much better performance characteristics.

- Improve visualization. At present, most of the information is presented in tabular form. It would be far better to display this information graphically, with a true indication of distance from instability, so that operators could quickly assess the dynamic security of the system.

Even though the DSA was only built as a proof of concept, and with its limitations, it has still proven itself an invaluable tool in NEMMCO's operational environment.

9.5 REFERENCES

[1] CIGRE, "CIGRE Technical Brochure on Review of Online Dynamic Security Assessment Tools and Techniques," Working Group 601 of Study Committee C4, January 2007.

[2] CIGRE, "CIGRE Technical Brochure on Wide Area monitoring and Control for Transmission Capability Enhancement," Working Group 601 of Study Committee C4, January 2007.

[3] Power Technologies Inc, PSS/E 29 Program Application Guide.

[4] Crisp, G. T. J., and Ledwich, G. F. "Advanced Tools to Manage Power System Stability in the National Electricity Market," presented at AUPEC 2004.

[5] Reinprecht, N. B., and Bones, D. "Data Management for the Electricity Market," *IEEE Computer Applications in Power,* Vol. 1, pp. 35–38, 2001.

[6] Reinprecht, N. B., and Loh, K., "Power Network Management Tool," presented at IPEC 2003, Singapore, 2003.

[7] P. Kundur, *Power System Stability and Control,* McGraw-Hill, New York, 1994.

[8] Savulescu, S. C., *Real-Time Stability in Power Systems: Techniques for the Early Detection of the Risk of Blackout,* Springer, Norwell, MA, 2006.

[9]   IEEE Task Force on Load Representation for Dynamic Performance, "Standard Load Models for Power Flow and Dynamic Performance Simulation," *IEEE Transactions on Power Systems,* Vol. 10, No. 3, pp 1302–1313, 1995.

[10]   Power System Dynamic Performance Committee of the Power Engineering Society, IEEE, "Techniques for the Power System Stability Limit Search," A Special Publication of the Power System Dynamic Performance Committee of the IEEE PES, TP-138-0, 1999.

[11]   NEMMCO, "Power System Incident Report, System Separation and Load Shedding," http://www.nemmco.com.au/marketandsystemevents/232-0052.pdf, January 2007.

[12]   NEMMCO "Dispatch of Contingency Frequency Control Ancillary Services," http://www.nemmco.com.au/ancillary_services/168-0044.htm.

10

ONLINE VOLTAGE SECURITY ASSESSMENT IN THE HELLENIC INTERCONNECTED SYSTEM

Costas Vournas, George Christoforidis,
and Thierry Van Cutsem

10.1 INTRODUCTION

10.1.1 Voltage Stability

According to IEEE and CIGRE recommended definitions [2], voltage stability deals with the ability of maintaining voltage magnitudes near their nominal values when the system is subject to disturbances. In normal operating conditions, the voltage drops caused by power transfers across the network are on the order of a few percent. However, in a unstable voltage situation this no longer holds, but instead voltages undergo a dramatic, generally monotonic decline in the seconds or minutes following a disturbance. When this decrease is too pronounced, the system integrity is endangered, mainly due to protecting devices that trip generation, transmission, or load equipment, not to mention the nuisance caused to customers by sustained voltage sags. This disruption process may eventually lead to a blackout in the form of a voltage collapse.

Voltage instability is due to the inability of the combined transmission and generation system to deliver the power requested by loads [4]. It is a dynamic phenomenon largely driven by the load response to voltage changes. As such, it has been much better understood recently [3,4] than it used to be in the mid 1980s. However, several of the theoretical and practical difficulties associated with the subject still remain [17]. In many power systems throughout the world, voltage instability is considered a major cause of blackouts, as important as thermal overloads and the associated risk of cascade line tripping.

There are two lines of defense against incidents likely to trigger such instability: preventive and corrective controls. Preventively, security margins must be evaluated with respect to contingencies and appropriate preventive actions must be taken to restore sufficient margins, where needed. In this respect, online voltage security assessment is essential. However, it would be extremely expensive, and most likely impossible, to protect a power system against any conceivable disturbance. As a trade-off between reliability and economy, power systems are usually operated in such a way that they can survive credible contingencies, that is, incidents with a reasonable probability of occurrence, whereas for more severe incidents the transmission system operator (TSO) relies on corrective controls. The latter should take on the form of automatic emergency actions through system protection schemes aimed at preserving operation of the largest possible part of the system by isolating the unstable part [1]. Corrective actions usually affect generators and/or loads, and, hence, are acceptable only in the presence of severe disturbances.

10.1.2 Voltage Security Assessment

Power system security in general refers to the ability of a system operating in a particular condition (operating point) to withstand any disturbance in a prespecified set. These disturbances are usually referred to as contingencies. In the case of voltage security, we are examining, in particular, voltage stability after a contingency. In this sense, an operating point is voltage secure if following any contingency in the specified set voltage stability is maintained and a new stable operating point is reached in each case. Usually, only long-term voltage stability is examined, but in some cases where induction machines dominate, the analysis can be extended to short-term voltage stability as well.

Since voltage security assessment (VSA) deals with voltage stability, it is part of dynamic security assessment (DSA) and is, thus, different from static security assessment (SSA), which concentrates on verifying that postcontingency voltages are within some limits and line currents below thermal capacity. However, it is relatively easy to combine VSA and SSA insofar as the tools for the former are also able to analyze the latter.

The contingencies that are relevant to VSA are in general equipment outages, such as generator, transformer, or transmission line tripping. All single contingencies are usually included in the specified contingency set, which is then referred to as the $N-1$ security set. An exception is made for radial links, whose opening will obviously result in isolated buses.

Apart from single contingencies, a number of selected credible double or even triple contingencies may be included in the specified contingency set. These contingencies are determined by considering common cause events, based upon the expertise of the VSA user and knowledge of system structure.

This chapter concentrates on the application of online VSA in the National Control Center of the Hellenic Interconnected System [10]. In this application, a relatively large number of $N-2$ contingencies was considered, as well as all single contingencies (excluding radial links). Use of the VSA tools presented in this chapter for arming system protection schemes, which involve load shedding in case an actual disturbance with insufficient security margin occurs, is also discussed.

10.2 THE CONTROL CENTER OF HTSO

10.2.1 The Hellenic Interconnected System

The Hellenic Interconnected System consists of the generation and transmission systems of mainland Greece and some adjacent islands. The main production center is located in the northwest part of the country close to the lignite mines, which is the basic fuel source, whereas the main center of consumption is in the southern part, mostly concentrated in the metropolitan area of Athens (see Figure 10-1). The installed generation capacity of the interconnected system in 2007 was 12,332 MW, powered by 43% lignite, 24.5% large hydro, 20.5% natural gas, 6% oil, and 6% small hydro, wind, and other renewable sources. The generation capacity at the southern part of the country is close to 3200 MW, installed mainly at the power plants of Lavrio, Ag. Georgios (both near Athens), Aliveri (in Central Greece), and Megalopolis (in the Peloponnese).

The transmission system operates mainly at the levels of 400 kV (EHV) and 150 kV (HV). There are five AC interconnection lines with the neighboring countries in the north (with a net transfer capability up to 900 MW) and a DC interconnection with Italy via a submarine cable (with a maximum capacity of 500 MW). Depending on topology, the model of the system used for voltage stability analysis consists on the average of 900 buses, 1500 branches and 90 generators. The geographical imbalance between generation and consumption leads to bulk transmission in the north-to-south direction. This transfer is continuously increased due to the high increase of the yearly peak loads, mainly in the south, and reached 3500 MW during the summer of 2007. The peak load occurs during the summer period and is largely due to air-conditioning and other cooling devices. The delays in the implementation of planned transmission projects (mainly due to public protest) make the situation even worse. Thus, occasionally the Hellenic System faces critical operational conditions with respect to voltage stability. Several such incidents have been experienced in the Hellenic System since 1996. The most severe and recent disruption occurred on July 12th, 2004, when the southern part of the system suffered a blackout lasting two hours, due to voltage collapse.

10.2.2 The Energy Management System

The Energy Management System (SCADA/EMS) operated by the Hellenic Transmission System Operator (HTSO) consists of four control centers: the National Control Center (NCC), the Southern Regional Control Center, the Northern Regional Control Center, and the Regional Control Center of Thessaloniki.

The NCC gathers information about the entire interconnected system of Greece. Field data are first retrieved at the regional control centers from remote terminal units and then are transmitted to the NCC as intersite data. The SCADA/EMS at the NCC consists of separate databases and applications for supervisory control and data acquisition (SCADA) and generation control and network analysis [13].

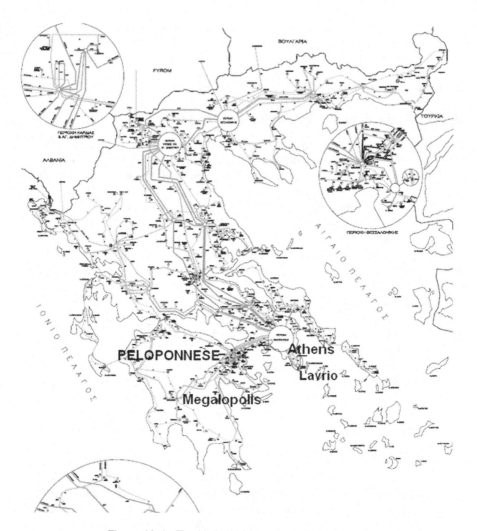

Figure 10-1. The Hellenic Interconnected System.

An offline application defines and maintains the network model database of the power system. Each network analysis application uses its own copy of the database to store the model and the state solution. The network analysis applications [14] are organized into two sequences: the real-time network analysis sequence and the study-mode network analysis sequence (Figure 10-2). The real-time sequence is executed periodically every few minutes and also by events or upon request, and consists of the state-estimator and contingency analysis applications. The study-mode sequence is executed upon request and consists of the power flow, optimal power flow, and another version of contingency analysis. The purpose of the real-time sequence is to maintain an up-to-date state estimate and perform contingency analysis for the power system network.

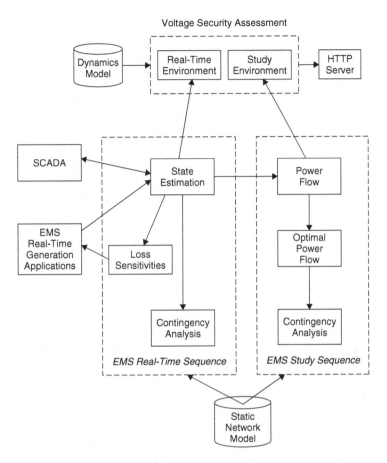

Figure 10-2. EMS network analysis applications and VSA.

The state estimation function consists of topology processing, observability analysis, state estimation, and error detection and identification. It uses measurements from SCADA and data from the real-time generation applications. It also feeds SCADA with estimated values and generation applications with unit-loss sensitivities. The study-mode sequence can be initialized from the real-time solution of the state estimator with the latest network model and system data.

10.3 ONLINE VSA IN THE HELLENIC SYSTEM

10.3.1 Application of Online VSA

In order to monitor closely voltage security in the Hellenic Interconnected System, an online VSA tool has been installed at the National Control Center of HTSO. The online VSA tool has been developed by the National Technical University of Athens and the University of Liège within the framework of the OMASES project coordinated by ALSTOM (now AREVA T&D) and sponsored by the European Union [5,10]. It was in experimental operation until June of 2003, when the OMASES project ended. Since January of 2005, the online VSA has been reactivated after implementation of user-recommended modifications and it is used continuously as an online and study tool at the HTSO control center [7]. The general architecture of the VSA application within the OMASES platform is shown in Figure 10-3.

The online VSA tool is fed with real-time system data from the real-time state estimator (in its online version) or the study-mode power-flow program (in its study version) available at the SCADA/EMS of NCC (Figure 10-2). In its online version, VSA runs periodically or on operator demand, whereas in its study version it runs only on demand.

The scope of VSA is the analysis of the impact of contingencies and the determination of security margins in terms of power transfers or power consumption in load areas. More precisely, the application involves:

1. Contingency filtering,
2. Determination of secure operation limits,
3. Computation of the system precontingency loadability limit, and
4. Computation of postcontingency loadability limits for the dangerous contingencies identified at item 2.

The main outcome of the loadability limit computations is a set of power–voltage curves (P–V curves) showing voltages at key buses versus (regional or system) load.

The quasi-steady-state (QSS) approach is used to determine the system response to both contingencies and the load demand increase. This technique is described in Annex 10-1 of this chapter.

10.3.2 Implementation Aspects

The online VSA application runs on a Windows server connected to a local area network (LAN) under the TCP/IP protocol (Figure 10-4). This server is separate from the mainframe computers (in an active/hot standby pair), which hosts the SCADA/EMS applications and databases. The communication with the SCADA/EMS is accomplished with text files loaded by the SCADA/EMS host onto the Windows server.

A SCADA/EMS process periodically creates 14 text files containing the latest network solution computed by the state estimator. Each file is a description of a class of SCADA/EMS components (e.g., substations, nodes, buses, lines, transformers, generators, and loads). These files are sent by the same SCADA/EMS process, via FTP, to a predefined real-time entry directory in the VSA Windows server, which triggers the VSA computation cycle. In addition, there is also a study VSA version operated in a

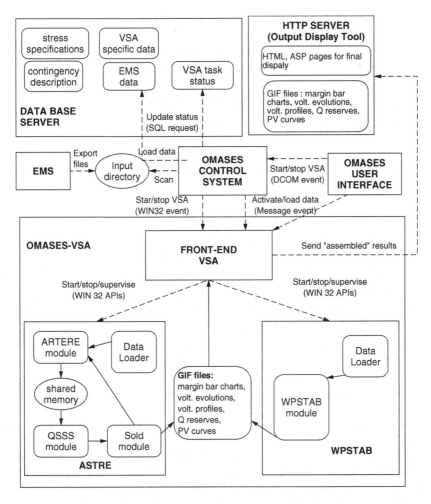

Figure 10-3. Architecture of the VSA application.

separate environment. It is fed with network data (14 text files) by the study power-flow application of SCADA/EMS.

The role of the OMASES control system shown in Figure 10-3 is to queue and dispatch messages to the VSA application, detect the presence of new SCADA/EMS files, load them into the database, and activate, synchronize, and supervise the VSA task automatically. The results are displayed to the operators through a Web interface using an HTTP server.

The VSA software itself is made up of two packages. The secure operation-limit computation and contingency filtering rely on the ASTRE software developed at the University of Liège [4,8,9,11]. ASTRE itself encompasses a load-flow module, which is used for stressing the system and filtering the contingencies, and a QSS simulator

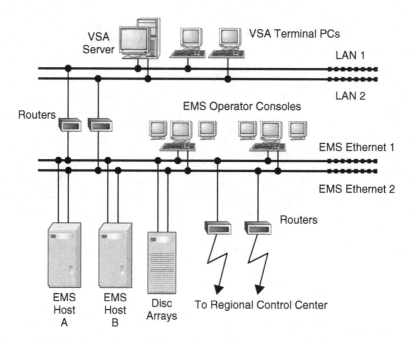

Figure 10-4. Incorporation of VSA application into the existing EMS configuration.

for simulating contingencies, both driven by a "conductor" module that collects diagnostics of QSS time simulations and infers the next stress levels to be tested. The former two executables communicate through a shared memory for the input data and through files for the outputs. The computation of postcontingency loadability limits and the preparation of P–V curves rely on the WPSTAB software developed at the National Technical University of Athens [12].

Data loaders connect to the relational database containing SCADA/EMS data, VSA dynamic data, stress specifications, and contingency descriptions. They produce the text files needed by ASTRE and WPSTAB.

A "front-end" program prepares, manages, and removes dynamically the working environment of a new VSA session, communicates (through messages) with the remaining of the OMASES environment, launches the data loaders, calls sequentially, and supervises the ASTRE and WPSTAB programs. It also collects the figures produced (in GIF format) and assembles them into HTML pages, together with ASP scripts, for Web-based display by the HTTP server.

10.3.3 Description of Online VSA Results

This subsection presents the VSA outputs available to operators at the HTSO control center. The results are illustrated with tables and figures produced in actual real-time operation.

10.3.3.1 *Secure Operation Limits*

In VSA, voltage security is analyzed in terms of power transfers, referred to as system stress. This stress is implemented by increasing the load and, consequently, the generation, leading to an increase of power transfers over weak corridors and/or depletion of reactive power reserves. For a given direction of stress, the secure operation limit (SOL) corresponds to the most stressed operating point such that the system can withstand any contingency from a specified list. This limit refers to the current, precontingency configuration of the system.

In the HTSO case, the contingencies resulting in the most constraining SOLs are computed and listed in a table format, as seen in Table 10-1, together with the corresponding margins (computed with a tolerance of 10 MW). In this case, the system stress corresponds to the increase of the total system load in proportion to the base case (state-estimator solution at the current operating point). Thus, the MW margins in the SOL table refer to the allowed system load increase before voltage instability is encountered after a contingency. Only contingencies with margins smaller than a maximum stress are shown (in the HTSO case this is 500 MW).

The SOLs of Table 10-1 correspond to the maximum system load of 2007, which reached 10,600 MW on July 23rd. Note that the first seven contingencies of Table 10-1 have a zero margin, meaning that the occurrence of any of these at the specific time that the snapshot was taken would result in unacceptable postcontingency conditions. However, the first five of these contingencies have very local effects (as will be shown in the next subsection) and refer to areas of the system where transmission upgrades are already under way. In particular, the first contingency refers to breaking the loop of cables feeding the island of Kerkyra, resulting in an abnormally long radial path to a

Table 10-1. Secure operation limits

Number	Contingency name	Margin (MW)
359	LINE_CON_MOURT-MESOG.1	0
289	LINE_CON_KTHES-SHOLAR.1	0
321	LINE_CON_KYL-ZAKYN.1	0
234	LINE_CON_PTOL-KOZ.1	0
177	LINE_CON_ARG2-ARG1.1	0
149	GEN_CON_MEGALO2.GEN4.UN	0
134	GEN_CON_K_LAVRIO.CC5.CC4.UN	0
50	DLINE_CON_CABLE_PELO-HPEI	109
143	GEN_CON_MEGALO1.GEN1.GEN2.UN	133
127	GEN_CON_THIS_KOMO.GFIC.UN	195
70	DLINE_CON_MEGA2_PYRG	203
57	DLINE_CON_KTHES-EYOSM	219
78	GEN_CON_AHSAG.GEN8-AHSAG.GEN9.UN	234
133	GEN_CON_K_LAVRIO.CC4.UN	289
122	GEN_CON_LAVRIO.GFIC.GEN2.UN	336
	All other contingencies	> 500

relatively large load and, hence, in an excessive voltage drop. The nature of the other local problems is similar.

The next two contingencies with zero margins (numbers 149 and 134) have a system-wide effect and are the most threatening ones. Contingency 149 involves the loss of a 300 MW unit (GEN4) at the Megalopolis Power Station in the Peloponnese area. This area was at the time the weakest in the system in terms of voltage security. Another unit of equal power capability (GEN3) at the same station was out of operation on the day the snapshot was taken, due to maintenance. Contingency 134 refers to the simultaneous loss of two combined-cycle plants (CC4 and CC5) at the Lavrio power station near Athens with a total capacity of 950 MW. These two power plants are connected to the same 400 kV bus that feeds a double circuit line; thus, a common cause for the simultaneous loss exists, even if the probability is relatively low. The rest of the contingencies in Table 10-1 have margins larger than 100 MW and are, therefore, less dangerous than the two contingencies mentioned above.

The extent of the problems caused to the system by a contingency is easily assessed using the "voltage profiles" automatically produced by the SOL computation and described in the next section.

10.3.3.2 *Voltage Profiles*

A voltage profile shows the number of buses with voltages below a certain level. The voltages are obtained from a snapshot of the collapsing system at the marginally refused level of stress during SOL computation. The snapshot is taken close to the end of the simulation interval, when voltages in the affected area are well below normal. This provides clear identification of the area in trouble and avoids running modal analysis as may be required in static tools, especially when the critical voltage (i.e., the voltage level at which instability starts) is high. The affected area is identified from the names of typical buses. For instance, Figure 10-5 shows the voltage profile corresponding to a contingency (number 289 in Table 10-1) whose effect is contained within a very limited area. As can be seen, very few buses with very low voltages exist in this case; almost all buses remain above 0.9 p.u. voltage. This is a typical profile for a local transmission problem without system-wide consequences.

In contrast, Figure 10-6 shows the voltage profile of a dangerous contingency (number 134 in Table 10-1) whose effect is system-wide. As can be seen, the disturbance effects are widespread, with the majority of the transmission system buses having a voltage lower than 0.9 p.u. From the same figure, it can be seen that the lowest voltages following this contingency are experienced in the Peloponnese, as well as in several substations in Metropolitan Athens. This picture is typical of voltage instability experienced by the southern part of the system.

Voltage profiles are computed and presented for all contingencies listed in the margins table. By examining such voltage profiles, it can be easily seen which contingencies would result in local and which in system-wide problems.

10.3.3.3 *Voltage Evolutions*

Voltage evolution plots show the postcontingency time evolution of voltage magnitude at a bus in the marginally accepted and refused cases, respectively. For each contin-

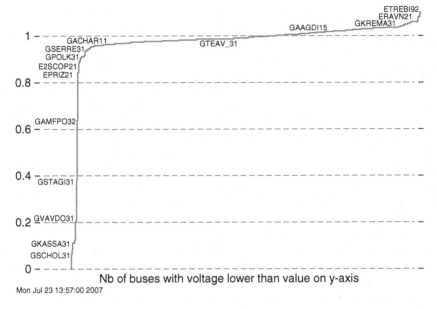

Figure 10-5. Voltage profile for a contingency with local effect.

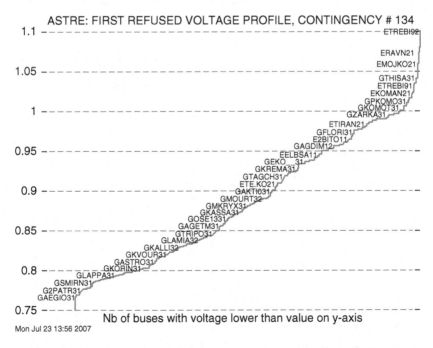

Figure 10-6. Voltage profile for a contingency with system-wide impact.

gency, the voltage evolution curve relates to the bus experiencing the largest voltage drop due to the contingency. The time horizon of the simulation is usually set to 1200 seconds. This value was chosen because it takes a long time to detect long-term voltage instability when the precontingency stress exceeds the SOL by only a small amount.

Voltage evolution curves allow one to distinguish between stability limits and low voltage limits. For instance, if the marginally refused simulation corresponds to a case in which some voltages finally settle to a value a little below a specified threshold, then the marginally refused case does not correspond to a voltage instability case. In this case, the SOL should not be interpreted as a stability limit and the value of the margin depends on the particular threshold chosen. On the other hand, when the marginally refused simulation shows voltages that drop well below the threshold without stabilizing to a low level, thereby causing the simulation to stop before the specified time of 1200 seconds, this signifies that the system has collapsed due to voltage instability. In this case, the SOL should be treated as a voltage stability limit.

A separate threshold is specified for generator buses, so as to refuse a situation in which the voltage of a field-current-limited generator drops to an unacceptable value that would trigger protective generator disconnection (for either undervoltage or auxiliaries overcurrent reasons), thus initiating a cascade of dangerous events.

Figure 10-7 shows the voltage evolution for a case of unacceptably low voltage (for contingency number 50 in Table 10-1). As can be seen, the marginally refused simulation does not correspond to voltage instability, but merely to final voltages that are a

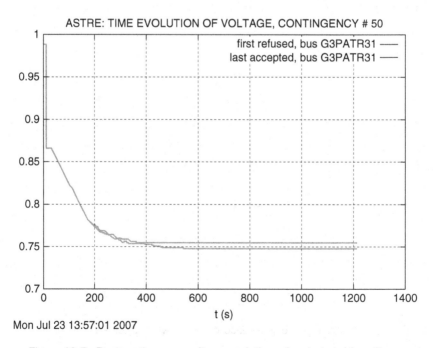

Mon Jul 23 13:57:01 2007

Figure 10-7. Postcontingency voltage evolutions: low but stable voltages.

little below the assumed threshold (0.75 p.u.). Hence, for this contingency, the SOL should not be interpreted strictly as a voltage stability limit.

A quite different situation is illustrated in Figure 10-8, which shows the voltage evolution corresponding to a contingency (number 134 in Table 10-1) leading to voltage instability and collapse. Here, the marginally unstable simulation ends up in a collapse (loss of equilibrium). This evolution is clearly very different from the one of Figure 10-7.

Voltage evolution curves are computed and presented for all contingencies in the margins table. By examining such curves, it can be easily seen which contingencies result in voltage collapse and which simply result in unacceptably low voltage levels.

10.3.3.4   Regional and National P–V curves

Finally, for each contingency with a margin below some predefined threshold, a set of P–V curves is produced, each relative to a particular area, or to the entire system. The abscissa shows the actual load power consumed in the area and the ordinate the voltage magnitude at a representative transmission bus in that area.

In the absence of any contingency, P–V curves are obtained through QSS simulation of a ramp increase in demand, as already mentioned. Compared to continuation power flow [19,4], this technique has similar computational efficiency and is also free from numerical problems when crossing maximum load power, but is more accurate

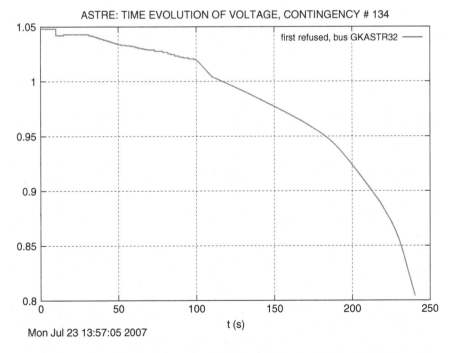

Figure 10-8. Postcontingency voltage evolution leading to instability.

since it relies on the more detailed system used for contingency analysis. An example of the national P–V curve for the July 23, 2007 snapshot without any contingency is shown in Figure 10-9. This result tells the operator how far the system could go without any incident. In this example, the load could increase by almost 700 MW (up to 11,300 MW) without collapsing.

P–V curves relative to a postcontingency situation, on the other hand, include the effect of the contingency followed by the demand increase. Figure 10-10 shows the national PV curve corresponding to the very severe contingency number 134 in Table 10-1 (loss of two combined-cycle plants at the Lavrio power station). The initial voltage drop (from 98% to 95%) and load power reduction (from 10,600 to 10,200 MW) are due to the immediate effect of the contingency, whereas the remaining part of the P–V curve stems form the effect of LTC controlled voltage restoration, as well as that of a slow load ramp. As seen in Figure 10-10, the total load is not able to return to its precontingency consumption even after the load ramp.

The regional P–V curve of the area of Attica (Metropolitan Athens region) for the same contingency is shown in Figure 10-11. As seen, the load power is severely reduced after the contingency and the demand cannot be restored to precontingency consumption. A mere inspection of the various P–V curves clearly reveals that the region affected by the specific contingency also includes the Peloponnese and central Greece.

10.3.4 Assessment of Transmission System Upgrades using VSA in Study Mode

The VSA tool is also used by HTSO in study mode, in order to assess the effect of various changes in the system on voltage security and the relative merit of different up-

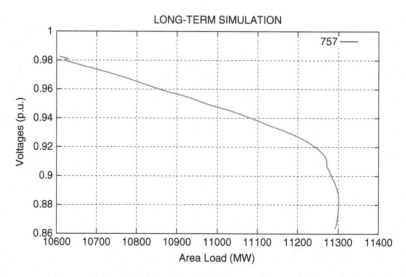

Figure 10-9. National P–V curve without contingency.

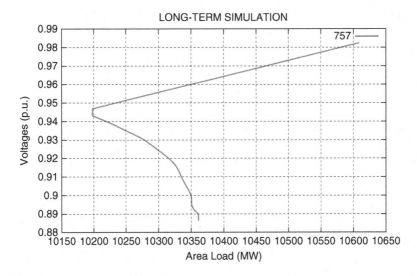

Figure 10-10. National P–V curve after severe contingency number 134.

grades. For instance, it was used to quantify the effect of the reinforcements made in the Hellenic System before and during the summer of 2004. These projects had been considered necessary since Athens was to host the 2004 Olympic Games. The reinforcements included the installation of additional autotransformers at EHV/HV substations in Metropolitan Athens and the adjacent area of central Greece, the installation of mechanically switched capacitor banks at some crucial substations (totalling 400

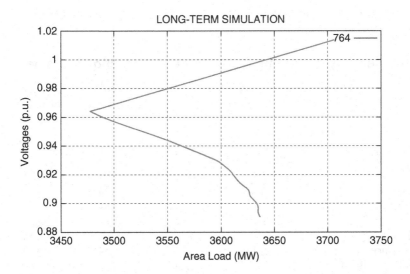

Figure 10-11. Attica regional P–V curve after severe contingency (number 134).

MVAr) on both medium- (MV) and high-voltage (HV) buses, and the reinforcement of the 150 kV network in the Athens region [6].

Results of this study are shown in Table 10-2, where SOLs for the most threatening contingencies at that time (all referring to generation loss) are shown in four cases, all with the same load and generation profile, that correspond to an actual snapshot on July 9th, 2004, namely:

- Case 1 corresponds to the actual July 9th, 2004 snapshot retrieved from the SCADA/EMS state estimator.
- Case 2 refers to the same system with the addition of the capacitor banks at the medium-voltage level of 20 kV.
- Case 3 was constructed from Case 2 with the addition of the capacitor banks at the 150 kV level.
- Case 4 was created by adding to Case 3 the autotransformers at the EHV/HV substations, thus including all the reinforcements in the system.

By looking at Table 10-2, one can see and quantify the effect of the individual reinforcements on system loadability margins, with the most important effect being that of the medium-voltage capacitors (Case 2) followed by that of the EHV/HV autotransformers (Case 4).

The national P–V curves for Cases 1 and 4 (before and after the complete system reinforcement) and for the most severe single contingency (LAVRION.CC4) are given in Figure 10-12. The critical point of the curve of Case 4 is at a higher load and voltage level than the one of Case 1, showing substantially improved loadability limits for the reinforced network.

10.4 USE OF ONLINE VSA FOR ARMING LOAD-SHEDDING PROTECTION

10.4.1 Security Assessment and System Protection Schemes

As power systems throughout the world are forced by economic and environmental considerations to "walk closer to the edge" of a possible catastrophic blackout, the

Table 10-2. Effect of system reinforcements on secure operation limits

Contingency name	Case 1	Case 2	Case 3	Case 4
LAVRIO.CC3.GEN2	0	295	388	555
LAVRIO.CC4	0	274	325	491
LAVRIO.CC3	318	>650	>800	>826
LAVRIO.GEN2	374	>650	>800	>826
MEGALOPOLIS.GEN4	261	284	325	529
MEGALOPOLIS.GEN3	>598	640	650	>826

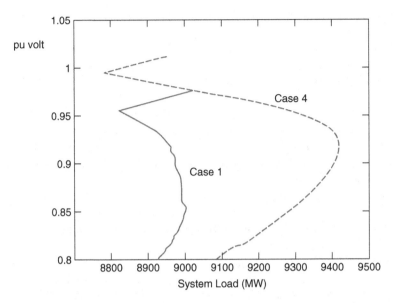

Figure 10-12. National P–V curves for contingency LAVRION.CC4.

need for effective means to identify an approaching critical condition, as well as to counteract an ongoing instability, is becoming more evident [16]. Online VSA can be used as the "yardstick" that measures the distance to the edge and provides information to take appropriate preventive actions and restore sufficient security margins, whenever needed. Preventive enhancement of security may lead the TSO to decrease the posted available transfer capabilities, to reschedule generation, to request some units to be kept in operation for voltage support, and, as a last resort, to shed load. Of course, any of these actions has a nonnegligible cost.

Coming back to the "walking close to the edge" paradigm, when everything else fails and the edge is reached or is even slightly overstepped, the system should have a "safety net" that will avoid the widespread catastrophe of a free fall. Clearly "falling to the net" has a cost, but this is minimal with respect to that of a generalized blackout.

As already mentioned, a system protection scheme (SPS) is a protection system designed to detect abnormal system conditions and take predetermined corrective actions (other than the isolation of faulted elements) to preserve as far as possible system integrity and regain acceptable performance.

System protection schemes should share some common design requirements with equipment protection schemes:

1. Dependability: the protection acts when needed,
2. Security: the protection does not act when not needed,
3. Reliability: the protection is both dependable and secure,
4. Selectivity: the size of the action fits the severity of the disturbance, and

5. Robustness: the protection can face a wide range of scenarios that could be encountered.

In recent years, several types of SPS have been proposed and/or implemented. They can be classified according to the following four criteria:

1. *Decentralized versus Wide-Area.* A decentralized SPS uses local measurements (e.g., instance collected in the same substation) and acts on local devices (e.g., located in the same substation). On the other hand, a wide-area SPS collects data from several (distant) locations and/or acts on devices located in several (distant) places of the system.

2. *Response-Based versus Event-Based.* A response-based SPS relies on measurements through which the consequences of system misoperation can be observed. On the contrary, an event-based SPS operates upon the recognition of a particular combination of events.

3. *Rule-Based versus Algorithmic Decision.* A rule-based SPS relies on "if . . . then . . ." rules (e.g., comparison of measurements with thresholds), whereas an algorithmic decision-based SPS relies on a more involved analysis of a model of the system.

4. *Closed-Loop versus Open-Loop.* An SPS is closed-loop when it is able to act several times, each action relying on the measured result of the previously taken actions.

Decentralized SPSs are more reliable since they do not rely on an extensive telecommunication system. On the other hand, they may lack the system view needed to coordinate various competing controls.

Event-based SPSs are expected to be faster than response-based SPSs, which have to wait for the system response to a specific event before acting. The former are appropriate in cases in which the threatening disturbances can be clearly identified. However, response-based SPSs are more robust, since they work by observing the consequences of disturbances without attempting to identify them.

A similar conclusion holds for algorithmic SPSs; since they rely on a model of the system, they are in principle better prepared to face unforeseen disturbances and adapt their action to the severity of the situation. On the other hand, the more detailed the model, the lower the robustness with respect to modeling and real-time data inaccuracies, and the higher the dependency upon the real-time information system. Rule-based SPSs are comparatively simpler, although the embedded rules have to be properly tuned.

When the counteracted phenomena make it possible, closed-loop SPSs combine selectivity and dependability; since they are allowed to operate as many times as needed, they automatically adapt their action to the severity of the disturbance. Furthermore, this increases the SPS robustness against the inevitable uncertainties in system behavior. This is particularly important in voltage instability, where loads play a central role; their composition changes with time and their behavior under large voltage drops is not known accurately.

10.4.2 Event-Based SPS in the Hellenic System

10.4.2.1. Considered Contingencies

As seen in the previous sections, the Hellenic Interconnected System is occasionally prone to voltage instability. A number of major system upgrades were performed to restore the security of the system; however, for certain specific and very severe contingencies, security margins may drastically drop at some point in the near future.

The contingencies in question are:

1. Tripping of a double circuit line serving radially the power station at Lavrio, close to Athens, in the Attica region, of nominally 950 MW (contingency number 134 in Table 10-1); and

2. Combined loss of two 300 MW units at the Megalopolis station in the southern Peloponnese peninsula.

At the last stage of the 2004 blackout, a manual load-shedding action was attempted, but it was incomplete since there was no time left for it to be fully applied. This led to the conclusion that for each of the above two contingencies, an automatic load-shedding scheme should be designed and implemented. Some basic design characteristics of the SPS that has been implemented in each of the above cases are outlined below [15,18].

Due to the long history of low voltages in the system, undervoltage was not deemed a sufficient indicator for load shedding. It was, thus, decided to select two separate event-driven load-shedding schemes that would recognize the above contingencies and act accordingly to shed prespecified loads.

On the other hand, transmission voltages below 0.88 p.u. at generator buses were seen from the blackout experience to cause disconnection of units (due to loss of auxiliaries), leading to a cascade of generator trips. This information was used to tune the amount of load shedding for each of the above schemes. Thus, the minimum amount to be shed was calculated so that no generator voltage would fall below the tripping threshold.

The final amount of shedding was determined by also examining available feeders for automatic disconnection. The feeders marked for the underfrequency protection system (used also to avoid overloading of the interconnection lines to the neighboring countries to the north) were used for this purpose. The amount of load shedding has been tested and decided by simulation.

10.4.2.2 Example of SPS Design in Attica

For the design of the SPS against contingency number 134 of Table 10-1, the voltage sensitivities to load rejection were used. In Table 10-3, a sample of four available load-tripping commands are ranked based on the quantity e_k, which is the increase of the weakest voltage ΔV_k divided by the amount of load shed ΔP_k (in p.u.). The most effective tripping command corresponds to the largest quantity e_k.

To compute the minimum amount of shedding required, the most effective tripping command (number 3) is tested first in the simulation. The simulation results are shown

Table 10-3. Most effective tripping commands

Tripping command k	e_k	ΔP_k (MW)	ΔV_k (p.u.)
3	0.02109	155.8	0.0329
1	0.01803	24.0	0.0049
6	0.01326	30.8	0.0041
2	0.01186	158.6	0.0188

in Figure 10-13, where the lowest 150 kV bus voltage is drawn for successive load shedding commands. It is assumed that the contingency occurs at $t = 200$ s, whereas load shedding is applied at $t = 250$ s. Since load shedding command number 3 is clearly not enough to prevent voltage instability, as seen in Figure 10-13, the simulation is repeated by executing simultaneously the two most effective tripping commands, numbers 3 and 1, and so on. The voltage collapse is finally prevented when the three most effective tripping commands (numbers 3, 1, and 6) are executed.

This corresponds to the minimum required load shedding, of approximately 210 MW. As seen in Figure 10-13, with the minimum load shedding the voltage remains quite depressed. The effect of adding tripping command 2 is also illustrated in Figure 10-13. In the latter case, all voltages are restored to acceptable values, but a total amount of 369 MW is shed.

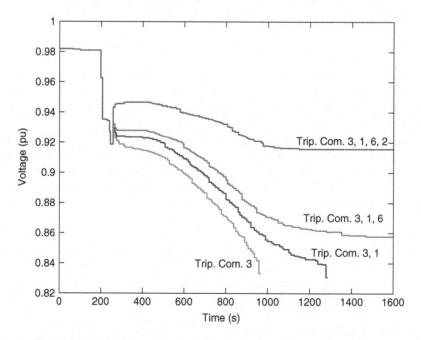

Figure 10-13. Voltage evolution at weakest bus after load shedding.

10.4.2.3 Design of SPS in the Peloponnese

The contingency considered for the design of the SPS in the Peloponnese is the successive loss of units 3 and 4 of the Megalopolis Power Station at $t = 100$ s and 200 s, respectively (total loss of 600 MW), followed by a slow proportional ramp on the load demand, which corresponds to a total increase in load of approximately 200 MW. The simulation is based on the 2006 system peak-load condition that occurred on August 21, 2006. The simulated voltage evolution of some weak 150 kV buses in the affected area of the Peloponnese for the above contingency is presented in Figure 10-14.

System degradation is observed after the disturbance, although voltage collapse is avoided because the LTCs in the affected region reach their tap limits and, thus, load restoration is stopped. Since, however, this condition cannot guarantee stability in the presence of downstream load restoration mechanisms (such as distribution system voltage regulators) this condition is not considered acceptable and a load shedding is required.

In the Peloponnese, load can be shed automatically only by the tripping of specific transmission lines. This immediate shedding action is only temporary, since a manual load-shedding program involving other substations in the Peloponnese area is scheduled to follow in order to reconnect the disconnected lines. The simulated voltage evolution of the two weakest 150 kV buses in the affected area is presented in Figure 10-15 with the protection system disconnecting two transmission lines after detecting the loss of the second unit of 300 MW (a 10 s time delay is assumed in the simulation). This action leads to a total shedding of approximately 183 MW. As seen, voltages re-

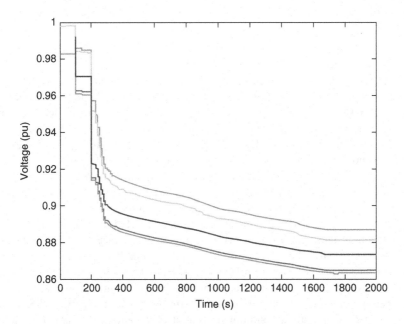

Figure 10-14. Voltage evolution at weak buses without load shedding.

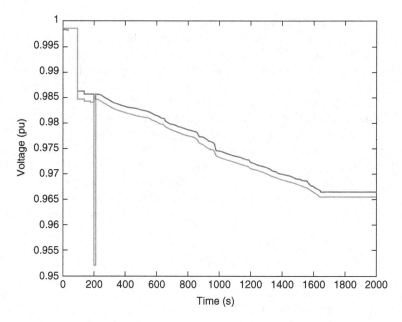

<u>Figure 10-15.</u> Voltage evolution of weakest buses in the Peloponnese after load shedding.

turn to normal after load shedding. The added load ramp ensures that a security margin is restored.

10.4.2.4 Arming SPS through Online VSA

One key consideration when deciding to commission an SPS in an actual system is the question of security—how to minimize the risk of its unintended operation. In order to avoid such risks in the Hellenic System, it was decided to arm the load-shedding schemes centrally from the National Control Center. The arming command is, thus, issued using the information provided by the online VSA application (when margins for the above contingencies are close to zero) as well as operator judgment.

In particular, the SPS for the Attica region is scheduled to be armed each time the specified double contingency has a zero, or very small positive margin (less than 20 MW). In the case of the Peloponnese, since there is no likely common cause for the double contingency considered (loss of two units connected to separate substations), the protection system is armed only when a single contingency results in the loss of both units, that is, only when one of the above units is already unavailable due to maintenance or whatever another reason. Thus, each load-shedding scheme is armed centrally, but it is triggered by local detection of the corresponding equipment outage.

The combination of local SPS with central arming increases security (SPS will not operate in normal operating conditions) without deteriorating dependability, as the SPS will, hopefully, be armed during all emergency conditions. Also, in this way the operators have control over the system (since they can arm or disarm the SPS) while load is shed automatically, only when really needed.

10.5 CONCLUSION

This chapter presented a voltage security assessment tool in operation at the control center of the Hellenic Transmission System Operator in real-time and study environments. Specifically, it described the application of the tool within the existing control center SCADA/EMS and some characteristic results in both real time and study mode.

The real-time application of VSA proved very useful during the daily operation of the system since it gives an updated indication to the operator, as the system changes with time, about possible voltage security threats to the system. The available tools indicate the nature of possible threats (voltage collapse or low voltages) and their geographic extent, as well as the corresponding security margin in terms of acceptable load increase. This allows system operators to make decisions and take actions based on unbiased information. In study mode, VSA can be used as a tool for planning system transmission and generation expansion and to assess the effect of reinforcements on voltage security.

The present state of the Hellenic Interconnected System appears quite secure due to the continuing upgrades. However, assuming reasonably weakened system conditions, an insecure condition may be encountered for two critical contingencies involving loss of generation in Attica and the Peloponnese.

Thus, HTSO has designed and set in operation, in cooperation with Public Power Corporation, two automatic load-shedding schemes in the areas of Attica and the Peloponnese, in order to protect the system against these critical contingencies. The protection systems are event-driven and are manually armed from the Control Center, when the system conditions are considered insecure for the specific contingencies based upon the level of the security margins calculated by the online VSA and operational experience. Thus, the risk of unintended operation of the system protection is minimized.

In order to enhance the ability of the Hellenic System to withstand voltage stability problems in the future, which could also stem from contingencies other than those considered, HTSO intends to revise periodically the existing load shedding schemes against voltage instability as system conditions evolve.

10.6 REFERENCES

[1] Karlsson, D. (convenor), CIGRE TF 38.02.19: *System Protection Schemes in Power Networks,* CIGRE Brochure, 2001.

[2] Kundur, P. (convener), Paserba, J. (secretary), Ajjarapu, V., Andersson, G., Bose, A., Canizares, C., Hatziargyriou, N., Hill, D., Stankovic, A., Taylor, C., Van Cutsem, T., and Vittal, V., "Definition and Classification of Power System Stability," *IEEE Trans. on Power Systems,* Vol. 19, pp. 1387–1401, 2004.

[3] Taylor, C., *Power System Voltage Stability,* EPRI/McGraw-Hill, New York, 1994.

[4]   Van Cutsem, T., and Vournas, C., *Voltage Stability of Electric Power System,* Kluwer Academic Publishers, Boston, 1998.

[5]   Bihain, A., Burt, G., Casamata, F., Koronides, T., Lopez, R., Massucco, S., Ruiz-Vega, D., and Vournas, C., "Advance Perspectives and Implementation of Dynamic Security Assessment in the Open Market Environment," paper 39–101, CIGRE, 2002.

[6]   Christoforidis G., Kabouris J., Vournas C., and Van Cutsem T., "Investigation of Parameters Affecting Voltage Security of the Hellenic Interconnected System," presented at Proceedings of 15th PSCC, Liege (Belgium), Aug. 2005.

[7]   Christoforidis, G., Vournas, C., Kabouris, J., and Van Cutsem, T., "Experience Gained by the Application of a Voltage Security Assessment Method at the Hellenic Transmission System Operator," presented at Proceedings of CIGRE Session, Paris, August 2006.

[8]   Van Cutsem, T., Jacquemart, Y., Marquet, J.-N., and Pruvot, P., "A Comprehensive Analysis of Mid-term Voltage Stability," *IEEE Transactions on Power Systems,* Vol. 10, pp. 1173–1182, 1995.

[9]   Van Cutsem, T., and Mailhot, R., "Validation of a Fast Voltage Stability Analysis Method on the Hydro-Québec System," *IEEE Transactions on Power Systems,* Vol. 12, pp. 282–292, 1997.

[10] Van Cutsem, T., Kabouris, J., Christoforidis, G., and Vournas, C. D., "Application of Real-Time Voltage Security Assessment to the Hellenic Interconnected System," *IEE Proceedings on Generation, Transmission and Distribution,* Vol. 152, No.1, pp. 123–131, 2005.

[11] Van Cutsem, T., Grenier, M.-E., and Lefebvre, D., "Combined Detailed and Quasi Steady-State Time Simulations for Large-Disturbance Analysis," *Electrical Power and Energy Systems,* Vol. 28, pp. 634–642, 2006.

[12] Vournas, C. (scientific coordinator), "Software Development for Voltage Stability Analysis," Project 96 SYN 95 Report, NTUA, Athens, 1999.

[13] Rigos, A., Koronides A., Charamoglis P., and Cachet J. M., "A Modern Energy Management System for the Control of the Hellenic Generation and Transmission System," in Proceedings of the Athens Power Technical Conference, Sept. 1993.

[14] Boardman, E., Christoforidis, G., and Koronides , T., "The Network Subsystem Implementation of the Energy Management System of the Public Power Corporation of Greece," in Proceedings of the Athens Power Technical Conference, Sept. 1993.

[15] Nikolaidis, V., Vournas, C., Fotopolulos, G., Christoforidis, G., Kalfaoglou, E., and Koronides, A., "Automatic Load Shedding Schemes against Voltage Instability in the Hellenic System," presented at IEEE PES General Meeting, June 2007.

[16] Van Cutsem, T., and Vournas, C., "Emergency Monitoring and Corrective Control of Voltage Instability," in S. C. Savulescu (Ed.), *Real-Time Stability in Power Systems,* Chapter 4, Springer, New York, 2005.

[17] Van Cutsem, T., and Vournas, C. D., "Emergency Voltage Stability Controls: an Overview," in Proceedings of IEEE/PES General Meeting, Tampa, FL, June 2007.

[18] Nikolaidis, V. C., and Vournas, C. D., "Design Strategies for Load-Shedding Schemes against Voltage Collapse in the Hellenic System," *IEEE Transactions on Power Systems,* Vol. 23, No. 2, May 2008.

[19] Ajjarapu, V., and Christy, C., "The Continuation Power Flow: A Tool for Steady State Voltage Stability Analysis," *IEEE Transactions on Power Systems,* Vol. 7, pp. 416–423, 1992.

[20]   Grenier, M.-E., Lefebvre, D., and Van Cutsem, T., "Quasi Steady-State Models for Long-Term Voltage and Frequency Dynamics Simulation," in *Proceedings of IEEE Power Tech conference,* St Petersburg (Russia), June 2005.

[21]   Gear, C. W., *Numerical Initial Value Problems in Ordinary Differential Equations,* Prentice-Hall, Englewood Cliffs, NJ, 1971.

ANNEX 10-1   QUASI-STEADY-STATE SIMULATION

The quasi-steady-state (QSS) approximation of long-term dynamics is a well-known and proven technique for fast, simplified simulation of power system long-term dynamics. Simply stated, this approximation relies on time-scale decomposition: faster phenomena are represented by their equilibrium conditions instead of their full dynamics [4]. This technique greatly reduces the complexity of the resulting model and, hence, provides the computational efficiency required for applications such as real-time contingency analysis or training simulators. In this annex, the principle and the formulation of QSS simulation are reviewed. The interested reader may find more details in [4, 8, 9, 11, 20].

Principle of the QSS Approximation

In stability studies, the general dynamic model of a power system takes the form:

$$0 = \begin{pmatrix} \mathbf{G} & -\mathbf{B} \\ \mathbf{B} & \mathbf{G} \end{pmatrix} \begin{pmatrix} \mathbf{v}_x \\ \mathbf{v}_y \end{pmatrix} - \begin{pmatrix} \mathbf{i}_x \\ \mathbf{i}_y \end{pmatrix} \tag{A.1}$$

$$0 = \mathbf{g}(\mathbf{v}_x, \mathbf{v}_y, \mathbf{i}_x, \mathbf{i}_y, \mathbf{x}, \mathbf{z}) \tag{A.2}$$

$$\dot{\mathbf{x}} = \mathbf{f}(\mathbf{v}_x, \mathbf{v}_y, \mathbf{i}_x, \mathbf{i}_y, \mathbf{x}, \mathbf{z}) \tag{A.3}$$

$$\mathbf{z}(t_k^+) = \mathbf{h}\big[\mathbf{x}, \mathbf{y}, \mathbf{z}(t_k^-)\big] \tag{A.4}$$

Equations (A.1) are the network relationships, where $\mathbf{G}$ and $\mathbf{B}$ are the real and imaginary parts of the bus admittance matrix, $\mathbf{v}_x$ and $\mathbf{v}_y$ are the projections of the voltage phasors on two rotating orthogonal axes denoted x and y, and similarly for $\mathbf{i}_x$ and $\mathbf{i}_y$. Equation (A.2) relates to the stator of synchronous and induction machines, to static loads, SVCs and similar FACTS devices, and so on. The differential equations (A.3) involve the vector of (continuous) state variables $\mathbf{x}$ and relate to a wide variety of phenomena and controls including:

- Short-term dynamics of generators, turbines, speed governors, automatic voltage regulators (AVRs), static Var compensators (SVCs), induction motors, HVDC links, and so on.

- Long-term dynamics of secondary frequency and voltage control, load self-restoration, and so on.

Finally, the discrete-time equations (A.4) capture discrete events that stem from:

- Controllers acting with various delays on shunt compensation, generator set-points, load tap changers (LTCs), and so on.
- Equipment protections such as overexcitation limiters (OELs).
- System protection schemes against short- and long-term instabilities, acting on loads and/or generators, for instance.

The corresponding (shunt susceptance, transformer ratio, etc.) variables are grouped into the vector, which undergoes discrete changes from $\mathbf{z}(t_k^-)$ to $\mathbf{z}(t_k^+)$ at some instants $t_k (k = 1, 2, \ldots)$. $\mathbf{G}$ and $\mathbf{B}$ often change with $\mathbf{z}$ but this dependency is not shown for clarity. Apart from digital controllers operating at constant sampling rate, the t_k instants are dictated by the system dynamics.

As indicated previously, the QSS approximation of long-term dynamics consists of representing faster phenomena by their equilibrium conditions instead of their full dynamics. Thus (A.3) is replaced by:

$$0 = \mathbf{f}_1(\mathbf{v}_x, \mathbf{v}_y, \mathbf{i}_x, \mathbf{i}_y, \mathbf{x}_1, \mathbf{x}_2, \mathbf{z}) \qquad (A.5)$$

$$\dot{\mathbf{x}}_2 = \mathbf{f}_2(\mathbf{v}_x, \mathbf{v}_y, \mathbf{i}_x, \mathbf{i}_y, \mathbf{x}_1, \mathbf{x}_2, \mathbf{z}) \qquad (A.6)$$

where $\mathbf{x}$ has been decomposed into (the fast/algebraic) $\mathbf{x}_1$ and (the slow/dynamic) $\mathbf{x}_2$ subvectors, with a corresponding decomposition of $\mathbf{f}(.)$ into $\mathbf{f}_1(.)$ and $\mathbf{f}_2(.)$. By way of illustration, a model of the synchronous generator is presented in detail later in this annex.

Handling of Frequency in QSS Simulation

When simulating long-term dynamics, system frequency changes must be appropriately accounted for. Two QSS models can be used, depending on how system frequency f_s is modeled:

Model 1. f_s is an algebraic variable of the type $\mathbf{x}_1$, assuming that speed governors and turbines react instantaneously.

Model 2. f_s is a dynamic state of the type $\mathbf{x}_2$, together with other states describing the turbines and speed governors.

All synchronous machines are assumed to rotate at the same electrical speed. By summing the swing equations of all machines, one has:

$$2\pi \left(\sum_{j=1}^{g} M_j \right) \dot{f}_s = \eta = \sum_{j=1}^{g} P_j^m - \sum_{j=1}^{g} P_j \qquad (A.7)$$

where g is the number of generators, M_j is the inertia of the jth machine, and P_j^m is its mechanical power, which is an output of the turbine model included in (A.6). The η variable represents the total imbalance between mechanical and electrical powers. Replacing $2\pi \dot{f_s}$ by its expression obtained from (A.7) in the swing equation of the ith machine, the latter becomes merely

$$P_i^m - P_i = \frac{M_i}{\sum\limits_{j=1}^{g} M_j} \eta \tag{A.8}$$

In both models, all phasors as well as the previously mentioned x and y reference axes are assumed to rotate at the angular frequency $2\pi f_s$.

QSS Model of the Synchronous Machine and its Regulations

Under the QSS approximation, the short-term dynamics of a generator and its regulators are neglected. In theory, the QSS model is obtained by setting $\dot{x_i}$ to zero in the corresponding equations of the complete model (A.3). In practice, however, the reduced model detailed hereafter is used as a good compromise between simplicity and accuracy.

Each synchronous machine is characterized by the electromotive force (emf) $\overline{E_q}$ whose magnitude is proportional to the machine field current, and the emf E_q^s behind saturated synchronous reactances. Assuming negligible armature resistance, the stator equations are written in the machine (d, q) reference frame as

$$\begin{pmatrix} 0 \\ E_q^s \end{pmatrix} = \begin{pmatrix} V_d \\ V_q \end{pmatrix} + \begin{pmatrix} 0 & X_q^s \\ -X_d^s & 0 \end{pmatrix} \begin{pmatrix} I_d \\ I_q \end{pmatrix} \tag{A.9}$$

and in the system (x, y) axes:

$$\begin{pmatrix} 0 \\ E_q^s \end{pmatrix} = \begin{pmatrix} -\sin \delta & \cos \delta \\ \cos \delta & \sin \delta \end{pmatrix} \begin{pmatrix} V_x \\ V_y \end{pmatrix} + \begin{pmatrix} 0 & X_q^s \\ -X_d^s & 0 \end{pmatrix} \begin{pmatrix} -\sin \delta & \cos \delta \\ \cos \delta & \sin \delta \end{pmatrix} \begin{pmatrix} I_x \\ I_y \end{pmatrix} \tag{A.10}$$

where δ is the phase angle of $\overline{E_q^s}$ in the x–y reference frame, and X_d^s and X_q^s are the saturated direct- and quadrature-axis synchronous reactances, respectively. They relate to their unsaturated values X_d and X_q through

$$X_d^s = X_1 + \frac{X_d - X_1}{k} \qquad X_q^s = X_1 + \frac{X_q - X_1}{k} \tag{A.11}$$

where X_1 is the leakage reactance and $k = (E_q/E_q^s)$ is a saturation coefficient. A widely used saturation model is

$$k = 1 + m(V_1)^n \tag{A.12}$$

where V_l is the magnitude of the voltage behind leakage reactance:

$$\overline{V}_1 = \overline{V} + jX_1\overline{I} \tag{A.13}$$

Expressing voltages and currents in terms of their components, (A.12) becomes

$$k - 1 - m[(V_x - X_1I_y)^2 + (V_y + X_1I_x)^2]^{n/2} \tag{A.14}$$

With X_d^s and X_q^s given by (A.11), the two equations in (A.10) make up the stator equations of the type (A.2). There are three algebraic variables of the type x_1, namely E_q^s, δ, and k, that must be balanced by three equations of the type (A.5). The first of these equations is (A.14). The remaining two are provided by voltage and speed regulations, as detailed next.

For voltage control, one of the following relations holds.

Under AVR control:

$$E_q = G(V^0 - V) \Leftrightarrow kE_q^s - G(V^0 - \sqrt{V_x^2 + V_y^2}) = 0 \tag{A.15}$$

where V^0 is the voltage setpoint and G the open-loop static gain.

Under error-free (integral) AVR control:

$$V = V^0 \Leftrightarrow \sqrt{V_x^2 + V_y^2} - V^0 = 0 \tag{A.16}$$

Under field current (lower or upper) limit:

$$E_q = E_q^{\text{lim}} \Leftrightarrow kE_q^s - E_q^{\text{lim}} = 0 \tag{A.17}$$

Under constant reactive power:

$$Q = Q^{\text{lim}} \Leftrightarrow V_yI_x - V_xI_y - Q^{\text{lim}} = 0 \tag{A.18}$$

For speed control, one of the following relations holds.

If frequency Model 1 is used, and under speed-governor control:

$$P = P^m = P^0 - \alpha f_s \Leftrightarrow V_xI_x + V_yI_y - P^0 + \alpha f_s = 0 \tag{A.19}$$

where P^0 is the power setpoint and α involves the permanent speed droop.

Under turbine limit:

$$P = P^{\text{lim}} \Leftrightarrow V_xI_x + V_yI_y - P^{\text{lim}} = 0 \tag{A.20}$$

If frequency Model 2 is used, according to (A.8):

$$P^m - P = \frac{M_i}{\sum\limits_{j=1}^{g} M_j} \eta \Leftrightarrow V_x I_x + V_y I_y - P^m + \frac{M_i}{\sum\limits_{j=1}^{g} M_j} \eta = 0 \qquad (A.21)$$

The QSS model includes an additional algebraic variable, namely f_s if Model 1 is used, or η if Model 2 is used. This variable is balanced by the phase angle reference equation:

$$V_{yr} = 0 \qquad (A.22)$$

where r is the number of the reference bus.

Numerical Integration of the QSS Model

Grouping all algebraic variables into the vector:

$$\mathbf{y} = [\mathbf{v}_x^T \mathbf{v}_y^T \mathbf{i}_x^T \mathbf{i}_y^T \mathbf{x}_1^T]^T \qquad (A.23)$$

the QSS model (A.1, A.2, A.4, A.5, A.6) can be rewritten in compact form as

$$\mathbf{0} = \mathbf{g}(\mathbf{y}, \mathbf{x}_2, \mathbf{z}) \qquad (A.24)$$

$$\dot{\mathbf{x}}_2 = \mathbf{f}_2(\mathbf{y}, \mathbf{x}_2, \mathbf{z}) \qquad (A.25)$$

$$\mathbf{z}(t_k^+) = \mathbf{h}[\mathbf{y}, \mathbf{x}_2, \mathbf{z}(t_k^-)] \qquad (A.26)$$

When $\mathbf{z}$ undergoes a discontinuity due to (A.26), so does $\mathbf{y}$ but not $\mathbf{x}_2$. The new value of $\mathbf{y}$ is obtained by solving (A.24) through a sequence of Newton iterations:

$$\mathbf{g}_y \Delta \mathbf{y} = -\mathbf{g} \qquad (A.27)$$

where $\mathbf{g}_y$ is the Jacobian matrix of $\mathbf{g}$ with respect to $\mathbf{y}$. The last available value of $\mathbf{y}$ is used as initial guess for this sequence of iterations.

The differential equations (A.25) can be integrated using the Trapezoidal rule:

$$\mathbf{x}_2(t + h) = \mathbf{x}_2(t) + \frac{h}{2} [\dot{\mathbf{x}}_2(t + h) + \dot{\mathbf{x}}_2(t)] \qquad (A.28)$$

where h is the time step. In practice, it ranges from 0.1 to 10 s, depending on the system.

The following partitioned scheme and functional iterations are used. Starting from a predicted value of $\mathbf{x}_2(t + h)$, (A.24) is solved using fully converged Newton iterations (A.27) to obtain $\mathbf{y}$. The latter is substituted into (A.25) to update the derivatives

$\dot{x}_2(t + h)$, which in turn are introduced into (A.28) to obtain a corrected value of $x_2(t + h)$. The procedure is repeated until the following conditions are satisfied:

- Under frequency Model 1, the last change in $x_2(t + h)$ does not trigger Newton iterations (A.27), that is, all components of g remain below a specified tolerance.
- Under frequency Model 2, the last change in $x_2(t + h)$ causes η to change by less than a specified tolerance.

Although functional iterations prevent increasing h beyond some value, for convergence reasons [21], they have been preferred because they allow using the same Jacobian g_y when integrating the differential equations and when solving the numerous discontinuities (A.26) typical of a long-term simulation. This Jacobian is updated as rarely as possible.

When Model 1 is used for frequency and the evolution is driven by LTCs and OELs, the QSS model has no differential equation (A.25); instead, it is purely discrete (A.26). In this case, the simulation only involves the Newton iterations (A.27).

11

THE REAL-TIME SUPERVISION OF TRANSMISSION CAPACITY IN THE SWEDISH GRID

Lars Sandberg and Klas Roudén

11.1 INTRODUCTION

11.1.1 Swedish Power System in the Context of the Nordel Interconnection

The Swedish National Grid encompasses 400 kV and 220 kV overhead transmission lines as well as HVDC sea cables. The grid, including its 158 substations, is owned and operated by Svenska Kraftnät (SvK, Swedish National Grid), a state owned transmission company that emerged in 1992 from the restructuring of Vattenfall, which today is the largest power producer in Sweden. The 130 kV and 70 kV regional grids are connected to the main grid but are owned and operated by distribution companies that are totally independent from SvK.

The Swedish National Grid is interconnected via AC transmission lines and sea cables with the transmission systems of the neighboring countries Finland, Norway, Denmark, Germany, and Poland. The integrated power system in Sweden, Norway, Finland, and Denmark is known as Nordel and is depicted in Figure 11-1.

A significant amount of hydro power is generated in the northern and central part of Sweden. Most of this power gets transferred to the southern part of the country where the load requirements are also covered with the output of three nuclear power stations with a total of 10 reactors. The largest part of the generation capacity is owned by

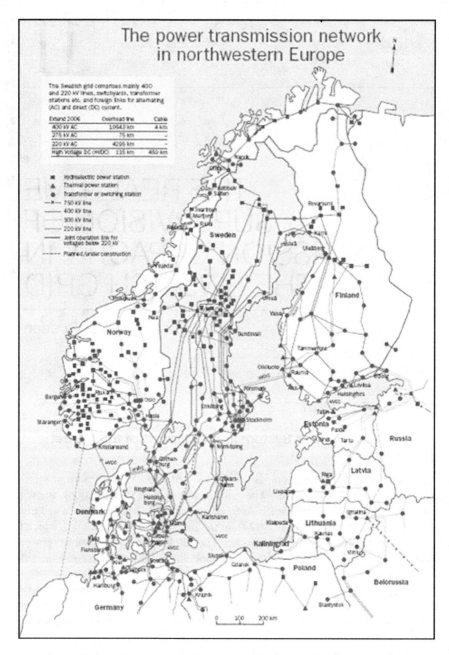

Figure 11-1. Nordel interconnected transmission system.

companies totally separated from SvK, which only owns 11 gas turbines. The peak load occurs during wintertime, has a typical value of approximately 27,000 MW, and is covered by a total average production capacity of approximately 28,500 MW at full availability of all 10 nuclear blocks.

11.1.2   Operational Characteristics of the Transmission Grid

11.1.2.1   *Physical Phenomena Caused by Topology, Pattern of MW Transfers, and Insufficient Reactive Compensation*

The internal transfer limits in the Swedish National Grid are most often set with respect to the risk of voltage collapse. Due to the longitudinal topology of the network and the limited reactive compensation resources in the middle part of the grid, the power transfers from north to south tend to result in high reactive losses, especially at high transfer rates and despite the installation of series capacitors with high compensation degree in all eight 400 kV lines from middle to southern Sweden. Furthermore, as shown in Figure 11-1, the cross-connections are rather few in the northern part of the grid, thus obstructing the reactive support from the large hydro units connected to adjacent parallel lines that might otherwise be available during transmission contingencies.

The topology of the southern part of the grid is more meshed, but here the reactive compensation resources consist basically of the MVAr generation in the nuclear power stations, which, in most cases, are situated relatively far from the location of critical contingencies. Accordingly, the reactive shortage in the south is often compensated by shunt capacitors installed both in the Swedish National Grid and in the regional subtransmission and distribution networks. In addition, a certain amount of reactive compensation in the southern part of the Swedish National Grid is provided by a few Static VAr Compensation (SVC) devices.

The voltage and reactive power management is rendered even more difficult by the normally large power transfer variations between day and night, which may cause bus voltage violations when the transmission lines are lightly loaded. At night, the voltages are kept within acceptable levels, for example, 420 kV, by switching on the shunt reactors in strategic substations. The total nominal capacity of shunt reactors installed at SvK is 5890 MVAr, with individual unit sizes in the range of 50–200 MVAr.

In addition to the limitations mentioned above, the power transfers toward Finland, in the north, and to Norway, in the south, are also restricted by the risk of voltage collapse. Depending upon the direction of power transfers, transient stability problems or poor damping may also occur, thus both creating further restrictions internally, within Sweden, and limiting the transfers to Finland, Norway, and eastern Denmark (Sealand). Finally, in a few cases during summertime, certain outages in the Swedish National Grid may also cause thermal violations of the transmission equipment.

11.1.2.2   *Concepts and Principles for Power System Control in Sweden*

TRANSMISSION BOTTLENECKS. The transmission capacity, expressed in MW, is normally related to certain *bottlenecks* in the Swedish National Grid. The main critical bottlenecks for the power direction from north to south are shown in Figure 11-2.

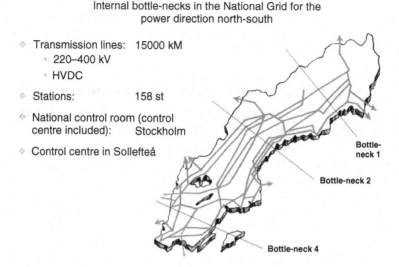

Figure 11-2. Internal bottlenecks within the Swedish National Grid.

The power flows across these bottlenecks are calculated by adding the power flows, telemetered by SCADA or computed by the state estimator, entering into the northern ends of all the transmission lines encompassed by the respective bottleneck. Then, for each bottleneck, a list of critical $N-1$ contingencies or *dimensioning faults* related to voltage security is built manually. Approximately 12–15 contingencies are listed for each bottleneck.

The contingencies in these lists can vary depending upon the current topology of the transmission grid and are identified by analysts in the operations department both as part of the process of planning outages and after postdisturbance transmission outages, which sometimes may cause contingencies that were not envisioned for normal operating conditions. Typical contingencies considered in the Swedish National Grid include bus faults that may cause the tripping of lines or of major generating units, especially in the south. Multiple events that consist of both line and generator trips are much more dangerous because, in addition to producing "a hole in the grid," they cause a frequency droop that is immediately compensated by an automatic increase (via the droop, as indicated below) of the hydro production in the north or southern Norway with the subsequent increase of power flows across the bottlenecks.

As shown subsequently in this chapter, the entire sequence of procedural steps when executing the voltage security assessment tool is referred to as a "job." A "job" begins with the preparation and execution of a complete suite of voltage security assessment calculations and ends with the presentation of the final voltage security limits corresponding to the current contingencies for the chosen bottlenecks. In order to account for potential inaccuracies in the network model as well as in the solution technique, the theoretical voltage security limits computed for the current contingencies in the respective bottleneck are decreased with a certain *operational margin,* which is specific for each bottleneck and most often consists of a standard figure. The lowest limit for all contingencies in a bottleneck after applying the operational margin correction is then assumed to represent the *operational limit* of that bottleneck.

PRIMARY AND SECONDARY VOLTAGE CONTROL. There are two types of *primary voltage control* procedures:

1. Automatic switching of shunt reactors and capacitors when certain bus voltages in the transmission system reach the upper or lower limits
2. Continuous (smooth) control of the bus voltages at the terminals of generating units, synchronous condensers, and SVCs by automatic voltage regulators

At present there is no *automatic secondary voltage control* with closed-loop control based on optimal power flow (OPF) or similar tool. The secondary voltage control is, therefore, manual. This type of voltage control at the generating units is ordered by phone from SvK's control centers to external operators in accordance with agreements concerning reactive power support, without costs for SvK. It is expected that SvK will begin to use the OPF tool for minimizing the network losses sometime in 2008, but still without closed-loop control. However, there is a plan to use closed-loop control for the HVDC link to Finland, which belongs to the same AC area as Sweden. Today, there is a coarse loss optimization function for this HVDC in the operational tool, which is briefly described in a subsequent section in this chapter.

PRIMARY AND SECONDARY FREQUENCY CONTROL. The *frequency control* is performed by *primary regulation (droop)* and *secondary regulation.* The primary regulation is executed automatically via the turbine governors of the units. The sensitivity in this regulation can be changed locally or remotely with various settings in the respective turbine governors.

The secondary regulation in Sweden is always carried out manually and encompasses the unit start-up, unit shutdown, and changing the set points for active power. There is no automatic generation control (AGC) but some producers use central computers (pilots) to automatically operate hydro units according to their plans. There are also some *system protections* installed in the Swedish National Grid. They act automatically at tripping of lines when certain lower voltage limits are reached, or when combinations of low voltage events take place, and attempt to compensate for decreasing voltage levels. System protections also act to counter the action of unit trippings, starting of gas turbines, injection of power from HVDC links, and so on. Tripping of units will, however, cause a frequency imbalance, which normally is handled via the frequency control procedures described above.

TAP CHANGING UNDER LOAD AND AUTOMATIC VOLTAGE REGULATION. There are no *tap changing under load (TCUL) transformers* in the SvK's 400/220 transmission grid, but TCUL transformers are installed at the interface with the distribution system at voltage levels from 130 kV and below. Most taps are installed at the low-voltage side, for example, on the 130 kV side of a 400/130 kV transformer. The two-winding aggregate transformers with connected units have no taps at all, but certain three-winding transformers with connected units are equipped with separate regulation for voltage control.

The automatic voltage regulators (AVR) of most generating units have integrated *field current limiters* that limit the MVAr output during disturbances. All the generators in the nuclear power stations and some large hydro units are also equipped with *armature current limiters.* From the voltage stability point of view, these limiters can cause significant problems during disturbances, as they only limit the reactive part of the current, which then will successively decrease the voltage and reduce the reactive capability of the machine, in the worst case down to zero. These and other voltage control processes are simulated in detail by the voltage security assessment tool described further down in this chapter.

SUPERVISORY CONTROL AND DATA ACQUISITION. Manual switching and control in Sweden is normally performed remotely from the control centers to substations, including hydro and gas turbine power plants, except for the nuclear and a few thermal power stations, which are manned.

The supervisory control in the Swedish National Grid is performed remotely from two dispatch centers, one in Stockholm and the other one with remote work sites in Sollefteå, which is located approximately 500 km north of Stockholm. The National Control Center (NCC) and the work site for the power balance are integrated in the Stockholm dispatch center. The configuration shown in Figure 11-3 depicts the principal solution for real-time supervision and control, program development, and dispatcher training in the XA/21 system from General Electric (GE).

In addition to online voltage stability assessment, the following SCADA/EMS functions are performed:

- Basic SCADA functions such as data acquisition, supervision, and control; dynamic network coloring; operator tools such as area operator roles, alarm filtering, tagging, and power view; and root-cause analysis;
- Custom SCADA functions including monitoring of thermal capacity in series branches and system transformers, monitoring of reactive generator reserves, calculation of corona losses, and handling of system protections;
- Load forecasting;
- Power network analysis, including state estimation, load flow, PNA forecasting of network bottleneck flows and losses; contingency analysis, and short-circuit calculations
- Voltage stability assessment (Spica);
- Optimal power flow (loss minimization); and
- Dispatcher training in the Aristo environment (the Aristo power system model is the power system behind a copy of the XA/21 system).

XA/21 at Svenska Kraftnät

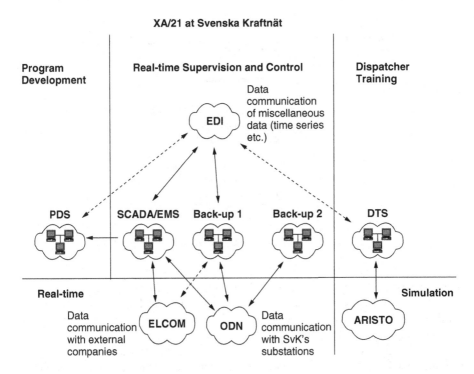

Figure 11-3. Principal solution for the main functions of the XA/21 system at SvK.

11.1.3 The Deregulated Nordel Electricity Market

All players in the electricity business in northern Europe today act in an open and deregulated electricity market. The market transactions are handled by NordPool, which is a power exchange owned by the Transmission System Operators (TSO) in the Nordel system. NordPool is responsible for both most of the internal trading within Nordel and with other deregulated electricity markets in Germany, Netherlands, and Poland. Figure 11-4 describes the relationships between the different players in the Swedish electricity market and how the generated power is handled.

NordPool incorporates a *spot market* for physical trading as well as a *financial market*. The physical trading is performed daily according to 24 hourly energy exchange plans agreed upon by the players. Of highest importance in electricity markets is, of course, the transmission capacity in the grids between various network areas. Every morning at 8.30 a.m. at the latest and after the operations planning and forecasting tasks have been completed, the network supervisor at SvK informs the TSOs in Norway, Denmark, Finland, Germany, and Poland of the hourly maximum capacity, in both directions, for the next twenty-four hours starting at midnight.

After the bids are received at NordPool from the market players, some capacity in certain network bottlenecks may still remain. These margins are then handled by Elbas, a special adjustment market that can be used by Finland, Sweden, Eastern Den-

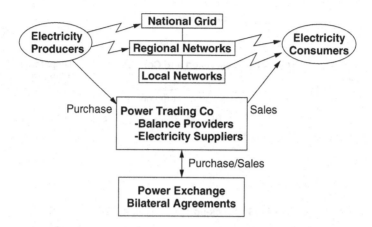

Figure 11-4. Overall structure of Nordel Electricity Market.

mark (Sealand), and Germany, and from late 2008 also by Norway and Western Denmark (Jutland). Elbas, which is an hourly power exchange and marketplace, makes it possible to increase the planned power exchanges up to the operational limits and allows for an optimal use of the transmission system.

So far, Sweden is just one single price area, whereas the electricity market in Norway often can be divided into separate price areas because of temporary network bottlenecks during outages, either planned or caused by disturbances, too high load levels, and so on.

11.2 PRIOR AND CURRENT APPLICATION DEVELOPMENT AT SVK

11.2.1 Earlier Methods for Computing Transmission Capacity

Prior to early 1994, the transmission capacities were first calculated by analysts in operations planning, and then used in real-time operation, where various outages in transmission and production with fixed capacity reductions were taken into account. A catalogue with reduction figures for various single outages or a combination thereof was used in the control room by the network supervisor. The current total measured active load in Sweden was also taken into consideration.

The final result consisted of presenting the bottleneck limit P_{max} to the supervisor as a slight stepwise line parallel with the load axis, where both the working point for the current transmission P across the bottleneck and the load was shown. All calculations were based on a load-flow balance as representative as possible for the current operational state. The calculations concerning voltage stability, for example, P–V curves, were performed with procedures developed in-house by using PSS/E and its internal language IPLAN.

The major limitation of this approach was the inability to continuously maintain a correct and actual load-flow balance to form a base case for the special applications mentioned above. As a result, the manual execution of such computational chores, for example, updating topology and load changes as they happen, became an impossible task.

11.2.2   State Estimation and Generator Reactive Capacity

State estimation has been fully operational since 1983, initially at Vattenfall in the old control center, and, from 2000, in the new control center. The earlier state estimator was written in-house by experts at Vattenfall. Today, the state estimation is performed within the real-time network analysis sequence in the XA/21 SCADA/ EMS system.

The state estimate is the "entrance ticket," so to speak, to all network applications and especially to those used for setting the capacity limits. Therefore, the high quality of the state estimation results is a "must" requirement. Poor quality state estimates would have economically catastrophic consequences for SvK, as the setting of the limits would have to be more conservative. In a worst-case scenario, SvK would have to revert to the manual methods from the past that were described earlier in this chapter. Fortunately, though, the quality of the results produced by the state estimator is outstanding both because of the implementation itself and due to a continuous and energetic internal effort consisting of maintenance, parameter tuning, database updates to reflect network expansion, and so on.

The state estimator uses a breaker-oriented model with approximately 2000 buses and represents the following grids where generating units are connected, as shown in Figure 11-1:

- 400–220 kV grid in Sweden—the Swedish National Grid,
- Western 400 kV grid in Finland,
- 400 kV grid in eastern Denmark (Sealand),
- All the 400–220 kV connections to Norway and the internal connections there,
- Practically the entire Swedish 130 kV and partly 70 kV regional networks, and
- A few 130 kV connections to Norway and eastern Denmark.,

The generator database for the units connected to the 70–400 kV grids has been developed by SvK in collaboration with the Swedish power utilities and includes, in addition to other parameters, current and accurate data concerning voltage control. Most of these data are based on protocols from field tests, but in some cases other data must also be accepted. This information is then used by programs developed in-house to create P–Q capability curves and to represent the occurring limiters for field and armature current. The resulting generator data are then imported into the XA/21 database and become available to the real-time and study-mode network applications and to special SCADA functions as well.

11.2.3 Spica—The SvK Voltage Security Assessment System

As discussed earlier in this chapter, the techniques used initially to determine the transmission capacity of the grid were hampered by serious limitations. The availability and the good quality of the state estimator, however, motivated an in-house software development effort aimed at developing a brand new operational tool that would use the continuously updated system model produced by the state estimator.

When the project began in 1991, the NCC was equipped with mainframe computers but SvK had already plans to replace the old machines with a modern, distributed computer system. It was, therefore, decided that the new tool would be designed for the upcoming SCADA/EMS solution and would not be integrated with the old system. Accordingly, the development and implementation of the new voltage stability assessment software were performed on SUN Microsystems servers and workstations. Following an earlier in-house tradition whereby SvK used to give star names to the main processors, the new server got the name Spica, which is the brightest star in the Virgin constellation, and this is how the new voltage security assessment application was named, too!

The development was predicated on operational and user interface requirements stemming from the SvK users, that is, the network supervisors at NCC and the analysts in the operations planning department. The first version of Spica retrieved its real-time data input from the state estimator via a PSS/E interface. At that time, the state estimator was running, and still runs, at 5 minute intervals. The cycle time for Spica calculations was set at, and still is, every 15 minutes based on every third state estimate. The first version of Spica consisted primarily of a conversion of the IPLAN programs from the earlier methods but with two significant differences:

1. The key input data were now retrieved from the state estimator.
2. The user interface was brand new and predicated on a workstation.

The initial system configuration is depicted in Figure 11-5. The field testing of the first Spica release was performed in late 1993. The program became fully operational and started to be used at NCC in early 1994. During the same year, Spica was presented at the CIGRE session in Paris as referential subject 39/11-03.

Since then, a comprehensive internal development effort was undertaken by SvK to enhance the user interface and add new computational capabilities, such as modeling the behavior of the armature current limiters and the ability to simulate the operational scheme directly on the screen. In 2000, Spica was loosely integrated with the XA/21 system and has been fully operational ever since.

11.2.4 Operational and User Interface Requirements

Spica is used in the control room, by personnel responsible for various aspects of the real-time operations, and in the operations planning office. Accordingly, the ability to perform calculations is provided to both these user categories. In the control room, however, SCADA alarms are also issued to alert the network supervisor to the risk of

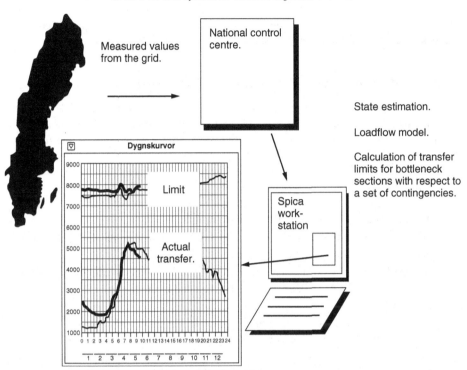

Figure 11-5. Initial Spica configuration.

violations of the current operational limit, that is, voltage collapse, in the respective network bottleneck. At the present time, the calculations and the corresponding presentation of results are performed cyclically every 15 minutes. There is a possibility that the cycle for automatic stability calculations might be reduced when the upgraded XA/21 system becomes operational in 2011.

An important user requirement that was taken into consideration was to present all the calculation results in engineering units, for example, MW and kV, rather than using percentages or p.u., as most of the thinking in a control room is in engineering units, for example, for security supervision and conveying capacity figures to the market. Incidentally, during the initial development, there was an attempt to represent a secure operating state with a point below the line $y = x$, whereas forbidden conditions would be represented with a point above this line. But immediately, the control room personnel raised the question: what is the transmission limit?

Therefore, the key Spica output intended for the network supervisor consisted of a display showing the daily load curve with the limit and the corresponding transfer in MW on the y-axis and the time on the x-axis. The presentation of results in a user-friendly format and the simplicity and ease of use of the tool have been appreciated by all dispatchers. This has been true both in the real-time supervision, where much of the work is simplified by using "soft" functional push buttons and interactions with operational displays, and in operations planning, for example, in the course of preparing contingency studies. The Spica user interface is further discussed later on in this chapter.

11.2.5 Integration and Performance Requirements

As indicated in the preceding sections, Spica has been entirely designed and developed in-house at SvK. A major objective was the ability to integrate it with the XA/21 SCADA/EMS platform "as seamlessly as possible," so to speak. Accordingly, the following design goals were established and subsequently met:

- The Spica user interface is started directly from the XA/21 native user interface. This is very important because it does not force the users to constantly switch between two different user interface environments.
- The Spica execution is triggered from XA/21 either automatically, at fifteen minute intervals, or manually on demand.
- Selected network contingencies that are evaluated by the contingency analysis functionality of XA/21 PNA can also be included in the file used in Spica calculations.
- The network protection system models used in XA/21 can also be used in Spica to the extent this is feasible due to the bus-oriented model in Spica.

The Spica–XA/21 data interface consists of an ASCII file that describes the system state and contains all the needed parameter data as well. Data related to the network protection systems are also included in this file and are used when interacting with the XA/21 program module that determines which network protection actions should be triggered, if any, given the characteristics of the current system state. The communication between Spica and this XA/21 program module is managed via message queues.

A significant effort has been made to accommodate the bus numbering and naming conventions used in the XA/21 database with those that are typically used in a load-flow program. Electrical buses in Spica are identified by a bus number and an eight character bus name. In most cases, bus names are unique, although they do not have to be uniquely defined. A Spica bus-naming table allows giving a name to each point in the bus switchyard. Since a bus can be split into one or more electrical buses, depending upon the actual state of breakers and switch disconnects, the name of a Spica electrical bus is determined by the named points included in the electrical bus. If only one name is included, that name will be used; otherwise, one of the available names is selected by a lexical comparison of the names.

Spica bus numbers are always unique and form the basis for checking the network connectivity. The bus number assigned to an electrical bus is derived by multiplying the unique switchyard number by 1000 and adding the number of the named point that determines the bus name. The point number is unique within the bus and always less than 1000.

Disconnected network objects can often be reconnected to more than one electrical bus by the operation of different switches. This is something of a problem since the information needed to determine those electrical buses is not supplied in the bus-oriented model. This has been solved by adding the possibility to relate each object of the network statically to a point in a switchyard. The disconnected object is then related to the electrical bus in which the related point at the moment is included.

The key performance requirement was the ability to execute Spica at least within 15 minute intervals. This was dictated by the fact that, in the current configuration, Spica shares the resources of a two processor UltraSparc servers running at 450 MHz with other network applications. The elapsed time to process one contingency is approximately 5 seconds for a network consisting of approximately 2800 electric nodes, and is affected by the "distance" in MW between the current state and the nose point; when the system is farther away from the nose, a heavier computational effort should be expected. A job containing 15 contingencies would require a total execution time of 75 seconds; a study encompassing three bottlenecks, that is, three jobs, would require approximately 4 minutes.

As indicated in Section 6.1, an important item on the list of future improvements is to reduce the Spica execution cycle to less than five minutes, or even to run it at one minute intervals. In order to reach this goal, a dedicated, and more powerful, Spica server would have to be incorporated in the SCADA/EMS LAN.

11.3 VOLTAGE SECURITY ASSESSMENT WITH SPICA

11.3.1 The Load-Flow Model

Spica's primary purpose is to support the dispatch center with the MW maximum limits for a number of network bottlenecks with respect to voltage collapse. For this purpose, Spica runs every 15 minutes and uses a load-flow program together with a steady-state load-flow model. The load-flow model is built from the state estimator's results. The network model is the same as the one used by the EMS system. Its state is determined by the EMS system state estimator. Therefore, the availability of a robust state estimator is fundamental since the system needs to deliver results when expected. In addition, the following provisions are considered of paramount importance for obtaining a good quality system model:

- The internal network is represented in detail and entails no equivalents.
- Adequate solutions have been adopted in order to maximize the system observability.
- Extensive and good quality communications that enable collecting network data and measurements.
- Thorough and continuous maintenance of data and measurements.

The EMS power system model encompasses the detailed representation of the national 400 kV and 220 kV grids as well as most of the national regional 130 kV networks. The neighboring countries are represented by limited parts of their highest voltage networks. Work is in progress to further extend the network model, for example, adding at least a partial representation of the 70 kV network. The load-flow network model is exported from the database into an ASCII file that uses a format derived from PSS/E specifically from Spica.

11.3.1.1 Network Modeling
The following system data are included in the load-flow model:

- Constant MW and MVAr loads;
- Generators, different types of MVAr limiters, P–Q capability curves in tabular format as a function of MW limiters, armature current limiters, and field current limiters;
- Shunts, with and without automatic switching at upper and lower voltage limits;
- SVCs;
- Transmission lines;
- Transformers.

11.3.1.2 Modeling the Primary Frequency Regulation
A frequency regulation increase or droop, in MW/Hz, together with maximum and minimum MW limits can be applied to generators. These values can be summed up for all frequency regulating generators to a resulting MW-to-Hz curve representing the entire network. During the load-flow computation, the algorithm keeps track of how much additional real power has to be injected into the system in order to maintain the real power balance. Then the frequency corresponding to this additional real power injection is read from the MW-to-Hz curve. By knowing the frequency, the actual production and the regulator strength of every frequency regulating generator can be determined.

11.3.1.3 External Network Equivalents
Network areas not found in the state-estimator network model can be represented in Spica through equivalents taken from other load-flow models that can be imported into Spica. Since the equivalents used are automatically determined for every new load-flow model that arrives to Spica, it is possible at any time to manually modify the load-flow model used by Spica to reflect changes in the external networks. The principle for connecting network equivalents is shown in Figure 11-6.

11.3.2 Calculations Performed by Spica

11.3.2.1 Procedure Overview
The overall computational process is as follows:

1. The state estimator sends a load-flow model file to Spica every 15 minutes. The process is initiated from the SCADA/EMS.

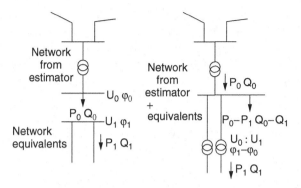

Figure 11-6. Use of external network equivalents.

2. Spica expands the load-flow model by connecting the equivalents and, in addition, performs several data processing tasks that are not directly related with the voltage stability studies. In order to compensate for the inevitable MW and MVAr mismatches at the points where the load-flow model is interconnected with equivalents, complex power injections are added in such a way that the tie-line flows remain unchanged. The resulting load-flow case is saved so it can be used in the subsequent computations.

3. Spica runs a number of "jobs" where each "job" encompasses a specific network bottleneck as well as the corresponding contingencies. For each contingency processed within a "job," the power transfer across the network is increased until the maximum network loadability has been found.

4. The calculation results are presented in graphical format, such as P–V as well as other types of curves.

11.3.2.2 A Job in Spica

The voltage security calculations are organized in the form of jobs. A job consists of running a number of contingencies for a network bottleneck for the purpose of determining the maximum loadability of the transmission corridor for each one of the contingencies that apply to that bottleneck. Physically, the job is defined by the contents of an ASCII file. Each file row contains the name of a parameter file and the name of a contingency file, and both files can be executed as scripts in the Spica load-flow module. A parameter file defines a network bottleneck by specifying its branches and/or transformers and the way in which the power flow across the bottleneck should be increased by identifying buses at which the production must be raised and buses at which the load must be increased. Usually, a single parameter file is common to all contingencies in a job. A job is run by executing each row of the job file. The calculation results are stored in ASCII files and, sometimes, in binary format, too. The contingency that yields the lowest limit for a job, that is, for a specific network bottleneck, is considered to be dimensioning if the contingency is among those considered to be taken in account.

A possibility, of course, would be to build just one single, and probably quite large, job file that would perform all the processing chores, since every contingency can have its own parameter file. However, this would be impractical since, in reality, the dispatcher determines that he has got a number of bottlenecks and may feel that different contingencies might have different relevance for different bottlenecks.

11.3.2.3 Contingency Calculations in Spica
The voltage stability calculations performed for a contingency encompass the following steps:

1. An initial evaluation of the power flow across the bottleneck is performed prior to processing the load-flow case. This initial load-flow case consists of the model retrieved from the state estimator expanded with the network equivalents but prior to making any further calculation.

2. The contingency is simulated on this load-flow model, the postcontingency load-flow case is obtained, and the *new* power flow across the bottleneck is examined. The solution entails two phases. During the first phase, transformer regulation is not performed in order to let the voltage reductions have an immediate impact on network areas where bus voltages are not regulated by TCUL transformers. However, the voltage-controlled shunts are enabled. In the second phase, the TCUL transformers are enabled, too.

3. The system generation and load are increased according to the parameter file in order to increase the power flow across the bottleneck.

4. The load flow is solved. A special technique is used to improve the stability of the load-flow calculations and find the maximum limit with greater accuracy.

5. Steps 3 and 4 are then repeated until either the maximum loadability has been found or a load-flow solution can no longer be obtained. The maximum loadability thus calculated is corrected with the difference between what has been noted in steps 1 and 2 just for the purpose of seeing what would have been the transfer at that point before simulating the contingency. It is this value that is regarded as the maximum possible transfer for the actual dimensioning contingency. The maximum transfer limit is then somewhat reduced to achieve an extra safety margin (see the end of 1.2.2.1 above).

Figure 11-7 depicts the power transfer from a generating area to a receiving load area across a network bottleneck.

11.3.2 Solution Technique

11.3.3.1 Identifying the State of Voltage Collapse
The Spica load-flow solution technique uses the Newton–Raphson method. In order to improve the numerical stability of the algorithm, the system state is not always updated all the way to the new state calculated by a new iteration. If between a previous

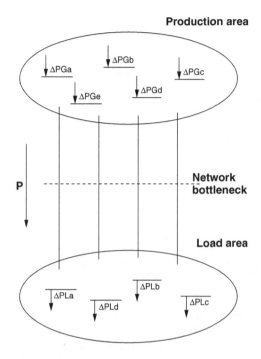

Figure 11-7. Power transfer from a generating area to a receiving load area across a bottleneck.

state and a new state a change occurs for some equipment, for example, a shunt getting connected or disconnected in order to keep the voltage within the specified range or a generator hitting the MVAr limit, the new system state is calculated by linear interpolation without performing a new iteration.

During the calculation of the maximum transfer limit, an additional network equation is used that creates a fictitious real load that depends on the voltage of the network at a number of selected buses and is allocated to some other buses at the other side of the network bottleneck. This additional equation provides for a strong coupling between voltages and the real power flow across the grid. The process is illustrated in Figure 11-8 and consists of considering that all the loads are voltage independent except for the P_u load that results from adding the additional fictitious load. The net effect is that instead of increasing the flow with ΔP, the actual increase is just ΔP_t.

It can also be seen that the voltage dependency of the load makes it possible to find intersection points at the lower side of the P–V curve, which improves the numerical stability of the calculations near the nose compared to what would had been the case if solely a voltage independent load were used. Note also that the intersection point comes close in the P_t direction around the nose. This means that the maximum can be found with great accuracy, even if a rather big ΔP step were used.

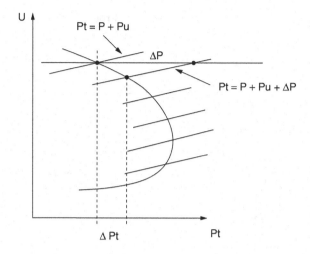

Figure 11-8. Spica load-flow procedure for identifying the limit.

11.3.3.2 Network Equations

A distributed swing bus approach is used in order to be able to simulate the response of frequency regulators when computing a postcontingency load flow. The distributed swing bus is also used to raise the generation at selected buses situated at one and the same side of the network bottleneck under evaluation when increasing the power flow towards the maximum transfer limit. The distributed swing bus principle is depicted in Figure 11-9 where the Jacobian matrix of the system of equations is schematically illustrated.

It can be seen that the swing bus has got branch (line or transformer) connections to buses a, b, and c, and that the calculated correction ΔP is propagated to buses i, j, and n by some allocation factors. These factors can be given such big values that the calcu-

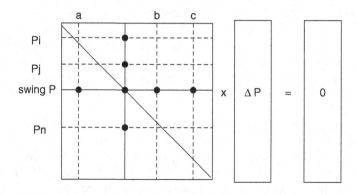

Figure 11-9. Distributed swing bus in the Jacobian matrix.

lated ΔP value will be almost zero. The principle of the fictitious voltage-sensitive load is illustrated in Figure 11-10 which shows that the voltages are measured at buses *d*, *e*, and *f* and that the fictitious load will get allocated to buses *l*, *m*, and *n*.

The off-diagonal elements marked can be of different sizes in order to have different weightings for voltages or different allocations of load between buses. If the factors are assigned large values, the coupling between voltage and real load will be strong. The choice of buses at which to place the voltage measurements should of course be those particular buses for which voltages are expected to be sensitive to the transfer of load.

11.3.4 Calculation Results

Figure 11-11 is identical to what can be seen in the dispatching center and shows how the calculation results are displayed. The time, in hours from 0 to 24, is shown on the *x*-axis; the *y*-axis corresponds to the real power in MW. The lowest curve is the actual transfer across the network bottleneck, whereas the curves above show the calculated limits for different contingencies. It can be seen that at 8 o'clock there is a disturbance somewhere in the power system since all the limits dropped to lower values some time after that the network is restored.

If the limits are lower than the actual transfer of a bottleneck for a contingency, the calculations reveal that the load flow is not solvable for that contingency. In that case, the algorithm reduces the transfer and attempts to start from a new point characterized by a lower transfer where the load flow hopefully would be solved when applying the contingency. In such a case, the diagram would show a limit curve *below* the actual transfer curve.

The actual system state will always move according to some more or less predictable path into the future. Eventually, this path will pass a point above which the voltages would collapse if a contingency would occur. Unless they would follow exactly the same path, which is virtually impossible, it is almost sure that the calculations

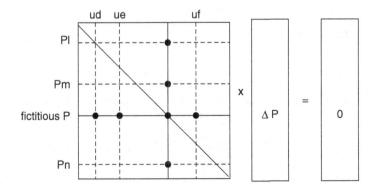

Figure 11-10. Fictitious voltage-sensitive loads in the Jacobian matrix.

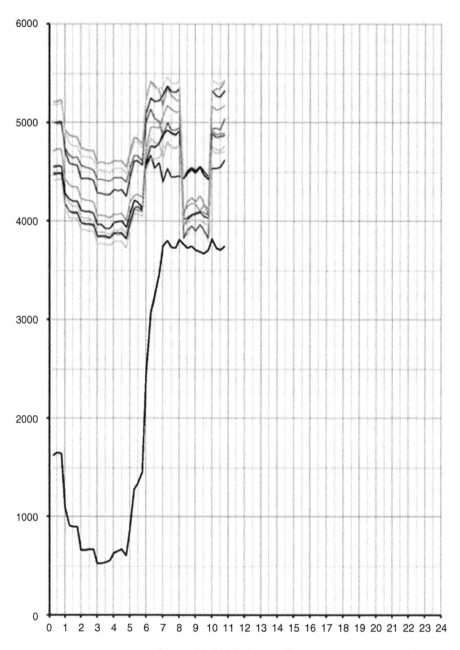

Figure 11-11. Spica results.

would not predict exactly the same limit. Therefore, the calculated limits are never constant and vary to a higher or lesser degree. However, Spica does not stress the system along a predicted path. Instead, it uses just a rough method determined by the contents in a parameter file of how to increase the power flow to reach the loadability limit.

As a result, the calculated limits are not constant and may vary significantly. Let us suppose that the state of the actual system is right at the limit with respect to a certain contingency. If we perform a calculation for this case, we will find out that the system loading cannot be increased at all, so the calculations will also tell us that we are at a limit. The consequence of this is that the algorithm will always tell us whether we are below, at, or above such a limit, provided that the model behaves correctly. It can also be inferred that the absolute error of the calculated limit approaches zero when the transfer approaches the limit.

The practical implication of this fact is that, in our opinion, it is more important to have an accurate system model rather than to focus on how the MW transfer is increased in order to reach the limit. Nevertheless, there may be other aspects for which the way the transfer is increased really matters. Different system-stressing procedures might have different impacts on network protection and might have consequences as well. In addition, there is also a risk that the system will not be dynamically stable for certain contingencies and network bottlenecks. These situations cannot be detected with Spica because the methodology only allows identifying bottlenecks where voltage collapse has been found to be the phenomenon behind major problems.

The P–V diagram shown in Figure 11-12 illustrates how some bus voltages might vary with the base-case power flow across a network bottleneck. The curves are not smooth because every time a limit that changes the behavior of a network object is reached or left, the curves change direction, too. The large jumps in the figure correspond to shunt devices automatically connected or disconnected when voltage limits are reached.

The more reactive compensation is added in the system and the more reactive power is injected when the power flow is increased to maintain the voltage profile and increase the transmission capability, the higher the voltage levels for the nose point of the P–V curves tend to be. In practical terms, this translates in the following rule of thumb: if the voltage profile does not seem to improve when adding more reactive compensation, it means that you are probably close to a transfer limit.

11.3.5 Working with Spica

Spica has a stand-alone graphical user interface that allows setting up the jobs, running them, and examining the calculation results. In addition, there is a graphical user interface for the Spica load-flow program, which allows reading the most recent load-flow case created for Spica calculations as well as cases from the past. The case is processed via conventional load-flow program commands and, once solved, it gets stored as a new case and can be used for Spica voltage security calculations in a separate study.

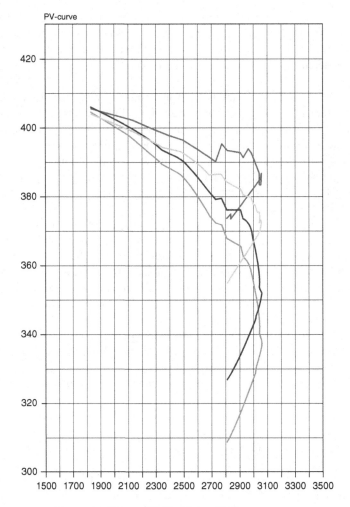

<u>Figure 11-12.</u> Spica P–V curves.

Such studies aim at answering questions such as what the limits would be if a certain line or transformer outaged. The Spica users encompass:

- NCC dispatchers responsible for checking the system security and the power balance between network areas so that the power system does not violate the limits,
- Dispatchers in NCC and Sollefteå responsible for performing the switching operations recommended by the studies for voltage control,
- Operations planning staff responsible for scheduling outages and determining future transfer limits, and
- Personnel involved in training, for example, instructors and dispatchers.

11.3.6   Limitations Implicit in Static Calculations

The static calculations performed by Spica are adequate for certain network bottle-necks but do not necessarily provide the correct answer for those bottlenecks in which the dynamics are the major problem. A load-flow solution in this context ba-sically consists of moving the system state from one point to another, still unknown, operating state out of a number of potential states that might be reached depending upon the actual behavior of the controls during this change. Different control actions may result in different trajectories that may not encompass the same objects and therefore do not necessarily lead to the same endpoint. Of course, a good load-flow program would always attempt to make a good choice from the control actions that are available.

The only way to verify whether a load-flow solution is correct is to perform two types of simulations: first a static solution, and then a dynamic calculation for verifica-tion. If both the static and the dynamic system models have the same origin for state and data, it is beneficial to reduce the differences in the common parts of the models. It is not practical to follow this approach each time a load-flow calculation is performed but it should at least be attempted now and again to ensure that the static calculation re-sults are reliable. This becomes more important when the load-flow program is used to calculate maximum loadability limits.

Since time is not represented in Spica, this can lead to difficulties to determine in which order and to what extent the power system control equipment should participate in building the system state. There is regulation for generators, transformers, SVCs, shunts and so on. All the regulation should preferably be performed in the right way. Again, dynamic verification might tell if the static calculations get close enough.

Another concern is that the quality of the calculated value of a limit strongly de-pends on the quality and completeness of the data used to perform the calculations. The need for correct data, including system state, is therefore very important. But this is true for static studies as well as for dynamic studies. One preliminary validation of the Spica method described in the following text has been done.

In a M.Sc. thesis at the Royal Institute of Technology, Stockholm, a comparison was made between calculation results obtained from Spica, on the one hand, and from SvK's fast dynamic system simulator ARISTO, on the other. As opposed to Spica, which uses static power system models, ARISTO uses dynamic models for load, induction motors, line protections, and so on. Accordingly, the results from ARISTO, with its extended and detailed dynamic modeling, were taken as reference. For evaluation purposes, the two studies used the same input cases, which were tak-en directly from the actual Nordel power system. The comparison between the results obtained with the static Spica and the dynamic ARISTO indicated a good and satis-factory similarity between the voltage collapse MW limits in the bottleneck scenar-ios, with slightly lower values in Spica, thus qualifying the Spica limits as more con-servative.

Nevertheless, and in spite of the above limitations of static calculations, Spica can be a very effective supervision tool when voltage instability is unambiguously identi-fied as the underlying physical phenomenon. The major benefit of performing static calculations to set the voltage security limits in a near real-time environment starting from the most recent state estimate is that this approach is much faster than methods that use dynamic models.

11.4 BENEFITING FROM THE KNOWLEDGE OF THE CURRENT TRANSMISSION CAPACITY

The transmission capacity is planned at SvK on a weekly basis, by operations analysts, and on a daily basis, by the network supervisor in the NCC. The planning procedure takes into account various factors, such as forecasted load, scheduled outages, and so on.

During the real-time operations, the limits for the three internal bottlenecks related to voltage stability are supervised by using the results from calculations performed every 15 minutes, whereas the thermal limits of all network series branches are continuously monitored at 30 seconds intervals based on the respective current ambient temperature, fixed limits, and so on. Transmission limits concerning the other physical phenomena mentioned earlier in this chapter are currently forecasted in real-time operation with manual methods. The limit calculated with good accuracy by the analysts in the weekly planning is decreased by using certain standard figures from operational planning when key transmission and generation equipment trips. During particularly stressed network conditions, either forecasted or due to disturbances, analysts can also assist the network supervisor.

The power transfers toward the neighboring countries in the south are often reduced because of restrictions in the internal bottlenecks in Sweden, especially in bottleneck 4 (see Figure 11.2). The first requirement is for the capacity to cover the net load, that is, load (including losses) minus production, in the current Swedish network areas. Then the remaining capacity up to the forecast limit for the bottleneck is divided between the neighboring countries according to agreements. This may often result in operational limits below the physical limits associated with the links to these countries.

If the limit violations occur in real-time operation, either because of network loadings that are higher than planned ones or because of disturbances, the network supervisor takes operational measures to lower the current power flow below the limit. Since, from the market perspective, Sweden is only one price area, a typical way to reduce the power flow is by "countertrade." The principle is illustrated in Figure 11-13.

Disturbances in southern Sweden may significantly raise countertrade costs for SvK, as this often means increasing the thermal power production in the disturbed area. As shown in Figure 11-13, the internal market players in Sweden see "an infinitely strong network," thus, the bottleneck limits are never known by them, neither in operations planning nor during real-time operation. During long-term outages of large nuclear blocks in southern Sweden, the loss of the corresponding energy is compensated by hydro power from north. In such cases, there can also be heavy costs for countertrade; otherwise, the bottleneck limits would be violated.

Quite obviously, SvK always tries to minimize these costs. The most economical first step is to bypass power via the HVDC link to Finland. The second step would be to inject hydro power from southern Norway, possibly into minor parts of southern

Constaints - Counter Trade

Market players see an infinitely strong network

Constaint (Bottlenecks)

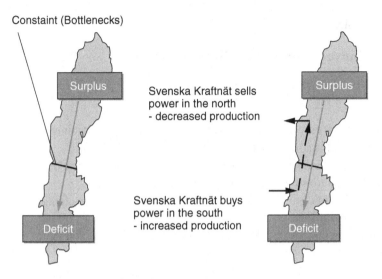

Svenska Kraftnät sells power in the north - decreased production

Svenska Kraftnät buys power in the south - increased production

Figure 11-13. Illustration of the countertrade principle.

Sweden. Finally, the third step consists of using conventional thermal power (oil- and coal-fired generators and gas turbines) in the southern part of the power system.

11.5 ADDITIONAL SPICA FUNCTIONALITY

In addition to the voltage security assessment calculations described above, Spica also performs a variety of other functions, such as:

- Determination of real power loss curves for different loadings of the DC connection between Sweden and Finland. These loss curves are used to identify the amount of power transmitted across the DC connection that minimizes the active power losses in the Swedish and Finnish National Grids.
- Calculation of the total MW losses in the Swedish National Grid for every state estimate that is sent to Spica. These values are then plotted as daily loss curves that are used when planning how much power Svenska Kraftnät must purchase to cover the losses in the Swedish National Grid.
- Loss calculations when dealing with the minimization of corona losses. In this case, Spica is used for simple load-flow studies aimed at identifying the sensitivities of bus voltages and MW losses with switching of transmission lines, shunt reactors, and so on.

- Standard load-flow studies during normal operation concerning voltage control, checking remedial actions to eliminate thermal overloads, and so on.

11.6 SUMMARY

11.6.1 Current and Future Enhancements

A number of functional and operational enhancements are currently being tested or have been already envisioned for Spica, as follows:

- Increase the execution frequency, for example, every other minute.
- Improve the quality of data that represent the generator limiters.
- Improve the modeling of network protection that was implemented in Spica but so far has not worked too well.
- Implement line protection schemes that would simulate the automatic tripping of heavily overloaded lines.
- Implement a procedure that would allow determining whether the MW limit for the network bottleneck has been reached if the voltages tend to descend below some lower limits (still to be specified).
- Improve the monitoring of branch and transformer overloads.
- Evaluate the possibility of performing Spica voltage security calculations directly in the ARISTO environment. The Spica load flow has already been implemented on ARISTO and, with the added ability to import the current state estimate in ARISTO and use the network equivalents directly from the ARISTO model, dynamic simulations and further studies become possible.
- A new load-flow solution technique has been developed and implemented in Spica. The new algorithm appears to be very stable and could be further adapted for the Spica voltage security calculations.

11.6.2 Conclusions

The material presented in this chapter can be summarized as follows:

- In the Swedish National Grid, with its long-distance transmission from north to south, the transfer limits are most often set with respect to the risk for voltage collapse.
- All the players in the electricity business in northern Europe today act in an open and deregulated market in which the setting of transmission capacity is of highest priority.
- Spica performs the real-time calculation and supervision of voltage security limits at internal bottlenecks in the Swedish grid. It was developed in-house and initially deployed in 1994. In 2000, it was loosely integrated with the GE XA/21 SCADA/EMS system used by SvK in the two control center sites. Spica's com-

putational engine consists of an in-house developed load-flow program and uses power system data imported from the state estimator. The voltage security calculations evaluate a number of contingencies for the network bottlenecks and determine the maximum corridor flow for each contingency across that bottleneck. The solution technique provides for a realistic simulation of the power system and assures the numerical stability during the search of the nose of the P–V curve.

- Spica is currently used by dispatchers for power system security supervision and for simple load-flow studies concerning switching and voltage control, by operation planning staff concerning outages and determining future limits, and by the personnel involved in operation training.

- The presentation of results in a user-friendly format and the simplicity and ease of use of the tool have been appreciated by all dispatchers. This has been true both in the real-time supervision, where much of the work is simplified by using "soft" functional push buttons and interactions with operational displays, and in operations planning, for example, in the course of preparing contingency studies.

- Spica was just built for static analysis and cannot be used in cases in which bottleneck dynamics can be the major problem. However, when voltage stability is the unambiguous physical phenomenon, performing calculations with static methods and models may result in effective supervision tools. The great advantage of static calculations when assessing voltage security limits in real time is that the approach is much faster than methods predicated on dynamic models, which normally are used in offline environments for system studies.

- Spica is supported and upgraded with new functionalities and user interfaces on a continuous basis in accordance with new requirements stated by SvK or originating in user wishes.

In a nutshell, SvK feels confident to continue using Spica, which is, and will be, the most important tool in the SvK control rooms in the future, and the importance of using this tool to make online decisions is fundamental for SvK in technical-economical terms.

DIMO'S APPROACH TO STEADY-STATE STABILITY ASSESSMENT: METHODOLOGY OVERVIEW, NUMERICAL EXAMPLE, AND ALGORITHM VALIDATION

Roberto D. Molina Mylius, Martín Cassano,
and Savu C. Savulescu

A.1 METHODOLOGY OVERVIEW

A.1.1 Steady-State Stability Revisited

The conventional method of small oscillations for estimating steady-state stability [1,5,25] consists of examining the eigenvalues of the characteristic equation associated with the linearized system of differential equations. Alternately, the Routh–Hurwitz criteria can be used. A condition for steady-state stability is obtained by evaluating the sign of the last term of the characteristic equation, which is the determinant of the dynamic Jacobian.

A significant hurdle for these approaches is the representation of the generators. Detailed analysis methods entail modeling the synchronous machines via transfer functions. The data requirements, the complexity of the ensuing algorithms, and the heavy computational burden render such techniques impractical for real-time implementation. But if the generator modeling is simplified, then it becomes possible to develop methodologies that are fast and can be applied in real time.

A trivial way to simplify the generator modeling is to get rid of generators altogether, that is, to neglect the internal reactances of the machines and assume that the generator terminal voltages are constant. This is the standard load-flow model, which is appropriate for performing load-flow calculations by increasing the total MW loading until the process diverges, or using continuation methods that identify "saddle-node bifurcations by looking only for singularities of the steady-state power flow Jacobian" [30]. But the singularity of the load-flow Jacobian is not a stability criterion [19,20,29] and if we really want to perform steady-state stability analysis, we need to keep the machines in the model. This can be achieved by expanding the network represented in the load-flow calculations with the internal nodes of the generators; in this context, "expanding the network" means connecting the external bus and the internal node of the machine with a reactance, such as x_d', x_d'', or even an arbitrarily small value as shown in Section A.1.2.3.2. If, in addition, certain modeling requirements are also met, the steady-state stability analysis can be greatly simplified by applying "practical steady-state stability criteria."

The practical steady-state stability criteria were developed in Russia [25]. They handle *aperiodic instability,* that is, they are not intended to detect instability due to self-sustained oscillations, and are derived from the mathematical conditions for the *dynamic* Jacobian to be singular. They are valid if:

- The generators are radially connected to a nodal point. This is not true in real-life systems, but, as we will show in the Section A.1.2.1, it is *always* the case if the short-circuit current transformation is applied to convert the meshed network to a radial scheme of short-circuit admittances.
- The system frequency is constant* during the short period of time associated with the transient process and, furthermore: either the voltage is constant at the nodal point, which leads to the *synchronizing-power criterion $dP/d\delta < 0$*; or the power balance is maintained at the nodal point, which leads to the *reactive-power steady-state (voltage) stability criterion $d\Delta Q/dV < 0$*.

The computation of the reactive power criterion $d\Delta Q/dV < 0$ instead of evaluating the eigenvalues of the dynamic Jacobian determinant results in an increase of the computational speed by at least one order of magnitude and is at the core of the fast and relatively accurate technique developed by Paul Dimo [6,7,8]. Dimo's method, which has been successfully implemented and is currently used in several SCADA/EMS installations to compute the system loadability limits in real time and to continuously monitor the distance to instability [12,22,23,27,28], is outlined in the following section.

A.1.2 Dimo's Approach to Steady-State Stability Assessment

Many references are available that describe the approach to steady-state stability assessment developed by Paul Dimo, starting, of course, with the books and papers writ-

*This assumption is valid for systems in which large amounts of MW are dispatched across a meshed network, but is not necessarily true in radial networks in which the short-circuit power is relatively small and special limits must de computed to guarantee that the system frequency remains stable.

ten by Dimo himself. In order to avoid overlap with already published material, in this section we briefly discuss just the key aspects of the technique and emphasize the assumptions that provide for computational speed while preserving precision and accuracy. The theoretically oriented reader is directed to review reference [22], in which the more subtle aspects of Dimo's methodology are addressed in detail, including the generalization of Dimo's formulation of the reactive-power steady-state stability criterion. In a nutshell, Dimo's approach is predicated on the:

- Short-circuit current network transformation;
- Zero power balance network method to aggregate the system loads into a fictitious load center;
- Simplified representation of generators;
- Reactive-power steady-state (voltage) stability criterion;
- Case-worsening procedure for stressing the system conditions instead of performing a sequence of load-flow computations.

These modeling and procedural steps are summarized in the following sections.

A.1.2.1 Short-Circuit Currents Transformation

This transformation is used to convert the power system network, which is highly meshed, to a scheme of short-circuit admittances connected radially to a nodal point. The radial network of short-circuit admittances thus obtained is known as the REI net. One of the key attributes of this transformation is that it allows "seeing" the generators from the nodal point. The transformation of a meshed power system network to an REI net can be applied to an *actual load bus,* to connect it radially with all the generators by means of short-circuit admittances, as shown in Figure A-1, or to the *fictitious load center* obtained by first introducing the zero power balance network, which is discussed in Section A.1.2.2 and illustrated in Figure A-2.

The steady-state conditions of the network shown in Figure A-1 are given by the standard load-flow equations written as

$$I = YV \tag{1}$$

$$\underline{S}_k = \underline{I}_k^* \underline{V}_k \tag{2}$$

where

$\underline{S}_k$ is the complex power injected into the bus k

$\underline{I}_k^*$ is the conjugate of the complex current injected into the bus k

$\underline{V}_k$ is the complex bus voltage of the bus k

I is the vector of complex bus-injected currents

V is the vector of complex bus voltages

Y is the matrix of complex nodal admittances

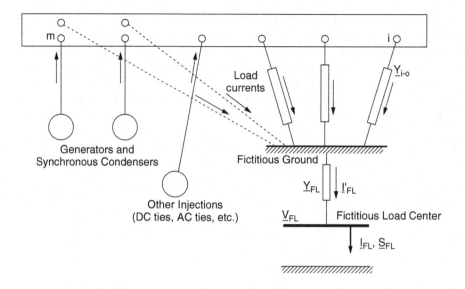

Figure A-1. Transition from the meshed power system network to the radial scheme of short-circuit admittances, also known as the REI net.

The buses are numbered sequentially as follows:

1 ... m ... G generator buses (either on the generator's terminals or on the high voltage side of the step-up transformer).

i ... N load buses where, for convenience, $G + 1$ was noted as i and corresponds to the load bus L_1 (Figure A-1).

Initially, we will illustrate the short-circuit current concept as used in [2, p. 689], where Barbier and Barret, like most technical authors in the field, assume that voltages are, or can be maintained constant at the G buses, and that the injected load currents at all the load buses except the study bus, say L_1, vary proportionally to the bus voltage magnitudes and can be replaced by constant admittances.* By changing the sign of these linearizing admittances, adding them to the diagonal elements in Y, and performing the Gauss–Seidel elimination of all the linear buses (except L_1), a new matrix Y' of order $(G + 1) \times (G + 1)$ is obtained, where the $(G + 1)$th row corresponds to the load bus L_1 and the following equations can be written:

$$I_{L_1} = \sum_{i=1}^{G} Y'_{iL_1} V_i + Y'_{L_1 L_1} V_{L_1} \qquad (3)$$

*Replacing the load with an admittance is referred to as *linearizing* the load. The admittance that replaces the load is called *linearizing admittance*. Buses without active injections are called *linear buses*.

where

$$Y'_{L_1L_1} = y'_{L_1L_1} - \sum_{i=1}^{G} Y'_{iL_1} \tag{4}$$

$y'_{L_1L_1}$ being the bus-to-ground admittance at the bus L_1.
    From (3) and (4) we obtain

$$I_{L_1} - y'_{L_1L_1}V_{L_1} = \sum_{i=1}^{G} Y'_{iL_1}V_i - \left(\sum_{i=1}^{G} Y'_{iL_1}\right)V_{L_1} \tag{5}$$

which can also be written as

$$I_{eq} = I_{cc} - Y_{cc}V_{L_1} \tag{6}$$

where, by using the same notations from [2]:

$$I_{eq} = I_{L_1} - y'_{L_1L_1}V_{L_1} \qquad \text{is the equivalent current at bus } L_1$$

$$I_{cc} = \sum_{i=1}^{G} Y'_{iL_1}V_i \qquad \text{is the short-circuit current at bus } L_1$$

$$Y_{cc} = \sum_{i=1}^{G} Y_{iL_1} \qquad \text{is the short-circuit admittance of the system seen from } L_1$$

From this background, Barbier and Barret developed the formula that gives the critical voltage at bus L_1 corresponding to the maximum power transferred toward L_1 from the generator buses [2, p. 690].
    Let us now look at these results from a *graph-theoretical* perspective. Equation (3) completely and uniquely defines the state of bus L_1 if the network admittances and the voltages V_1 through V_G are constant. The innovation introduced by Paul Dimo consists of *expanding the original network* with the internal nodes of the generators, which causes the voltages V_1 through V_G to be replaced with the electromotive force (emf) E_1, ..., E_m, ..., E_G, and the $Y_{cc} = \sum_{i=1}^{G} Y'_{iL_1}$ obtained after the Gaussian elimination to incorporate the internal reactances of the machines—synchronous, transient, subtransient, or even an arbitrarily small value as shown further down in Section A.1.2.3.2, depending upon the physical phenomenon that is being investigated. The resulting radial network of short-circuit admittances Y_{mL_1} (the prime ' symbol was dropped in order to simplify the notation) becomes the "REI net" in Dimo's terminology and the current injected into the load bus L_1 can now be rewritten as

$$I_{L_1} = \sum_{m=1}^{G} Y'_{mL_1}E_{mL_1} - \left(\sum_{m=1}^{G} Y_m + Y_{L_1-o}\right)V_{L_1} \tag{7}$$

thus telling us that the load current at bus L_1 can be expressed as the vectorial difference between $\sum_{m=1}^{G} Y_{mL_1}E_{mL_1}$, which is the symmetric three-phase short-circuit current flowing into the bus L_1, (for $V_{L1} = 0$), and $(\Sigma_{m=1}^{G} Y_m + Y_{L1-o})V_{L1}$, which is the "no-load short-circuit current" of the bus L_1 and corresponds to the short-circuit current at L_1 if the load current was equal to zero before the short circuit. This result was known in France from the 1930s under the name of "Fallou's theorem" [6,7] and lays at the foundation of Paul Dimo's REI approach to analyzing power system states and performing stability assessment. The algorithm is known as the short-circuit currents transformation. Figure A-1 shows how the power system network is "seen" from the load buses L_1 across the short-circuit admittances of the REI net. The REI net is built for a reference state, or base case, for which a load-flow computation has converged. The base case may depict a peak-load case or some other system state involving certain particular features, for example, high percentage use of hydraulic power, line, and/or generator contingency case, and so on.

The study bus retained in the model may be actual or fictitious. If the study bus is an actual load bus connected to generators through a typical transmission network, the generator buses are "seen" via short-circuit currents flowing across short-circuit admittances between generators and the load bus. If the study bus is a *fictitious load center,* that is, an equivalent bus in which all the system loads have been aggregated by inserting a zero power balance network, as shown in Figure A-2, and then reducing the system by eliminating all the linear buses, the machines are "seen" via the short-circuit currents that flow across short-circuit admittances between generators and the fictitious load bus. The generator buses correspond to *actual* synchronous machines and SVCs, but may also designate *virtual generators* introduced to model tie-line imports and DC ties.

A.1.2.2   Zero Power Balance Network

Paul Dimo introduced the zero power balance network for the purpose of aggregating the system loads into a fictitious single-load center while preserving the properties and the power balance of the base case. This method is known in the industry as "REI equivalencing" and has been demonstrated to be accurate if the individual bus loads vary conformingly with the total system load [30]. This is shown in Figure A-2. A detailed numerical example illustrating step by step the process of building the zero power balance network is provided in Section A.3.1.1.

It must be emphasized that in the context of evaluating stability, the machines, either real, such as generators and synchronous condensers, or "virtual," such as tie-line injections, are *not* equivalized and the zero power balance network is used only for the purpose of building the fictitious load center.* If the short-circuit current transformation is applied *after* having introduced the zero power balance network, the REI net thus obtained will connect the system generators to the fictitious load center, rather than to an actual load bus. The two different types of REI nets are shown side by side in Figure A-3 for comparison.

---

*This is quite different from the procedure known in the industry as "REI equivalencing," in which the generators are also equivalized and the final result is an "REI equivalent" that is built for the purpose of replacing external networks in static security assessment [31,32,33,34].

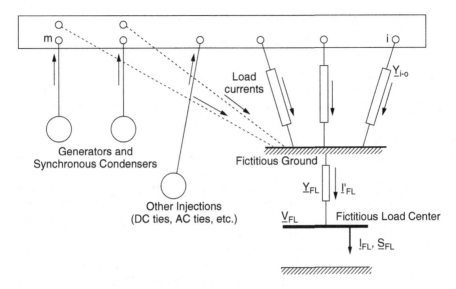

Figure A-2. Zero power balance network.

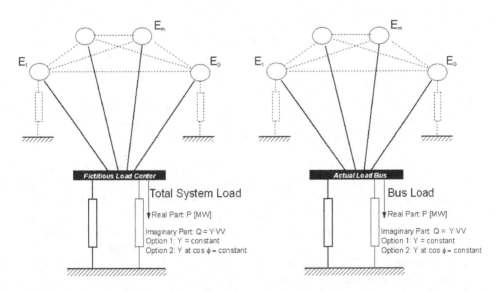

Figure A-3. REI nets for a fictitious load center (left) and for an actual load bus (right).

A.1.2.3 *Simplified Representation of Generators*

BACKGROUND. In order to understand why the old fashioned simple synchronous machine models are still useful today when steady-state stability must be assessed in real time quickly yet with reasonable accuracy, we must step back and recall how the machine models have evolved in time. Stability analysis was of paramount importance from the early industry stages when power systems started to incorporate long transmission lines and the electrical distances between generators and loads started to increase. At that time, the computations were performed manually, the generator models were simple and conservative, and modeling inaccuracies were compensated with large security margins.

Things changed completely with the advent of modern computing technologies that made possible the introduction of complex generator models representing in detail the excitation control systems, turbine governors, power system stabilizers, and so on. As a result, the stability studies became much more precise but, at the same time, the need to incorporate and update the continuously increasing populations of sophisticated, computationally intensive models relegated such studies to off-line environments. Those were the "golden years" of the industry, when power systems were operated more or less as planned, and when off-line operating guidelines tended to provide adequate protection, even for multiple cascading events. But the "golden years" are over.

In the current power system operating environment, the knowledge of stability limits computed offline, though extremely useful, is not enough for real-time decision making. Modern system operators must be ready to accommodate MW transfers that can be quite different from those for which the transmission system was originally planned and, under certain deteriorated operating conditions, may result in blackouts due to instability. Accordingly, simple models and tools are once again needed, this time to monitor the real-time evolution of certain stability indicators either for the system as a whole or across critical transmission corridors or both. As shown in Chapter 2, a useful stability indicator is the steady-state stability limit (SSSL). But can the SSSL be computed by deploying simplified models *without* introducing excessive security margins, that is, without producing excessively pessimistic results?

GENERATOR MODELING FOR THE COMPUTATION OF SSSL. During electromechanical oscillations, typically between 0.3 and 2.5 Hz, that take place after a small perturbation that causes a machine to change its MW output and to settle into a new stable state, the generator appears to have an internal reactance smaller than the steady-state reactance; this is the transient reactance x'_d. Accordingly, the "classic" generator model consists of representing the machine as seen from the stator as a constant emf E'_d behind the transient reactance x'_d. This model is very attractive due to its simplicity, but what is the true emf of the generator during this process?

Prior to answering this question, we need to briefly anticipate the "case-worsening" procedure used to compute the SSSL. Initially, the base case is verified to be stable by applying a steady-state stability criterion (as shown in A.1.2.4). Then, the system conditions are "stressed" by increasing the total system load in small increments which, if they were to happen in real life, would correspond to 5–10 sec time intervals. In order to meet the increasing load requirements while keeping the voltages constant at the synchronous machine terminals, each unit will generate more MW and, also, will increase the MVAr output by raising its internal emf. During this process, the machine is seen as a reactance, and a potential simplifying hypothesis for computing the SSSL is to assume that the internal emf of the generator is constant and equal to the value accepted for the base case. This assumption, however, is too conservative and may lead to pessimistic results because:

- In steady-state operation, the reactive generation Q_g of generators and synchronous condensers is maintained at or below the value Q_{max} imposed by the overexcitation limiter (OEL), with most units set to generate $Q_g < Q_{max}$ in order to provide a margin for contingencies when the need for MVArs may exceed the scheduled amounts. As a result, under normal operating conditions the emf is smaller than the maximum allowed by the OEL.

- When the system is stressed, that is, when the system load is increased, the fast automatic voltage controllers (AVRs) of today, with typical droops smaller than 0.25%, are capable of raising the emf to maintain the voltage at the generator's terminals practically constant until both P_{max} and Q_{max} limits are reached, in which case the OEL would stop controlling the voltage.

- On the other hand, if the case-worsening procedure *maintains a constant emf* when increasing the load, the terminal voltage of the generator will decrease, but this is true only if the machine is not equipped with a fast AVR or if both the generated reactive power and the MW output have already reached the Q_{max} and P_{max} limits. Therefore, maintaining the emf constant throughout the entire case-worsening procedure will produce conservative results.

The difference between the base-case operation and a "stressed" operating scenario is illustrated in Figure A-4.

The generator's emf increases in direct proportion the field current I_{fd} when the AVR raises the I_{fd}. If the generator is underexcited [field current I_{fd_1} in Figure A-4(a)] it absorbs reactive power from the network. When the generated MW reaches P_{max}, the field current becomes $I_{fd_2} > I_{fd_1}$. During the case worsening scenario [Figure A-4(b)], that is, when the MW transfer across the transmission network increases, the amount of reactive power needed to be shipped from the active buses (generators, synchronous condensers, SVCs, etc.) toward loads will increase and, as a result, the AVR will raise the I_{fd} in order to maintain the nominal voltage at the machine's terminals until the reactive generation would reach the Q_{max} value allowed by OEL for $I_{fd_3} = I_{fd_{max}} > I_{fd_2} > I_{fd_1}$. Beyond that point, the voltage at the machine's terminals will start to decay and the unit will be seen as a constant emf behind the synchronous reactance x_d.

Therefore, if all the internal generator nodes are retained in the computational model and if, in addition, the algorithm is capable of monitoring the real and reactive output of the machines, and detecting those units that have reached both P_{max} and Q_{max} during the case worsening process, one of the following approaches can be adopted in

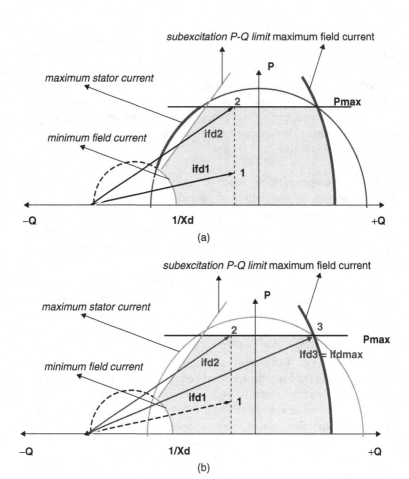

Figure A-4. (a) basic operating scenario; (b) case-worsening scenario.

conjunction with the "classic" generator model such that the SSSL value would be less pessimistic, yet still conservative:

- *Basic Approach.* Represent the machines through x_d' or x_d'' to build the basic model except for those generators equipped with fast AVRs, for which an arbitrarily small internal reactance shall be used instead. Calculate $d\Delta Q/dV$ for the reference case, and then maintain constant the emf behind the internal reactance $(x_d', x_d'',$ or the very small value, whichever applies) when worsening the case. This way, for those units for which the initial reactance value was arbitrarily small, the voltage on the machine terminals will indirectly be maintained approximately constant. If, while worsening the case, a generator reached both P_{max} and Q_{max}, replace the internal reactance used initially with the synchronous reactance x_d. As a result, the voltage on the unit's terminals will quickly drop during the subsequent steps of case worsening, which is consistent with the findings of [2, p. 681] and [13], among other references.

- *Alternate Approach.* Use the classic model for all the machines and represent them through x_d' or x_d'' to build the basic model. Compute $d\Delta Q/dV$ for the reference case. Then, prior to starting the case-worsening procedure and only for those units equipped with fast AVRs, replace the initial internal reactance with an arbitrarily small value. When worsening the case, assume that the emf behind the internal reactance (either x_d' or x_d'' or the very small value, whichever applies) is constant, and if a generator reached both $P_{\max}$ and $Q_{\max}$, replace the internal reactance used initially with the synchronous reactance x_d.

Calculation results obtained by applying the basic approach with different values of the "internal" generator reactances have clearly indicated that the SSSL tends to increase, that is, to become less conservative, when the values of the reactances used to represent the machines get smaller.

One word of caution, though: "arbitrarily small" is a matter of interpretation; therefore, a good software design should allow the user to calibrate the "arbitrarily small" value as per his/her vision of "how conservative" the simulation should be. As a rule of thumb, acceptable values for the "internal" generator reactances are x_d'', x_d', or somewhere between 0.0001 p.u. and x_d''.

A.1.2.4 Reactive Power Steady-State Stability Criterion dΔQ/dV

The radial nature of the REI net meets one of the rules of applicability of the reactive power steady-state stability criterion (voltage stability) that "the generators be connected to a central node." The other rule, which stipulates that the MW power balance be maintained at the central node, is met with a small approximation but can be reinforced by adjusting appropriately the values of the REI net reactances. For a system of $1, \ldots, i, \ldots, G$ generators, synchronous condensers and active injections such as DC or AC ties, connected radially to either a fictitious or an actual load bus through admittances $Y_1, \ldots, Y_i, \ldots, Y_G$, Dimo developed the following formula:

$$\frac{d\Delta Q}{dV} = \sum_m \frac{Y_m E_m}{\cos \delta_m} - 2\left(\sum_m Y_m + Y_{\text{load}}\right)V \tag{8}$$

where

E_m = internal voltages of the machines (assumed to remain constant, unaffected by small adjustments made under steady-state stability conditions)

δ_m = internal angles of the machines with reference to the voltage V on the load bus (either fictitious or actual)

In this approach, the real part of the load is represented by its MW value, whereas the reactive part varies with the square of the voltage in accordance with $Y_{\text{load}} = Q_{\text{load}}/V^2$, where the value of Q_{load} is either taken from the base case or recalculated at each step of case worsening (see below) by considering that the structure of the load is constant, that is, cos ϕ remains constant, as shown in Figure A-3. Dimo's formula (8) entails a certain approximation because it neglects the angles $\gamma_1, \gamma_2, \ldots, \gamma_n$ of the REI net susceptances. The general formulation of this criterion has been developed in [22] and is given in (9):

$$\frac{d\Delta Q}{dV} = \sum_m \frac{Y_m E_m}{\cos(\delta_m + \gamma_m)} - 2\left(\sum_m Y_m \cos \gamma_m + Y_{\text{load}} + Y^\gamma\right)V \qquad (9)$$

Let us also note that formula (7) can be derived from (8) by setting $\gamma_1 = \gamma_2 = \ldots = \gamma_n = 0$.

A.1.2.5   Case-Worsening Procedure—Distance to Instability

In voltage and steady-state stability problems it is not the base case, which presumably comes from a fully converged load-flow or state estimate, that is of primary importance, since, in most cases, the base case is stable. What really counts is the ability to characterize the system state by its "distance" from an unstable one. The stability calculations per se, either via simplified techniques such as practical stability criteria, or based on detailed simulation; for example, evaluating the eigenvalues of the Jacobian associated with the system of dynamic equations, do not give such information.

In order to compute the steady-state stability limit (SSSL) and, implicitly, to find the distance to instability, the steady-state stability calculation must be combined with a "system-stressing" or "case-worsening" procedure whereby various system parameters are changed in a direction that is unfavorable to stability. This is achieved by the case-worsening procedure that is used, instead of a succession of load-flow computations, to stress the system until it becomes unstable. The procedure is extremely fast, entails certain approximations which, if not applied carefully, may lead to very conservative SSSL values, and encompasses the following:

- Model the generators and synchronous condensers as shown in the Section A.1.3.2.3.
- Increase the total system generation to meet successively higher load levels by raising *coherently* the MW produced by each generator while observing the maximum MW limit of the machine.
- Represent the real part of the loads as MW, rather than as impedances, and model the reactive part of the load as a susceptance that either has a fixed value, derived from the base case, or varies proportionally with the real part of the load, the proportionality factor being the cos ϕ of the load in the base case.
- Simulate the sudden change of the operating conditions of generators that have reached their reactive power limits. If this happens under light load conditions, replacing the initial value of the internal reactance with the synchronous reactance will cause the steady-state stability to decrease but not to be destroyed. If it happens under high loadings, the same model change may destabilize the system and precipitate a voltage collapse [10].

Throughout the case-worsening process the system model remains constant. If major topology or other changes need to be simulated, for example, line and generator contingencies, a new load-flow solution is required. Once the base case has been recalculated, the REI net and the nodal image are updated, the case-worsening procedure is performed, and the SSSL, steady-state stability reserve, and security margin for the new system state are obtained. The SSSL thus calculated tends to be conservative at low system loadings, but the prediction of the distance to instability becomes more accurate when the total MW system loading increases and additional reactive compensation resources get committed.

This apparent paradox can be explained if we note that operating policies typically call for raising TCUL taps, removing shunt reactors, adding shunt capacitors, and bringing online synchronous condensers when the system is approaching peak-load conditions. At medium and light load levels, capacitors are removed, shunt reactors are reconnected, and synchronous condensers and/or units that were running essentially for generating MVArs are taken offline. Such operating procedures push the network's maximum loadability to values much higher than those at medium and light load levels. A fine example that illustrates this situation is described in [27].

A.1.3 Validating the Methodology

In accordance with long-accepted software engineering practices, the testing and validation of a computer application encompasses essentially the following steps:

- Functional validation, which entails verifying that the application behaves as per the stated specifications.
- Testing the algorithm implementation, that is, demonstrating that the results produced by the program being tested are the same as those obtained by implementing the same algorithm in a different way, for example, executing the calculations by hand.
- Benchmarking the underlying methodology, that is, verifying that the application, already demonstrated to have been implemented correctly, produces results that are similar, or reasonably close, to results obtained with a different technology.

The functional validation of the computer application that embodies the steady-state stability assessment method developed by Dimo was performed extensively over a period of time by running the program both with good and with bad data, comparing its behavior with the stated requirements, and verifying that the software fully meets the design specifications. Albeit extremely important in the software implementation process, this step is not relevant in the context of this book and will not be addressed herein.

Thorough numerical testing of the method has also been extensively performed during the early stages of software implementation, but a custom request to numerically illustrate the algorithm for the purpose of helping to understand the key details of Dimo's methodology offered a unique opportunity to test the program output against results of computations performed independently by a third party. This was done by developing a small-scale model, manually executing the full array of calculations by two project teams at two different locations, and comparing the results produced by the software with those calculated manually. The results of this extremely useful and interesting exercise are presented in Section A.2.

The benchmarking of the underlying methodology was performed in the past by using a small-signal stability program, as shown in [31], or by using the simple procedure described in [27]. A novel, and more accurate, benchmarking approach was developed by Roberto D. Molina Mylius and Martin Cassano at CAMMESA (Compañía Administradora del Mercado Mayorista Eléctrico Sociedad Anónima de la República Argentina) in conjunction with the numerical example described in this Appendix. It aims at verifying that the SSSL computed with Dimo's technique can be found, with a minor approximation, by applying the maximum power transfer theorem to a series of successively degraded operating states. The benchmarking calculations executed by Molina and Cassano are addressed in Section A.3.

A.2 NUMERICAL EXAMPLE—INDEPENDENT TESTING OF ALGORITHM IMPLEMENTATION

A.2.1 Six-Bus Network Used for Numerical Testing

In this section, we will use a simple network example to illustrate how to:

- Represent the generators,
- Build the zero-power balance network,
- Apply the short-circuit current transformation to obtain the REI net, and
- Calculate the $d\Delta Q/dV$ criterion.

The numerical example, which will serve both to illustrate the methodology and to perform the independent testing of Dimo's stability assessment algorithm used by QuickStab® [35], is based on a six-bus power system model (Figures 5(a) and 5(b)] developed by Molina and Cassano, with minor approximations, directly from the 500 kV Argentinean Interconnected Transmission System. The generation, load and bus voltage data are shown in Table A-1.

In accordance with the widespread load-flow convention, the MW and MVAr values of the loads are expressed as positive quantities if they represent true loads, as shown in Table A-1, whereas for abnormal cases, such as "negative loads" introduced to artificially depict a generation or an import, the signs must be reversed. However, when performing the actual computations, we will take into account the physical nature of the phenomena and, accordingly, use negative values for the MW and MVAr loads.

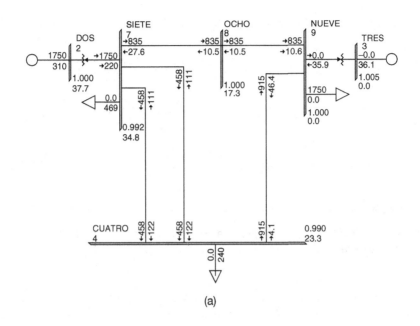

(a)

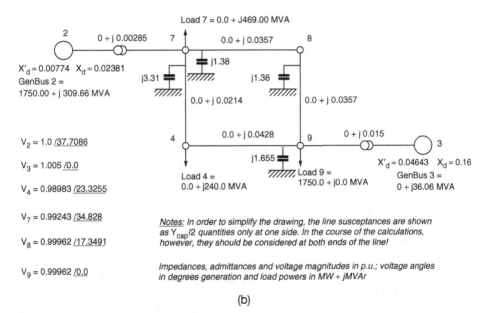

(b)

<u>Figure A-5</u> Six-bus network used for illustrating the technique. (a) Overview; (b) details.

Table A-1. Generation, load, and bus voltage data

Bus no.	Gen MW	Gen MVAr	Load MW	Load MVAr	V (p.u.)	Angle (°)	Q_{min} (MVAr)	Q_{max} (MVAr)
2	1750	309.66	—	—	1	37.7086	−1500	3000
3	—	36.06	—	—	1.005	—	−500	500
4	—	—	—	240	0.98983	23.3255	—	—
7	—	—	—	469	0.99243	34.828	—	—
8	—	—	—	—	0.99962	17.3491	—	—
9	—	—	1750	—	0.99962	—	—	—

A.2.2   Calculation Procedure

A.2.2.1   Adding the Internal Generator Buses and the Zero Power Balance Network

At the outset, we expand the original network with the internal nodes of the generators. For this example, we are using the E'_d behind the transient reactances x'_d. Then we replace the all the loads with "linearizing impedances" and connect them to a fictitious ground as shown in Figure A-6. The impedances that linearize buses 4, 7, and 9, that

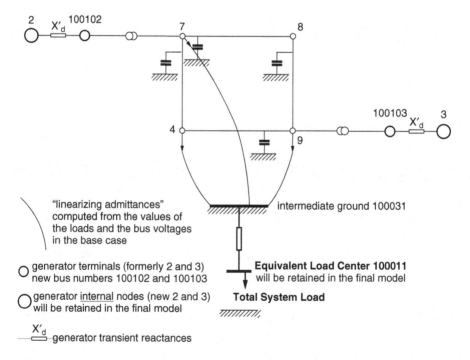

Figure A-6. Six-bus network extended with the internal nodes of the generators and the zero power balance network.

is, convert them to "passive buses," are connected between the buses 4-100031, 7-100031 and 9-100031, and are computed as follows:

$$G = P/V^2 \quad \text{and} \quad B = -Q/V^2$$

For example, the linearizing conductance G and susceptance B at bus 4 are given by

$$G_{4\text{-}100031} = (0.0)/(0.98983)^2 = 0.0 \text{ p.u.} \quad \text{and} \quad B_{4\text{-}100031} = -(2.4/(0.98983)^2) = -2.45 \text{ p.u.}$$

The current $\underline{I}_{4\text{-}100031}$ is computed as follows:

$$\underline{I}_{4\text{-}100031} = \left(\frac{S_4}{\underline{V}_4} \right)^* = \frac{\underline{S}_4^*}{E_4 - jF_4} = \frac{0.0 + j2.4}{0.90893 - j0.39193} = \frac{-0.940632 + j2.181432}{0.979763}$$

$$= -0.96006 + j2.22649 \text{ p.u.}$$

Similarly, we get

$$\underline{I}_{7\text{-}100031} = -2.698958 + j3.879269 \text{ p.u.} \quad \text{and} \quad \underline{I}_{9\text{-}100031} = -17.506652 \text{ p.u.}$$

Then we "delinearize" the total load and bring it on the bus of the fictitious load center by applying the Kirchhoff's first theorem, $\Sigma_{100031}^k \underline{I}_{100031}^k = 0$, and the power conservation theorem, $\Sigma_{100031}^k \underline{S}_{100031}^k = 0$. Since the currents that flow from the load buses into the fictitious ground 100031 are "load currents," the total current that "gets out" of the fictitious ground 100031 will be a "load current" as well. Accordingly, the current between the fictitious ground 100031 and the fictitious load center 100011 is calculated as follows:

$$\underline{I}_{100031\text{-}100011} = -21.16567 + j6.105759 \text{ p.u.}$$

and, by applying the power conservation theorem, we get

$$\underline{S}_{100031\text{-}100011} = -17.5 - j7.09 \text{ p.u.}$$

Now, from the complex current and power that circulate between 100031 and 100011, we calculate the impedance $\underline{S} = \underline{V}\underline{I}^* = \underline{Z}\underline{I}\underline{I}^* = \underline{Z}I^2$; thus, $R + jX = (P + jQ)/I^2$:

$$R_{100031\text{-}100011} = -0.0360627 \text{ p.u.}$$

$$X_{100031\text{-}100011} = -0.0146105 \text{ p.u.}$$

$$G_{100031\text{-}100011} = R/(R^2 + X^2) = -23.819698 \text{ p.u.}$$

$$B_{100031\text{-}100011} = -X/(R^2 + X^2) = 9.6503508 \text{ p.u.}$$

The complex voltage of the fictitious load bus 100011 is calculated as:

$$\underline{V}_{100011} = \frac{\underline{S}_{100011}}{\underline{I}^*_{100011}} = \frac{-17.5 - j7.09}{-21.16567 - j6.105759} = 0.8525002 + j0.0890507$$

$$= 0.857139 \underline{|5.985}$$

It must be clear by now that the zero power balance network (ZPBN) is the lossless network formed by the branches 4-100031, 7-100031, 9-100031, and 100031-100011. In spite of the fact that most of its branches have a resistive (conductive) part, this network is lossless because the power that gets in is equal to the power that gets out, which is why we applied the power conservation theorem. Therefore, by replacing the loads in the original system with the ZPBN, we obtain an extended network that has exactly the same energy balance as the original network. The line data for the extended network are summarized in Table A-2.

A.2.2.2   The REI Net

The REI net is obtained by applying the short-circuit current transformation, that is, by performing the Gaussian elimination of all the passive buses in the extended network and retaining only the internal nodes of the generators and the fictitious load bus. The Y-matrix of the extended network before applying the short-circuit current transformation is shown in Figure A-7. The row and column numbering reflects the fact that the buses that became "linear" because they have no injections will be eliminated and only the buses 2 and 3 (internal nodes of the generators) and 100011 (fictitious load bus) will be retained. Please note that the complex voltage of the fictitious load bus 100011 calculated previously will appear in the system of "nodal" equations that depict the equilibrium of the network obtained *after* having eliminated the passive buses.

The Gauss–Seidel elimination is a trivial linear algebra algorithm that can be applied without introducing any error because the nodes to be eliminated have no injec-

Table A-2. Line data for the extended network

From bus	To bus	R (p.u.)	X (p.u.)	$Y_{cap}/2$ (p.u.)	Tap (p.u.)
7	2	—	0.00285	—	1
9	3	—	0.015	—	1
4	7	—	0.0214	3.31	—
4	9	—	0.0428	1.655	—
7	8	—	0.0357	1.38	—
8	9	—	0.0357	1.38	—
4	100031	—	0.408235	—	—
7	100031	—	0.210004	—	—
9	100031	0.057099	—	—	—
100011	100031	−0.03606	−0.01461	—	—
3	100101	—	0.04643	—	—
2	100102	—	0.00774	—	—

	2	3	100011	100102	100103	100031	4	7	8	9
2	-j129.199			j129.199						
3		-j21.538			j21.538					
100011			-23.819 +j9.650			23.819 -j9.650				
100102				-j480.076				j350.877		
100103		j21.538			-j88.205					j66.667
100031			23.819 -j9.650			-6.306 +j2.439	j2.449	j4.762		-17.513
4						j2.449	-j67.577	j46.729		j23.364
7				j350.877		j4.762	j46.729	-j425.689	j28.011	
8								j28.011	-j53.262	j28.011
9					j66.667	-17.513	j23.364		j28.011	17.513 -j115.007

Figure A-7. Y-matrix associated with the extended six-bus network. The first three lines and columns correspond, respectively, to the internal nodes of the generators and the fictitious load bus, and will be retained after performing the Gaussian elimination of the passive buses.

tions (no currents), that is, they are *linear,* and they are linear precisely due to the way we expanded the original system by adding the ZPBN. In pseudocode, and *without* considering sparsity, the algorithm is as follows:

Renumber sequentially, from 1 to 10, the buses 2, 3, 100011, 100102, 100103, 100031, 4, 7, 8, and 9 (i.e., consider the original bus numbers as bus "names"). Then:

For k = 10 (corresponding to bus "9") to 3 (corresponding to bus "100011")

For $[(i, j = 1, 2, \ldots) \; \&\& \; (i < k) \; \&\& \; (j < k)]$

$$y_{ij}^{new} = y_{ij}^{old} - \frac{y_{ik}^{old} y_{jk}^{old}}{y_{kk}^{old}}$$

where

y_{ij}^{old} designates the complex entry i–j in the Y-matrix of nodal admittances of the extended network

y_{kk}^{old} designates the diagonal element k–k at the elimination step k

y_{ij}^{new} the new value of the i–j entry after the node k has been eliminated

Let us note in passing that a different type of reduction can be performed if, instead of expanding the original system with the ZPBN, we replace the loads at the load buses with impedances except for the load bus that we want to study. If, for example, we wanted to study the load bus 4, the Y-matrix would *not* contain lines and rows for the nodes 100031 and 100011; first lines and rows in the matrix would be 2, 3 (internal generator nodes), and 4; Gaussian reduction would eliminate the buses 7, 8, 9, 100102, and 100103; and the reduced network would contain just the nodes 2, 3, and 4.

After the Gauss–Seidel elimination has been completed, the:

- Complex admittances of the REI net are retrieved from the bottom row of the Y-matrix associated with the system of "nodal" equations that depict the equilibrium of the network obtained by eliminating the passive buses.
- Voltage phasors associated with the REI net are rotated with an angle equal to the angle of the complex voltage of the fictitious load center, thus converting its voltage phasor into a voltage angle reference.

Shown in Figure A-8 are the Y-matrix of nodal admittances after having applied the short-circuit current transformation and the REI net computed with QuickStab. The admittances, bus voltages, generated powers, and total system load in the REI net are shown in Figure A-9.

For verification and algorithm testing purposes, the entire suite of calculations was performed independently at CAMMESA. The results, which are practically identical, are shown in Table A-3. The network synthesized from the Y-matrix given in Table A-

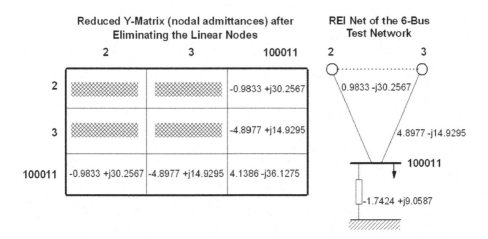

Figure A-8. Y-matrix after Gaussian elimination (left) and the resulting REI net (right).

Bus No	Bus Name	Yre G [p.u]	Yim B [p.u]	REI MW	REI MVAr	Vmod [p.u]	Vang [°]
3	TRES	4.89770	−14.92950	0.00	36.66	1.021660	0.0000
2	DOS	0.98330	−30.25670	1750.00	554.12	1.032890	45.2400
100011	REILOAD	1.74240	−9.05870	−1750.00	−709.00	0.857140	5.9600

QuickStab Professional TextDisplay: REI Net Parameters Report
Wed Oct 24 10:22:38 2007   *Case: g:\cammelfs\sixbuscase\base/F41P*

Figure A-9. REI net parameters report for the six-bus network.

3 is depicted in Figure A-10. A further analysis of this network is provided in Section A.2.3.

In addition, let us note that:

- The branches of the REI net are called "short-circuit admittances" because if a three-phase symmetrical short-circuit were applied at the bus 100011, thus making $V_{100011} = 0$, the currents in these branches would be "short-circuit currents."

Table A-3. Reduced Y-matrix after Gaussian elimination obtained at CAMMESA

Bus Number	2	3	100011
2	−1.1188 −j23.3588	2.9004 j0.3379	−0.9823 j30.2555
3	2.9004 j0.3379	3.5669 −j14.7053	−4.8966 j14.9309
100011	−0,9823 j30.2555	−4.8966 j14.9309	4.1439 −j36.1224

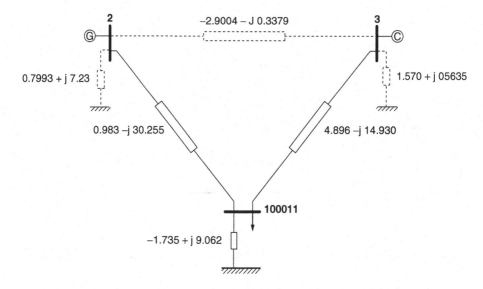

Figure A-10. Reduced network of admittances out of which the REI net has been extracted (p.u. on 100 MVA base).

- The currents across the branches of the REI net are exactly the currents flowing from the synchronous machines to the bus 100011. This can be seen very clearly from the lowest row (corresponding to bus 100011) in the reduced matrix of nodal admittances.

- The bus voltages at the nodes 2 and 3 are the internal E'_d emf of the generator (bus 2) and synchronous condenser (bus 3) because the units were represented through their x'_d; if they had been represented through their subsynchronous reactances x''_d, the bus voltages at the nodes 2 and 3 would have been the internal internal E''_d emf.

- The MW input (generation) in the REI net is shown in the column *REI MW* in the display "REI Net Parameters Report." These values are the same as in Table A-1 because the machines were represented by x'_d (no resistance).

- The MVAr input in the REI Net is not identical to the MVAr generation as shown in Table A-1 precisely because, internally, the machines produce additional MVAr to get over the x'_d.

- Physically, the radial branches of the REI Net look like regular impedances where the imaginary part is much higher, at least one order of magnitude, than the real part. This is very different from actual (typical) R/X line ratios, which tend to be in the range of one-third to one-fifth. Further analysis of this aspect is provided in Section A.2.3.

A.2.2.3 Computation of dΔQ/dV

At the outset, let us remember that the reduced network from which the REI net has been extracted also contains links between generators. These links correspond to nondiagonal elements in the reduced Y-matrix of nodal admittances and *do not* partici-pate in the bottom equation that expresses the current I_{100011}. Thus, when applying the formulae (9) or (10) to calculate $d\Delta Q/dV$, we must be aware of the fact that the amount of power that goes from generators to the equivalent load center is not 100% equal to the total amount of generated MW. This is because some power will in-evitably flow between the machines across the lateral links, which, actually, corre-sponds to the small, or very small, amount of power that may circulate between gen-erators across the power system network. Two scenarios for solving this difficulty are possible:

- *Scenario A.* Use the REI net admittances exactly as they appear in the reduced Y-matrix when computing $d\Delta Q/dV$. This process entails a small error (negligible in this case but a little bit larger in big systems, perhaps on the order of magni-tude of the losses in the entire network) because the power that flows across the REI net admittances to reach the REI load bus, which is the "nodal point" re-ferred to by Venikov in the conditions of applicability of the $d\Delta Q/dV$ criterion, is slightly smaller than the total generated MW.
- *Scenario B.* Adjust the admittances of the REI net such that 100% of the generat-ed MW will flow toward the equivalent load center, thus also matching perfectly the conditions of applicability of the $d\Delta Q/dV$ criterion.

With this clarification, three sets of separate calculations of the reactive-power steady-state (voltage) stability criterion $d\Delta Q/dV$ have been performed as follows:

- Manual calculation of $d\Delta Q/dV$ by applying Dimo's original formula given in (8) (calculation executed in New York by Savu Savulescu)
- Manual calculation of $d\Delta Q/dV$ by applying the exact formula given in (9) (calcu-lation was executed in Rosario by Roberto Molina Mylius and Martin Cassano)
- Computer calculation of $d\Delta Q/dV$ by using QuickStab®

MANUAL CALCULATION OF $d\Delta Q/dV$ WITH DIMO'S FORMULA. The various steps en-tailed in the manual calculation of $d\Delta Q/dV$ with the original Dimo's formula are as follows:

$$\frac{Y_2 E_2}{\cos \delta_2} = \frac{1.039289\sqrt{0.9833^2 + 30.2567^2}}{\cos(45.24 - 5.96)} = \frac{1.039289 \times 30.2567}{\cos(45.24 - 5.96)} = 40.6240$$

$$\frac{Y_3 E_3}{\cos \delta_3} = \frac{1.02165\sqrt{4.8977^2 + 14.9295^2}}{\cos(0.0 - 5.96)} = \frac{1.02165 \times 15.71233}{\cos(0.0 - 5.96)} = 16.1397$$

$$\sum \frac{Y_m E_m}{\cos \delta_m} = 56.7637$$

$$2\left(\sum Y_m + Y_{\text{load}}\right)V = 2[(30.2567 + 15.71233) + 9.058 - 9.73519] \times 0.857 = 77.63021376$$

$$d\Delta Q/dV_{\text{manual calculation by Savu Savulescu}} = \sum \frac{Y_m E_m}{\cos \delta_m} - 2\left(\sum_m Y_m + Y_{\text{load}}\right) = -20.867 \text{ p.u.}$$

MANUAL CALCULATION OF $d\Delta Q/dV$ WITH THE EXACT FORMULA. Roberto Molina and Martin Cassano calculated the reactive power criterion for the base-case conditions of the six-bus example by taking into account the angles γ_m of the complex admittances Y_m and applying the exact $d\Delta Q/dV$ formula (9):

$$\frac{d\Delta Q}{dV} = \sum_m \frac{Y_m \times E_m}{\cos(\delta_m + \gamma_m)} - 2 \times \left[\sum_m Y_m \times \cos(\gamma_m) + Y_{\text{load}} + Y^\gamma\right] \times V$$

where m is the number of synchronous machines in the REI net and V is the magnitude of the complex voltage at the fictitious load bus 100011. The calculations produced the following results:

$$Y_{1m} = \sqrt{30.255^2 + 0.9823^2} = 30.27; \qquad \gamma_1 = 90° - 88.14° = 1.86°;$$
$$E_1 = 1.0328 \angle 45.237°$$

$$Y_{2m} = \sqrt{4.8966^2 + 14.93^2} = 15.71; \qquad \gamma_2 = 90° - 71.84° = 18.16°;$$
$$E_2 = 1.02166 \angle 0°$$

$$V_{100011} = 0.857 \angle 5.983° \text{ p.u.}$$

$$Y_{\text{shunt\_100011\_0}} = \sqrt{1.73^2 + 9.062^2} = 9.22 \text{ p.u.}$$

$$\sum_m \frac{Y_m \times E_m}{\cos(\delta_m + \gamma_m)} = \frac{30.27 \times 1.0328}{\cos(45.237 + 1.86 - 5.983)} + \frac{15.71 \times 1.02166}{\cos(0 + 18.16 - 5.983)}$$
$$= 41.51 + 16.42 = 57.93$$

$Y^\gamma = Y_1 \times \sin(\gamma_1) \times \tan(\delta_1 + \gamma_1) + Y_2 \times \sin(\gamma_2) \times \tan(\delta_2 + \gamma_2)$

$\Rightarrow Y^\gamma = 30.27 \times \sin(1.86) \times \tan(45.237 + 1.86 - 5.983) + 15.71 \times \sin(18.16)$
$\times \tan(0 - 5.983 + 18.16)$

$\Rightarrow Y^\gamma = 0.857 + 1.056 = 1.91 \text{ p.u.}$

By subtracting the susceptance corresponding to the reactive part of the system load injected into the fictitious load bus 100011

$$B = \frac{Q_{\text{load}}}{V^2} = -\frac{7.15}{0.857^3} = -9.73 \text{ p.u.}$$

from the imaginary part of the shunt-to-ground at the bus 100011 $B_{\text{shunt\_100011-0}} = 9.062$ p.u., the following admittance of the reactive load is obtained: $Y_{\text{load}} = 9.062 - 9.73 = -0.668$ p.u. After making the necessary substitutions,

$$2 \times \left(\sum_m Y_m \cos(\gamma_m) + Y_{\text{load}} + Y^\gamma \right) \times V =$$

$$2 \times (30.27 \times \cos(1.86°) + 15.71 \times \cos(18.16°) + (+9.062 - 9.73) + 1.91) \times 0.857$$

$$= 79.56$$

and by applying the formula (9),

$$\frac{d\Delta Q}{dV} = \sum_m \frac{Y_m \times E_m}{\cos(\delta_m + \gamma_m)} - 2 \times \left(\sum_m Y_m \times \cos(\gamma_m) + Y_{\text{load}} + Y^\gamma \right) \times V = 57.93 - 79.56$$

$$= -21.63$$

Molina and Cassano obtained the following value for the criterion:

$$d\Delta Q/dV_{\text{exact calculation by Molina \& Cassano}} = \sum \frac{Y_m E_m}{\cos \delta_m} - 2 \left(\sum_m Y_m + Y_{\text{load}} \right) = -21.63 \text{ p.u.}$$

COMPUTER CALCULATION OF $d\Delta Q/dV$. For comparison, we show in Figure A-11 the value of $d\Delta Q/dV$, called StabIndex in the summary results display, computed with QuickStab® under the Scenario B, that is, by first adjusting the admittances of the REI net such that 100% of the generated MW will flow toward the equivalent load center, and then applying Dimo's formula.

As shown in Figure A-11, the differences between the three sets of calculations are practically insignificant and, as anticipated, the results of this numerical example fully confirm both the theory and the accurate computer implementation of the algorithm.

A.2.2.4 Case Worsening—Computer Calculation of SSSL

As emphasized earlier, both in this Appendix and in Chapter 2, the most important result, and, as a matter of fact, the ultimate goal of Dimo's steady-state stability analysis methodology is to compute the SSSL, that is, to determine how far are the current system conditions from a hypothetical state in which voltages may collapse or units may loose synchronism.

QuickStab Professional TextDisplay: Summary Results					_ □ ☒
Wed Oct 24 10:22:38 2007				Case: g:\cammelfs\sixbuscase\base/F21P	
State	**GridUtilMW**	**AvgSysVolt**	**SysAngle**	**StabIndex**	**StabReserve**
Actual	1750.00	0.8571	27.4900	−21.54	5.44
SecMargin	1565.45	0.9367	21.6990	−31.38	15.41
Critical	1850.70	0.7443	35.8000	−1.57	0.00
MaxStability	1750.00	0.8571	27.4900	−21.54	5.44

Figure A-11. $d\Delta Q/dV$ (StabIndex) computed by QuickStab for the six-bus network.

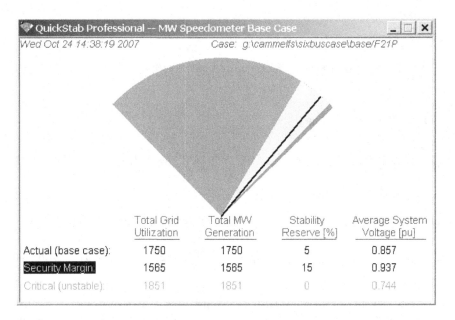

Figure A-12. Distance to instability shown on a speedometer chart. The needle corresponds to the *actual* system state (1,750 MW) and the right-most speedometer edge corresponds to *critical* state, that is, SSSL (1,851 MW).

After the model was built from the actual system conditions (base case) and the computation of the $d\Delta Q/dV$ reactive power steady-state (voltage) stability criterion confirmed that the system was stable, the next step consisted of performing case worsening* to stress the system until it becomes unstable. The SSSL thus computed for the six-bus example has the value 1850.70 (approximately 1851) MW; it is shown in tabular format in Figure A-11 where it is identified as Critical State under the column heading GridUtilMW (Total Grid Utilization MW, which is equal to the total system MW generation plus the total MW imports), and it is depicted in Figure A-12 on a speedometer chart that visualizes the distance to instability.

Figure A-13(a) shows how the system is stressed by the case-worsening procedure from 1750 MW (base-case value) to 1851 MW (highest system grid utilization MW before becoming unstable). The case-worsening results can then be plotted as a P–V curve as shown in Figure A-13(b).

A.2.3   Further Analysis of the Reduced Network

If we examine the reduced network obtained by applying Dimo's procedure (Figure A-10), out of which the REI net has been extracted and the power flows circulating in this

---

*The theoretical aspects of the case-worsening procedure have been briefly addressed in Section A.1.2.5. The implementation details go beyond the scope of this book and will not be illustrated in detail herein.

QuickStab Professional TextDisplay: Case Worsening Report ___ □ ✕

Wed Oct 24 14:38:19 2007   Case: g:\cammelfs\sixbuscase\base/F26P

Step No	GridUtilMW	AvgSysVolt	SysAngle	StabIndex
0	1750.00	0.857	27.49	−21.54
1	1768.59	0.849	28.18	−20.18
2	1781.68	0.839	28.86	−18.77
3	1794.39	0.830	29.55	−17.31
4	1805.64	0.820	30.24	−15.80
5	1815.66	0.811	30.93	−14.24
6	1824.41	0.801	31.62	−12.62
7	1831.90	0.792	32.31	−10.95
8	1838.14	0.782	33.01	−9.21
9	1843.13	0.773	33.70	−7.41
10	1846.88	0.763	34.40	−5.53
11	1849.40	0.754	35.10	−3.59
12	1850.70	0.744	35.80	−1.57
13	1850.78	0.735	36.50	0.54

(a)

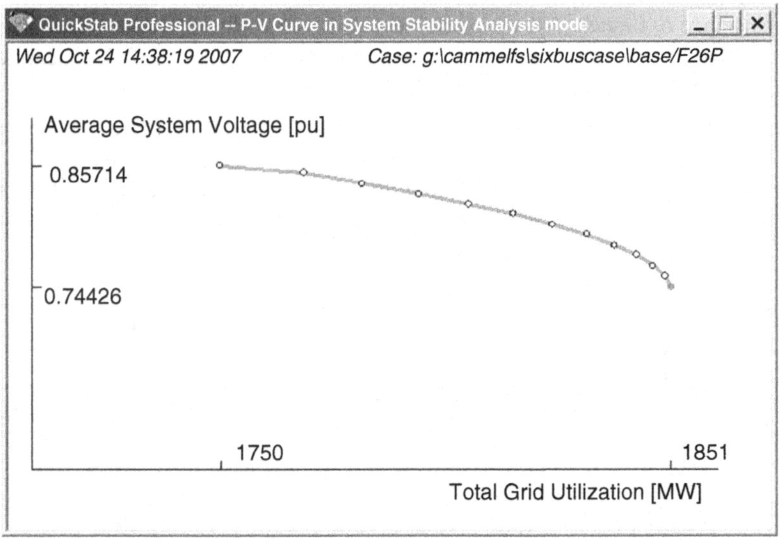

QuickStab Professional -- P-V Curve in System Stability Analysis mode   _ □ ✕

Wed Oct 24 14:38:19 2007   Case: g:\cammelfs\sixbuscase\base/F26P

(b)

Figure A-13. (a) numerical values computed by the case worsening algorithm; (b) P–V curve showing how the system is stressed from the base case to SSSL.

network have been determined, we can make a number of interesting observations as follows:

- *Power transfer from generators toward the equivalent load center.* The real power generated at bus 1 arrives at the equivalent load center across three paths as follows:

□ From the generator directly toward the fictitious load center (bus 100011)

□ From the generator toward the synchronous condenser (bus 3)

□ From both active buses through ground

The network of impedances is shown in Figure A-14. A comparison of the moduli of the branch and shunt impedances

$$|Z_{2\text{-}3}| \cong 0.34 \qquad |Z_{2\text{-}100011}| \cong 0.033 \qquad |Z_{3\text{-}100011}| \cong 0.063$$

$$|Z_{2\text{-ground}}| \cong 0.13 \qquad |Z_{3\text{-ground}}| \cong 0.59 \qquad |Z_{100011\text{-ground}}| \cong 0.10$$

indicates that the impedance $Z_{2\text{-}100011}$ of the direct link between the generator and the equivalent load center is approximately 10 times smaller than the impedance of the lateral link $Z_{2\text{-}3}$. Since the impedance $Z_{3\text{-}100011}$ of the link between the synchronous condenser and the fictitious load center is also small, approximately five times smaller than $Z_{2\text{-}3}$, it is obvious that the MW flow in the lateral link can be disregarded. The power flows in this network are shown in Figure A-15 and confirm the excellent precision of the calculations. The negative value of the resistance of branch 2-3, which is one of the "side effects" of the network reduction, has no physical meaning but explains why the branch 2-3 appears to be generating a very small amount of MW.

• *Fictitious (equivalent) branches and shunts.* The short-circuit transformation (Gaussian elimination) produces a network containing fictitious branches and shunts that do not exist in the actual network. A significant amount of the reac-

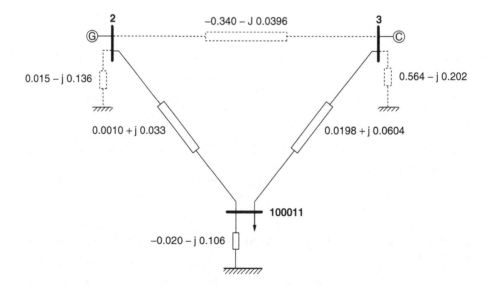

Figure A-14. Reduced network represented as a scheme of impedances (p.u. on 100 MVA base).

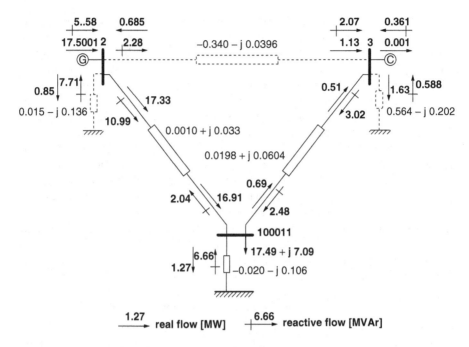

Figure A-15. Power flows in the network depicted in Figure A-14.

tive load at the equivalent load center 100011 is supplied by the fictitious capac-
itive shunt connected between bus 100011 and the ground (665 MVAr). This
complex shunt also injects approximately 127 MW to supply the power lost in
the "fictitious" resistances of the REI net, thus acceptably preserving the power
balance and the operating conditions at the buses retained in the model.

- *Power balance accuracy at the buses retained in the model.* The REI net extracted from the reduced network reproduces the state of buses 2 (generator) and 3
  (synchronous condenser) with acceptable accuracy. As shown in Figure A-14,
  the generator (bus 2) supplies the full amount of real power absorbed in the fictitious load center, 17.5001 p.u. ≈ 1,750 MW, and generates reactive power as
  well: 5.558 p.u. = 555.8 MVAr. The synchronous condenser absorbs a negligible
  amount of real power and generates 0.361 p.u. = 36.1 MVAr.

A.3   BENCHMARKING THE METHODOLOGY

A.3.1   Validating the SSSL Computed with Dimo's Method

A.3.1.1   Approach
Benchmarking an engineering application assumes in the first place that the underlying
method and its approximations, if any, on the one hand, and the alternate solution tech-

nique, on the other, are theoretically sound. Another important assumption is that the alternate technical approach used for reference is similar in purpose and intent to the methodology that is being benchmarked. When it comes to stability, this step is inherently difficult because the assessment depends upon the type of stability evaluation that has to be conducted. For example, in order to avoid comparing apples with oranges, a steady-state stability program that performs "case worsening" can not be benchmarked against an application that aims just at verifying that the system state is stable but does not stress the network to find the steady-state stability limit. Therefore, when assessing the validity of a stability analysis technique, the purpose should be, as a minimum, to demonstrate that both the program being benchmarked and the alternate technology give the same result, that is, stable or unstable, for the base case and, most importantly, predict the same system loadability for the state in which voltages may collapse and units may loose synchronism.

In this section, we will show how the SSSL computed by using Dimo's technique for the six-bus network, as shown in Section A.2.2.4, has been benchmarked at CAMMESA by Molina and Cassano by performing accurate calculations based on the *maximum power transfer theorem*. The procedure is "accurate" in the sense that it:

- Entails no aggregation of the system loads into a fictitious load center,
- Does not use the $d\Delta Q/dV$ steady-state stability criterion, and
- Stresses the system toward its steady-state stability limit by applying conventional network calculations rather then performing the "case-worsening algorithm."

A.3.1.2 Building the Model

In order for the results of the two different sets of SSSL computations to be comparable, the same six-bus network described in Section A.2 was used for benchmarking. In this network, the real and reactive powers are transferred from the synchronous machines to the bus where the 1750 MW load is located. The branches 7-8, 8-9, 7-4, and 9-4 correspond to 500 kV lines and are represented as pure reactances. Since the resistances have been disregarded, the network has no MW losses. The starting conditions are the same as those given in Figures 5(a) and 5(b) and Table A-1.

The generator connected to bus 2 through a step-up transformer actually represents the equivalent of 12 machines as shown below:

$$S_n = 280 \times 12 = 3360 \text{ MVA}; \qquad P_{max} = 250 \times 12 = 3000 \text{ MW}$$

$$-1500 \leq Q \leq 3000 \text{ MVAr}; \qquad X'_{deq} = \frac{X'_d}{12} = 0.00773 \text{ p.u. on 100 MVA base}$$

The synchronous condenser connected to bus 3 through a step-up transformer is also the equivalent of two machines:

$$Q_{max} = 500 \text{ MVAr}; \qquad X'_d = \frac{0.0928}{2} = 0.0464 \text{ p.u. on 100 MVA base}$$

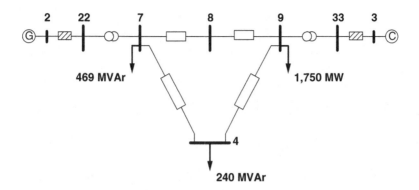

| internal machine node where the e.m.f. is applied

—▨— internal machine reactances —◯— transformers —▢— transmission lines

Figure A-16. Six-bus study network extended with the internal nodes of the machines.

In order to represent the generator and the synchronous condenser in the ensuing stability calculations,* the first step consisted of expanding the six-bus network to include the internal nodes of the machines, as shown in Figure A-16. Let us note that this is similar to the extended network used to build the REI net, with the major difference that all the buses are represented with their actual identity and no zero power balance network has been introduced.

The internal voltages (emf) behind the transient reactance of the generator and the synchronous condenser were calculated manually by using the well-known phasor diagram and related equations (Figure A-17):

$$\text{emf}_{\text{generator}} = E'_G = 1.0328 \angle 45.237° \text{ p.u.} \tag{10a}$$

$$\text{emf}_{\text{synchronous\_condenser}} = E'_C = 1.02166 \angle 0° \text{ p.u.} \tag{10b}$$

The values (10a) and (10b) of the generator and the synchronous condenser internal voltages were corroborated with a power-flow calculation performed with the PSS/E load-flow program as shown in Figure A-18, and are identical to those computed with QuickStab (Figure A-8).

The next step consisted of applying the short-circuit current transformation to convert the six-bus network to a model containing just the load bus and the active buses, in this case a generator and a synchronous condenser, as shown in [2]. Initially, the reactive loads at buses 7 and 4 were converted to shunts by using the bus voltages from the base case. Then, the star-delta, or Gaussian, elimination was performed. The resulting

---

*Please refer to Chapter 1, Section 1.2.3.2, which provides a detailed discussion of the need to represent the synchronous machines in voltage stability calculations, and to references [2,15,16,23].

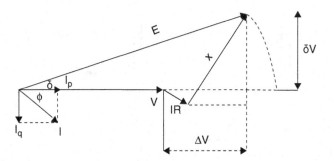

$$E^2 = (V + \Delta V)^2 + \delta V^2 = (V + RI\cos(\phi) + XI\,sen(\phi))^2\,(XI\cos(\phi) - RI\,sen(\phi))^2$$

$$E^2 = \left(V + \frac{RP}{V} + \frac{XQ}{V}\right)^2 + \left(\frac{XP}{V} - \frac{RQ}{V}\right)^2$$

$$\Delta V = \frac{RP + XQ}{V} \qquad \delta V = \frac{XP - RQ}{V} \qquad \delta = arcsen\left(\frac{\delta V}{E}\right)$$

Figure A-17. Synchronous machine representation in steady state.

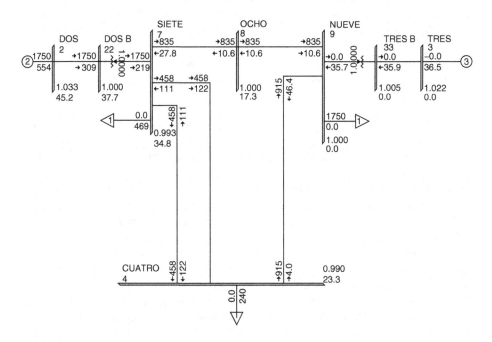

Figure A-18. Power flows in the six-bus network extended with the internal nodes of the machines.

network is shown in Figure A-19; its Y-matrix is part of equation (11) where P_{load} is the real part of the load at bus 9, initially 1750 MW, and V_{load} is the magnitude of the complex voltage at bus 9, initially 0.99962 p.u.

$$
\begin{bmatrix} I_2 \\ I_3 \\ 0 \end{bmatrix} = \begin{bmatrix} -j21.45 & 0 & j23.86 \\ 0 & -j16.28 & j16.28 \\ j23.86 & j16.28 & \dfrac{P_{load}}{V^2_{load}} - j34.014 \end{bmatrix} \begin{bmatrix} V_2 \\ V_3 \\ V_9 \end{bmatrix} \tag{11}
$$

Several interesting observations can be made if we compare the reduced network shown in Figure A-19 with the reduced network obtained by applying the fictitious load center version of Dimo's methodology (Figure A-10):

- Bus 9 is an actual bus and its load is the same as it was in the original six-bus network. This is quite different from what happens when applying Dimo's technique in conjunction with a zero power balance network, where the fictitious load center aggregates all the loads in the network (by coincidence, though, the equivalent system load in the six-bus example is identical with the load on bus 9 because there are no other loads in the network).

- Since the bus 9 is an actual bus, its complex voltage in the reduced network is identical to the value it had prior to performing the star-delta elimination. Again, this is different from the fictitious load bus in the REI net where the bus voltage V_{100011} is calculated as shown in Section A. 2.2.1

- There is no link between the generator (bus 2) and the synchronous condenser (bus 3) because neither the passive buses in the extended six-bus network nor the

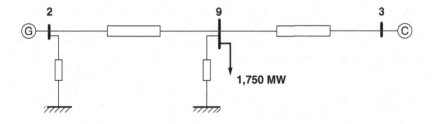

actual load bus   internal machine node where the e.m.f. is applied

equivalent impedances (admittances) after the star-delta elimination

Figure A-19. Reduced network after eliminating the passive buses from the extended six-bus system.

new buses created during the star-delta reduction were connected to both bus 2 and bus 3. Again, this is very different from Dimo's reduction which, due to the introduction of the zero power balance network, will almost certainly introduce links between the active buses retained in the final model.

A.3.1.3 Maximum Power Transfer Limit—Stressing the System

The SSSL calculation procedure consists of gradually increasing the load P_L at bus 9 and assuming that the moduli of the internal complex voltages of the machines are maintained constant.* When raising P_L, the

- Generator will augment its MW output and the angle $\delta_{2\text{-}9}$ will increase as well;
- Synchronous condenser will produce additional MVArs while the bus voltage magnitude V_9 at the load bus will decrease.

The MW value of the current power transfer angle $P_{2\text{-}9}$ and its maximum value $P_{\max 2\text{-}9}$ are given by the formulae (12a) and (12b) respectively:

$$P_{2\text{-}9} = \frac{V_2 \times V_9 \sin \delta_{2\text{-}9}}{X_{2\text{-}9}} \tag{12a}$$

$$P_{\max 2\text{-}9} = \frac{V_2 \times V_9}{X_{2\text{-}9}} = SSSL \tag{12b}$$

It is important to note that the SSSL value computed with (12b) is purely theoretical and is valid if, and only if, the voltage V_9 can be maintained constant regardless of the value of P_L. If there are no structural changes in the network, that is, $X_{1\text{-}9}$ is constant, and since V_2 is also constant, the steady-state stability limit SSSL will depend only upon the value of V_9, that is, $SSSL = f(V_9)$ and, when P_L increases and V_9 decreases, SSSL will decrease as well, as shown in Figure A-20.

Therefore, the problem consists of recalculating V_9 at each step of load increase. From the last row of matrix equation (11) we get

$$0 = \overline{Y}_{31} \times \overline{V}_1 + \overline{Y}_{32} \times \overline{V}_2 + \left[\left(\frac{P_L}{V_3^2} \right) + \overline{Y}'_{33} \right] \times V_3 \Rightarrow - V_3 \times \left(\frac{P_L}{V_3^2} + \overline{Y}'_{33} \right)$$

$$= \overline{Y}_{31} \times \overline{V}_1 + \overline{Y}_{32} \times \overline{V}_2$$

We take $\underline{V}_9$ for angle reference and express $\underline{V}_2$ and $\underline{V}_3$ in rectangular coordinates:

$$-V_9 \times \left(\frac{P_L}{V_9^2} + jY'_{99} \right) = jY_{9\text{-}2} \times (H_2 + jF_2)\, jY_{9\text{-}3} \times (H_3)$$

---

*The same assumption is used in case worsening (both basic and alternate approaches in Section 1.2.3.2).

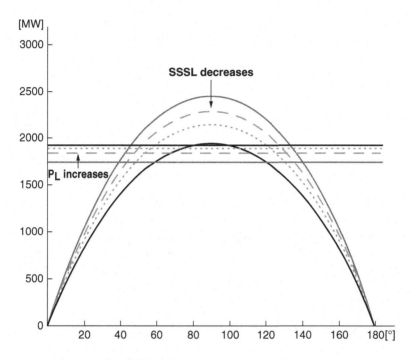

Figure A-20. SSSL and V_9 decrease when P_L increases.

After performing the complex algebra manipulations, we obtain the formulae (13a) and (13b) for the real and imaginary components of $\underline{V}_9$:

$$H_2 = -\frac{V_9 \times Y'_{9\text{-}9} + Y_{9\text{-}3} \times H_3}{Y_{9\text{-}1}} \tag{13a}$$

$$F_2 = \frac{V_9}{Y_{9\text{-}2}} \times \left(\frac{P_L}{V_9^2}\right) \tag{13b}$$

By substituting (13.a) and (13.b) in $V_{22} = H_{22} + F_{22}$ we obtain

$$V_{29}\left[Y'^2_{9\text{-}9} + \left(\frac{P_L}{V_9^2}\right)^2\right] + V_9 \times (2 \times Y'_{9\text{-}9} \times Y_{9\text{-}3} \times H_3) + (Y^2_{9\text{-}3} \times H_{32} - V_{22}Y^2_{9\text{-}2}) = 0 \tag{14}$$

Since P_L is known and both V_2 and $H_3 = V_3$ are maintained constant, the relationship (14) takes the form of the second-degree equation

$$a \times V_{29} + {}^*bV_9 + c = 0$$

with $a = Y'^2_{9\text{-}9} + (P_L/V_9^2)^2$, $b = (2 \times Y'_{9\text{-}9} \times Y_{9\text{-}3} \times V_3)$, and $c = (Y^2_{9\text{-}3} \times V_3^2 - V_2^2 Y^2_{9\text{-}2})$, where the coefficients b and c are constant (because V_2 and V_3 were assumed to be constant) and constant a varies as a function of P_L and V_9. The system stressing procedure entails computing a new V_9 for $P_L^{\text{new}} = P_L^{\text{old}} + \Delta P_L$ and replacing P_L^{new} with $Y_{\text{load}}^{\text{new}}$ as follows:

$$Y_{\text{load}}^{\text{new}} = \left[\frac{P_L^{\text{old}} + \Delta P_L}{(V_9^{\text{old}})^2} \right] > Y_{\text{load}}^{\text{old}} = \left[\frac{P_L^{\text{old}}}{(V_9^{\text{old}})^2} \right]$$

For each new value of P_L there are two possible values of V_9:

$$V_9^{1,2} = \frac{-(2 \times Y'_{9\text{-}9} \times Y_{9\text{-}3} \times V_3) \pm \sqrt{(2 \times Y'_{9\text{-}9} \times Y_{9\text{-}3} \times V_3)^2 - 4 \times \left\{ Y'^2_{9\text{-}9} + *\left[\frac{P_L}{(V_9^{\text{old}})^2} \right] \right\} \times (Y^2_{9\text{-}3} \times V^2_3 - V^2_2 Y^2_{9\text{-}2})}}{2 \times \left\{ Y'^2_{9\text{-}9} + \left[\frac{P_L}{(V_9^{\text{old}})^2} \right] \right\}}$$

(15)

Depending upon whether we use plus or minus in the above formula, we obtain two sets of values for the voltage V_9. Since the basic operating scenario was stable, we will use only the plus sign to get the higher values of V_9, thus disregarding the lower part of the parabola depicting the "PV–Nose" curve.* The calculations consist of the following iterative process:

- Step 1. Increase P_L by ΔP_L and calculate V_9 by solving the quadratic equation (15).
- Step 2. Verify whether the first Kirchhoff theorem is met within a small tolerance by using the value V_9 computed in Step 1.
- If the mismatch is larger than the acceptable tolerance, reduce ΔP_L in half and repeat the process.
- Stop when the mismatch is within the tolerance limits.

In addition, at each step of this iterative process, which is illustrated in Figure A-21, we calculate the angular steady-state stability limit by applying formula (12b) and verify that the reactive powers generated by the machines do not exceed the Q_{max} limits. The results are summarized in Table A-4, where the value of the angular steady-state stability margin is given by [SSSL (angular) $- P_L$] × 100, and indicate that:

- The maximum amount of power that can be transferred from the generator to the load bus is equal to 1854 MW and, as will be shown in Section A.3.1.3, corresponds to a *voltage stability limit.*
- The angular steady-state stability limit computed by maintaining constant the emf behind the transient reactances of the machines is in the neighborhood of 1900 MW (note, however, that the tolerance margin adopted in the iterative procedure was not sufficiently tight and the current mismatch at bus 9 increases simultaneously with P_L).

---

*The existence of two voltage values for one value of the load MW corresponds to the "dual states" noticed by Ionescu and Ungureanu [36] when the load is represented as an impedance.

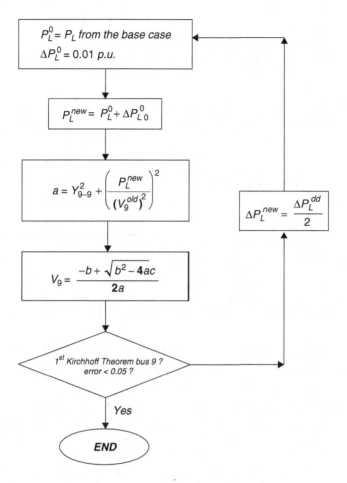

Figure A-21. Iterative procedure for calculating the SSSL for the six-bus numerical example.

Table A-4. Summary results of the benchmarking calculations

P_L (MW)	V_9 (p.u.)	SSSL(angular) (MW)	Margin (%)	Error (ΣI_3), bus 9
1750	1.00	2463	28.95	0.03
1800	0.96	2369	24.02	0.05
1825	0.93	2290	20.31	0.11
1840	0.89	2204	16.52	0.27
1850	0.84	2064	10.37	1.04
1854	0.77	1909	2.88	4.03

A.3.1.4 The P–V Curve

The P–V curve corresponding to the series of values P_L (P_{load}) and V_9 (V_{load}) from Table A-4 is shown in Figure A-22. A closer look at the points that define the "nose" region of this P–V curve provides for some interesting observations. When the system sends 1850 MW to load bus 9, the voltage V_9 takes the value 0.84 p.u., (420 kV), as shown in Table A-5. When the system is in this operating state, a 1% increase of the load will correspond to a voltage drop of 55%, which is a clear indication that we are near the "tip of the nose" of the P–V curve. If the generator would ship 1854 MW to the load bus, the voltage V_9 would drop to 0.77 p.u. (385 kV!), which means that, in this operating state, a 1% increase of load would produce a 175% drop of the bus voltage, thus causing the voltage to collapse.

At each step of the iterative procedure used to stress the system conditions, the MVAr flows in the network and the balance of reactive powers have been calculated for the purpose of verifying that the MVArs generated by the synchronous machines do not exceed their respective Q_{max} limits. The reactive power produced by the generator was within the limits during the entire process, but the reactive generation of the synchronous condenser was found to exceed the maximum limit for P_{load} = 1854 MW, which further corroborated the pattern of the P–V curve and induced us to infer that, for the six-bus example evaluated, the SSSL corresponds to voltage instability.

A.3.2   Physical Interpretation of the Reactive Power Steady-State Stability Criterion $d\Delta Q/dV$—The ΔQ–V Curve

Section A.2.2.3 illustrated the manual computation of the $d\Delta Q/dV$ criterion for the test network with Dimo's original formula, which neglects the resistances of the REI net branches, and with the exact formula developed in [22]. The value of $d\Delta Q/dV$ calculated with a computer program has also been provided for comparison. Now we will offer a physical interpretation of the $d\Delta Q/dV$ reactive power steady-state stability criterion by using ΔQ–V *curves* developed for two different operating states of the six-bus numerical example: a starting scenario, which is simply the base case used in Section 2

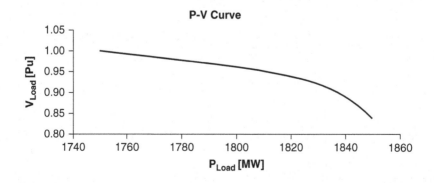

Figure A-22. P–V curve obtained for the six-bus example.

Table A-5. Neighborhood of the "tip of the nose" of the P–V Curve

P_{Load} (p.u.)	V_{load}	ΔV_{load}	$(\Delta V_{load}/\Delta P_{load}) \times 100$
18.40	0.895		
18.50	0.84	−0.055	−55%
18.54	0.84	−0.055	−175%

for testing; and a second scenario, in which the operating conditions of the same network have been stressed to move the system in the direction of the SSSL. The same modeling assumptions adopted previously will be considered: the machines are represented through their x_d' reactances; the internal voltages behind these reactances are constant, provided that the maximum reactive limits Q_{max} have not been reached; and the bus voltage V of the equivalent load center (bus 100011) and, implicitly, the reactive power injected in this bus, are allowed to vary when the system load is increasing.

Both the base case, with 1750 MW load, and the reduced network obtained by applying Dimo's procedure (Figure A-10), are in a stable steady state when the bus voltages are constant. Accordingly, if the bus voltage of a bus does not change, the net reactive power ΔQ of that bus should be equal to zero. For the study node 100011, the reactive power balance is given by

$$Q_G + Q_C + Q_{shunt\text{-}REI} + Q_{load} = \Delta Q \qquad (16)$$

where

Q_G is the reactive power sent from the generator (bus 2) to the load

Q_C is the reactive power sent from the synchronous condenser (bus 3) to the load

$Q_{shunt\text{-}REI}$ is the reactive power injected by the ground-to-bus$_{100011}$ shunt

Q_{load} is the reactive power absorbed from the load bus 100011

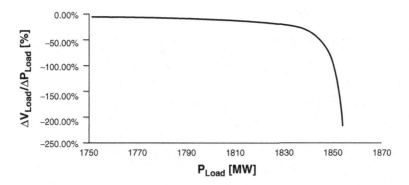

Figure A-23. Variation of $\Delta V_{load}/\Delta P_{load}$ when P_{load} increases in the proximity of the P–V nose tip.

The generic formula for the reactive power that travels from a source, such as Q_G and Q_C, toward the load bus is given in (17):

$$Q = YEV \cos(\gamma + \delta) - YV^2 \cos(\gamma) \qquad (17)$$

where Q and V are variables, whereas E, Y, δ, and γ are known and constant.

In order to simulate a ΔQ change of the reactive power at the bus 100011 while maintaining the MW load constant and equal to its base-case value (1750 MW), we first connect a source of reactive power Q_{comp}, and then we allow the amount of reactive power generated or absorbed to change and, for each new value of Q_{comp}, we calculate the new value of the load bus voltage V. These points can then be plotted to build a ΔQ–V curve. Accordingly, formula (16) can be re-written as

$$Q_G(V) + Q_C(V) + Q_{shunt\text{-}REI}(V) + Q_{load}(V) = \Delta Q(V) = Q_{comp}(V) = Q_{net}(V) \quad (18)$$

By substituting in formula (18) the numerical values for the reduced network obtained by applying Dimo's procedure (Figure A-10),

$$Q_{net} = 30.3 \times [1.0328 \times \cos(40.994) - V^2 \times \cos(1.74)] + 15.87 \times [1.02166 \times \cos(12.16) - V^2 \times \cos(18.15)] + V^2 \times 6.65 - V^2 \times 7.15$$

we get

$$Q_{net} = \Delta Q = 39.46 - V^2 \times (2.44) \qquad (19)$$

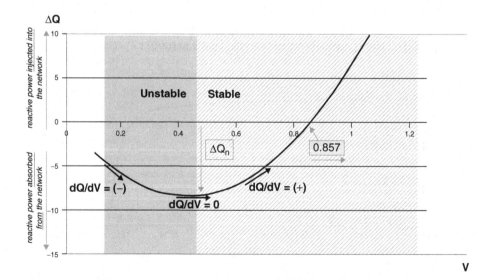

Figure A-24. ΔQ–V curve for the six-bus network base case.

which allows drawing the ΔQ–V curve (Figure A-24) under the assumption that P and δ are constant. This curve shows how the voltage of the load bus 100011 will change when we increase or decrease the net reactive power of that bus and, also, how much we can increase the net reactive power of the load bus 100011 without a voltage collapse. Please note that for the base case, the voltage of the load bus (equivalent load center) is equal to 0.857 p.u.

From geometrical considerations, the ΔQ–V curve reaches the minimum value when $d\Delta Q/dV = 0$, which divides the plane ΔQ–V into two zones: stable and unstable.

The curve clearly shows that in the stable zone, an increase of the net reactive power injected into the bus causes the bus voltage to increase, whereas in the unstable zone, an increase of the net reactive power of the bus would cause a voltage reduction. Let us emphasize, however, that the unstable area is just a mathematical representation without any basis in reality because a power system cannot operate stably with voltages smaller than 0.75 p.u. If now we revert the sign convention used above for ΔQ and consider that the net consumption of reactive power is positive (like an inductive shunt) and the net injection of reactive power is negative (like a capacitor) we obtain the "mirror" curve ΔQ–V shown in Figure A-25.

Therefore, the $d\Delta Q/dV$, which corresponds to the slope of the curve ΔQ–V, is negative for the stable operating points. In other words, during stable steady-state conditions, a reduction of the net reactive power absorbed by the load bus will cause an increase of the bus voltage and vice versa. Although less known that other curves, such as the P–V curve and Q–V curve, this mirrored version of the ΔQ–V curve is extremely useful for evaluating the steady-state stability conditions of a load bus for which there are no additional reactive sources and in which, if the absorbed real power increases, the absorbed reactive power increases as well. An interesting characteristic of this curve is due to the

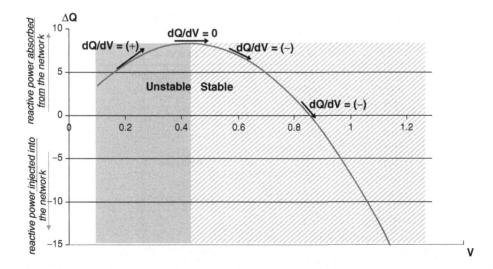

Figure A-25. Reversed ΔQ–V curve for the six-bus network base case.

parabolic form of equation (19); when the operating point is far from the steady-state stability limit, the slope of the ΔQ–V curve is practically constant.

It must be strongly emphasized, however, that the $d\Delta Q/dV$ criterion per se is just a stability index and does not say how far from instability is the current operating state.

The analysis developed in the preceding paragraphs can be further extended if we relax the assumption that both P and δ are constant when the net reactive power of the load bus is changing. Accordingly, if we again apply equation (18) to the steady state of the six-bus example but this time we allow δ to vary,

$$Q_{net} = 30.3 \times [1.0328 \times \cos(\delta + 1.74) - V^2 \times \cos(1.74)] + 15.87 \times [1.02166 \times \cos(\delta_c + 18.15) - V^2 \times \cos(18.15)] + V^2 \times 6.65 - V^2 \times 7.15$$

and, after making the calculations, we obtain the expression of Q_{net} as a function of both δ and V:

$$Q_{net} = 31.29 \times \cos(\delta + 1.74) + 16.21 \times \cos(\delta_c + 18.15) - V^2 \times (45.86) \quad (20)$$

Therefore, in order to build the curve ΔQ–V, now we need to correlate V and δ. The real power received at the load bus 100011 from the generator is given by

$$P = YEV \times \sin(\delta + \gamma) - YV^2 \times \sin \gamma \quad (21)$$

where we apply the standard case-worsening assumptions E = const, Y = const, and γ = const. Equation (20) allows correlating δ and V as shown in equation (22) and Table A-6.

Table A-6. Relationship of δ to V when the generated power increases (case worsening)

V	δ	Comments
0.55	85.27	angular stability limit
0.65	56.51	
0.7	50.68	
0.75	46.207	
0.8	42.60	
0.85	39.6	
0.857	39.23	base case
0.9	37.07	
0.95	34.89	
1	32.99	
1.05	31.31	
1.1	29.83	
1.15	28.50	maximum operation voltage limit
1.2	27.31	
1.25	26.23	

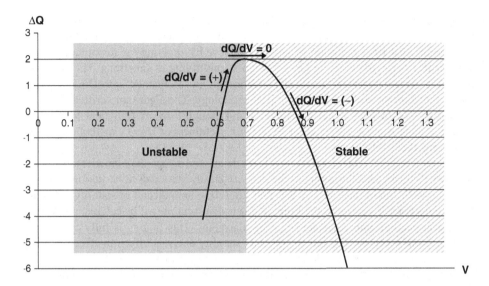

<u>Figure A-26.</u> Reversed ΔQ–V curve for the six-bus network base case with $\delta = f(V)$.

$$\delta = \arcsin\left\{\frac{1}{E} \times \left[\frac{P}{VY} + V \times \sin(\gamma)\right]\right\} - \gamma \qquad (22)$$

The values from Table A-6 will now be used in formula (20) to create the new ΔQ–V curve, which is shown in Figure A-26.

Under these assumptions, the $d\Delta Q/dV$ derivative vanishes for $V = 0.7$ p.u., whereas angular instability would occur for $V \leq 0.55$ p.u. In other words, for the six-bus example considered for this analysis, the SSSL corresponds to a point of voltage collapse rather than angular instability.

A.4 CONCLUSIONS

This appendix reviewed the technique introduced by Paul Dimo in the late 1950s and early 1960s to determine the steady-state stability limit of a power system. Several theoretical aspects that are fundamental for understanding and applying correctly the methodology have been addressed in detail, for example, the representation of generators through a constant emf behind a reactance and the $d\Delta Q/dV$ steady-state stability criterion. In order to shed light on implementation details that otherwise may go unnoticed, a small-size yet realistic network model that illustrates all the algorithm steps was developed by a project team at the Argentinean System and Market Operator (CAMMESA) in Rosario.

The calculations were performed independently in Rosario and in New York, and offered an excellent opportunity to test the precision of the computer program that embodies the REI–Dimo technology. In addition, the Argentinean project team developed a procedure that finds the point of voltage collapse by applying the maximum power transfer theorem without using any of Dimo's simplifying assumptions. The stability limits thus calculated were practically identical to those obtained with Dimo's technique, thus further validating the overall methodology. The algorithm verification performed at CAMMESA by using standard and straightforward analytical tools also helped to understand the foundation, extent and usefulness of Dimo's methodology for computing the SSSL.

In addition, the CAMMESA project team developed a separate numerical example that reveals the physical meaning of the $d\Delta Q/dV$ steady-state stability criterion. An important conclusion of this exercise was that $d\Delta Q/dV$ is an indicator of stability or instability, does not tell *how far* is the limit, and is truly useful only if its calculation is performed in parallel with a system-stressing algorithm. This finding is fully consistent with the long accepted fact that steady-state stability analysis offers an effective way of determining the maximum loadability of a power system provided that it is used in conjunction with a case-worsening procedure to stress the operating conditions until the system becomes unstable.

Last, but not least, the analysis performed at CAMMESA resulted in the recommendation to assess, prior to using Dimo's method for a particular power system, the sensitivity of the load model with voltage changes and the modeling of generators depending upon the behavior of their AVRs. These model fine-tuning steps are needed in order to avoid obtaining SSSL values that may be too conservative.

A.5 REFERENCES

[1] Anderson, P. M., and Fouad A. A., *Power System Control and Stability,* Iowa University Press, Ames, Iowa, 1990.

[2] Barbier, C., and Barret, J. P., "An Analysis of Phenomena of Voltage Collapse on a Transmission System," *RGE,* special edition, CIGRE, July 1980, pp. 3–21.

[3] Campeanu, H. S., L'Helguen, E., Assef, Y., Vidal, N., and Savulescu, S. C., "Real-Time Stability Monitoring at Transelectrica," Paper PSCE06-1288, presented at the "Real-Time Stability Applications in Modern SCADA/EMS" Panel, IEEE PSCE 2006 Atlanta, GA, October 29–November 2, 2006.

[4] Canizares, C., and Alvarado, F., "Point of Collapse and Continuation Methods for Large AC/DC Systems," *IEEE Transactions on Power Systems,* Vol. 8, No. 1, February 1993.

[5] Crary, S. B., 1945, *Power System Stability,* General Electric Series, Schenectady, New York, Copyright 1945, Third Printing October 1955.

[6] Dimo, P., "Etude de la Stabilité Statique et du Réglage de Tension," R.G.E., Paris, Vol. 70, No. 11, pp. 552–556, 1961.

[7] Dimo, P., "L'Analyse des Réseaux d'Energie par la Méthode Nodale des Courants de Court-Circuit. L'Image des Nœuds," R.G.E., Paris, 1962, Vol. 7, pp., 151–175.

[8] Dimo, P., *Nodal Analysis of Power Systems,* Abacus Press, Kent, England, 1975.

[9] Dimo, P., Manolescu, G., Iordanescu, I., Groza, L., Ionescu, S., Albert, H., Moraite, G., and Ungureanu, B., *Computation and Design of Electrical Energy Systems* (Romanian edition), Editura Tehnica, Bucharest, Romania, 1971.

[10] Dobson, I., and L. W. Liu, "Immediate Change in Stability and Voltage Collapse when Generator Reactive Power Limits are Encountered," in *Proceedings of International Seminar on Bulk Power System Voltage Phenomena II,* edited by L. H. Fink, pp. 65–74, 1993.

[11] DyLiacco, T. E., Savulescu, S. C., and Ramarao, K. V., "An Online Topological Equivalent for a Power System," *IEEE Transactions on PAS,* Vol. PAS-97, No. 9, pp. 1550–1563, 1978.

[12] Erwin, S. R., Oatts, M. L., and Savulescu, S. C., "Predicting Steady-State Instability," *IEEE Computer Applications in Power,* July, pp. 15–22, 1994.

[13] Ionescu, S., and Ungureanu, B., "The Dual Power States and Voltage Collapse Phenomena," Rev. Roum. Sc. Tech., Série Electrotechnique et Energétique, Vol. 26, No. 4, pp. 545–562.

[14] Pomarleanu, M., "The Selection and Ranking of Power System Dangerous Cuts," presented at Symposium on Bulk Power System Dynamics and Control, IV—Restructuring, Santorini, 1998.

[15] Pomarleanu, M., and Bejuscu, L., "Selectarea Operativa a Sec.tiunilor Unui Sistem Electroenergetic cu Rezervele de Stabilitate Statica Cele Mai Mici," *Energetica,* Vol. 47, No. 2, 1999.

[16] Magnien, M., Rapport Spécial du Groupe 32 Conception et Fonctionnement des Réseaux, Conférence Internationale des Grands Réseaux Electriques à Haute Tension, CIGRE Session, 1964.

[17] Moraite, G., Ionescu, S., Feldmann, S., and Chenzbraun, I., "Problèmes Soulevés par la Stabilité Statique des Réseaux Bouclés," presented at Conférence Internationale des Grands Réseaux Electriques à Haute Tension, CIGRE Session, 1966.

[18] Oatts, M. L., Erwin, S. R., and Hart, J. L., 1990, "Application of the REI Equivalent for Operations Planning Analysis of Interchange Schedules," *IEEE Transactions on PAS,* Vol. PAS-109, No. 5, pp. 547–555, 1990.

[19] Sauer, P. W., and Pai, M. A., "Power System Steady-State Stability and the Load-Flow Jacobian," *IEEE Transactions in Power Systems,* Vol. 5 T-PWRS, No. 4, pp. 1374–1381, 1990.

[20] Sauer, W. P., and Pai, M. A., "Relationships between Power System Dynamic Equilibrium, Load-Flow, and Operating Point Stability," in *Real Time Stability in Power Systems,* pp. 1–30, Springer Verlag, Norwell, MA, 2006.

[21] Savulescu, S. C., 1981, "Equivalents for Security Analysis of Power Systems," *IEEE Transactions on PAS,* Vol. PAS-100, No. 5, pp. 2672–2682, 1981.

[22] Savulescu, S. C., "Fast Assessment of the Distance to Instability. Theory and Implementation," in *Real Time Stability in Power Systems,* pp. 31–64, Springer Verlag, Norwell, MA, 2006.

[23] Savulescu, S. C., Oatts, M. L., Pruitt, J. G., Williamson, F., and Adapa, R., "Fast Steady-State Stability Assessment for Real-Time and Operations Planning," *IEEE Transactions on Power Systems,* Vol. 8 T-PWRS, No. 4, pp. 1557–1569, 1993.

[24] Tinney, W. F., and Powell, W. I., 1977, "The REI Approach to Power Network Equivalents," presented at PICA'77 Conference, May 1977, Toronto, Canada.

[25]   Venikov, V. A., *Transient Processes in Electrical Power Systems,* Edited by V. A. Stroyev, English Translation, MIR Publishers, Moscow, 1977.

[26]   Venikov, V. A., Stroev, V. A., Idelchick, V. I., and Tarasov, V. I., "Estimation of Electrical Power System Steady-State Stability," *IEEE Transactions on PAS,* Vol. PAS-94, No. 3, pp. 1034–1041, 1975.

[27]   Vergara, J. S., Thai, T. A., Cuong, N. D., Nam, N. T., Campeanu, H. S., and Savulescu, S. C., "Accuracy Testing and Real-Time Implementation of Dimo's Stability Analysis Technique," in *Real Time Stability in Power Systems,* pp. 65–92, Springer Verlag, Norwell, MA.

[28]   Vickovic, D., Eichler, R., and Savulescu, S. C., "Real-Time System Stability Monitoring in the Transmission Network of Bosnia and Herzegovina," paper presented at the Power-Grid Europe Conference and Exhibition, Feria de Madrid, Madrid, Spain, 26–28 June 2007.

[29]   Vournas, C. D., Sauer, P. W., and Pai, M. A., "Relationships between Voltage and Angle Stability of Power Systems," *Electrical Power and Energy Systems,* Vol. 18, No. 8, pp. 493–500, 1996.

[30]   Wu, F. F., and Narasimhamurti, N., "Necessary Conditions for REI Reduction to be Exact," IEEE PES Winter Meeting 1979, Paper A 79 065-4, 1979.

[31]   EPRI, "Power System Steady-State Stability Monitor Prototype," Final Report EPRI TR-100799, July 1992; and "Power System Steady-State Stability Monitor," Final Report EPRI TR-103169, December 1993.

[32]   IEEE PES Task Force on Terms and Definitions, "Proposed Terms and Definitions for Power System Stability," *IEEE Transactions on PAS,* Vol. PAS-101, No. 7, 1982.

[33]   NERC, 1996, Available Transfer Capability Definitions and Determination, North American Electric Reliability Council, June 996.

[34]   CIGRE WG C4.6.01, "Review of Online Dynamic Security Assessment Tools and Techniques," Final Report, Draft 3, August 2006.

[35]   http://www.quickstab.com.

SIME: A COMPREHENSIVE APPROACH TO TRANSIENT STABILITY

Mania Pavella, Daniel Ruiz-Vega, and Mevludin Glavic

B.1 INTRODUCTION

Transient stability of electric power systems is a challenging issue both mathematically and physically: mathematically because it involves a strongly nonlinear and highly dimensional formulation; physically because the transient instability phenomena develop very quickly and may have disastrous consequences.

Transient stability encompasses two main aspects: analysis and control. Analysis, which in the old days was the main concern, is addressed today quite satisfactorily, thanks to impressive progress in computer performance and the development of ad hoc methods. More recently, control started becoming increasingly important and today is a main concern. Indeed, the deregulation of the electric industry increases the economical pressure and leads to intensified transactions, larger interconnections, shorter time horizons for power system monitoring, and operation closer to security limits. This makes transient stability control an extremely important and challenging issue, especially when market requirements must be considered.

The increased emphasis on transient stability explains the advances in transient stability methods along the years. To address analysis aspects, the equal-area criterion was proposed in the mid-1920s for the study of simple one-machine systems [14], whereas multimachine system analysis was tackled by time-domain numerical integration methods [24]. Later on, various "direct" methods started being developed: energy-type methods in the late 1940s and early 1950s [2,21], Lyapunov-like approaches in the mid-1960s (e.g., [8,25,26], and the many references therein) and, finally, the extended equal-area criterion proposed in the late 1980s, whereby multimachine system dynamics were studied through a suitable one-machine infinite bus (OMIB) system [40,41]. In spite of the extensive and innovative research efforts, however, the direct methods were not able to fully overcome some inherent difficulties.

During all these years, the time-domain numerical integration approach has persistently remained the universal method for transient stability analysis, thanks to its ability to handle any power system size and model, and any stability scenario. The spectacular advances in computing technology substantially increased the speed of the involved and quite complex computations and helped the time-domain method become (almost) compatible with online requirements. Yet, this method remains essentially an analysis technique, unable to properly address control issues.

Attempting to circumvent the above shortcomings has led to the genesis of the method called SIME, a hybrid temporal–direct transient stability method. SIME (for SIngle Machine Equivalent) initially aimed to combine the advantages of time-domain and direct methods. The initial outcome has thus been the preventive SIME, which uses time-domain simulations. Later on, emergency SIME techniques were developed, wherein the temporal information is provided by data originating in the real-time environment. Preventive and emergency SIME techniques are described in Sections B.3 and B.4, after the general formulation introduced below.

B.2   BASIC FORMULATION*

B.2.1   Fundamentals

The general SIME methodology relies on the fact that, no matter how complex, the mechanism of loss of synchronism originates from the irrevocable separation of the system machines into two groups: critical machines, which comprise the units responsible for the loss of synchronism; and noncritical generators, that is, the remaining machines.

Accordingly, when analyzing an unstable case (defined by the prefault system operating conditions and the contingency scenario), SIME starts with a search aimed at identifying the relevant critical machines as soon as the system enters its postfault phase. To this end, SIME gathers temporal information on the multimachine system dynamics and considers, at each new incoming set of temporal information, a few candidate decomposition patterns[†] and related candidate one-machine infinite bus (OMIB)

---

*This section revisits fundamentals of SIME and introduces the main notation. The illustrations are obtained from simulations described in subsequent sections. The material is mainly borrowed from [28,34]

[†]A *decomposition pattern* divides the system machines into two groups, separated by a (generally) "large" angular distance between adjacent machine angles; see Figures B2(a) and B2(b). For a given pattern, SIME aggregates the machines of each group into an equivalent machine, and replaces the multimachine system dynamics by the dynamics of the resulting two-machine equivalent and, further, by the dynamics of the corresponding OMIB equivalent.

equivalents, to which it applies the equal-area criterion (EAC).* SIME stops the process as soon as a candidate OMIB reaches the EAC *instability conditions* expressed by

$$P_a(t_u) = 0; \qquad \dot{P}_a(t_u) > 0 \tag{1}$$

and declares that this is the *relevant or actual OMIB; P$_a$* is its accelerating power ($P_a = P_m - P_e$) and t_u is the *time to instability,* that is, the time at which the system loses synchronism and its machines split irrevocably into the group of *critical machines* (CMs). The group of the remaining units are called the *noncritical machines* (NMs). Thus, at t_u SIME provides the critical twofold information:

- Identification of the CMs, responsible for the system loss of synchronism
- Assessment of the stability (negative) margin, defined by the closed-form equation,

$$\eta_u = A_{\text{dec}} - A_{\text{acc}} = -\tfrac{1}{2} M \omega_u^2 \tag{2}$$

Figure B-1 illustrates the principle of the above definitions. Part (a) refers to an unstable scenario and part (b) to a stable state governed by the *stability conditions,* where

$$\omega(t_r) = 0 \qquad \text{and} \qquad P_a(t_r) < 0 \tag{3}$$

Stable and unstable cases and corresponding parameters are further addressed in Section B.2.2, items 2 through 5. Time t_r denotes the "return" time, where the OMIB angle stops increasing and "returns back" (starts decreasing); at this time, the (positive) stability margin can be computed by

$$\eta_{st} = \int_{\delta_r}^{\delta_u} P_a d\delta. \tag{4}$$

Figures B.2 through B.4 illustrate instability and stability conditions for an actual stability case simulated on the EPRI 88-machine system described in Section B.3. More precisely:

- Figure B-2 depicts the multimachine swing curves provided by the ETMSP program, stopped respectively at the "time to instability," t_u [unstable case as shown in Figure B-2(a)], and at the maximum integration period [5 s, stable case, Figure B-2(b)]. The symbol P_C used in the legends denotes the total generation power of the group of CMs.

---

*An OMIB is defined by its angle δ, speed φ, mechanical power P_m, electrical power P_e, and inertia coefficient M. All these parameters are derived from the multimachine system parameters.

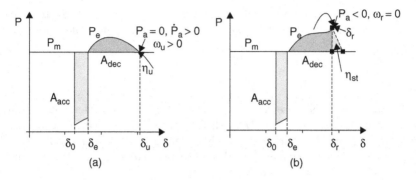

Figure B-1. Principle of SIME stability assessment via the EAC.

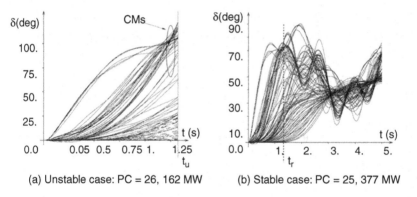

(a) Unstable case: PC = 26, 162 MW   (b) Stable case: PC = 25, 377 MW

Figure B-2. Multimachine swing curves, computed by ETMSP on a stability case applied to the EPRI 88-machine system. Critical clearing time: 69 ms; clearing time t_e = 95 ms. Adapted from [34].

- Figure B-3 portrays the equivalent OMIB swing curves.
- Figure B-4 represents, in the δ–P plane, the evolution with δ of powers P_e and P_m.* To simplify the representation, Figure B-4(b) only displays part of the curve P_e versus δ.

---

*$P_m(\delta)$ may not be a horizontal straight line, since it expresses a combination of the system machines' mechanical powers, which are not necessarily constant.

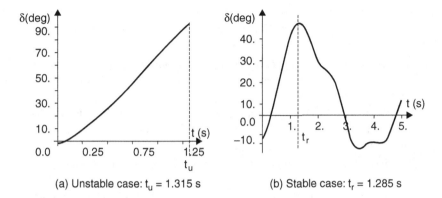

(a) Unstable case: t_u = 1.315 s   (b) Stable case: t_r = 1.285 s

<u>Figure B-3.</u> OMIB swing curves, equivalent to the multimachine curves of Figure B-2. Adapted from [34].

B.2.2   Salient Parameters and Properties

The salient parameters and properties of SIME methodology can be summarized as follows:

1. The OMIB dynamics is a faithful replica of multimachine system dynamics, since OMIB parameters are continuously refreshed by new incoming temporal multimachine system information. The SIME transient stability assessment is, thus, as accurate as the assessment provided by the multimachine temporal information. In addition, it is fully unbiased, since it relies on fully objective stopping criteria expressed by closed-form analytical expressions, whereas the assessment of multimachine temporal information uses pragmatic criteria.

2. The OMIBs are built based on unstable scenarios, since they are defined at the very instant (t_u) at which the system machines irrevocably split into CMs and

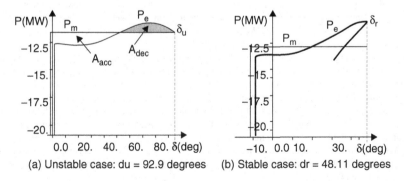

(a) Unstable case: du = 92.9 degrees   (b) Stable case: dr = 48.11 degrees

<u>Figure B-4.</u> OMIB-P curves, corresponding to the stability conditions of Figures 2 and 3. Adapted from [34].

NMs. Subsequently, however, the OMIB of a "borderline unstable case" can also be used for a "borderline stable case", that is, a stable case close to the "least unstable" one.

3. Thus, strictly speaking, only instability conditions and margins are analytically expressed and precisely computed by SIME, whereas stable conditions and margins are approximate.

4. The "time to instability," t_u, is a very important parameter. It indicates the very instant the system loses stability irrevocably, and is an effective measure of the instability "size".*

5. Similarly, the "time to first-swing stability," t_r, indicates the time at which the system is identified as first-swing stable. If multiswing instabilities are not of concern, it can be used as an early termination criterion for stable simulations.

6. The margin expressed by equation (2) is often "normalized" by the OMIB inertia coefficient. We will call it the "standard margin."

7. Under very unstable conditions,† it may happen that the standard margin does not exist (the OMIB P_m and P_e curves do not intersect; there is no postfault equilibrium solution). A convenient substitute is then the "minimum distance" between postfault P_m and P_e curves. Figure B-5 illustrates this substitute margin under the particularly stressed conditions described in Section B.3. Note that, here, the "time to instability" is the time at which this minimum distance is reached and the simulations are stopped. To simplify, we will still denote it "t_u."

8. A very interesting general property of the stability margins (standard as well as substitute ones) is their quasilinear variation with the stability conditions. Figures B-6 and B-7 illustrate the margin variation with the fault-clearing time and with the total generation power of the group of CMs, P_C. The SIME-based control techniques benefit greatly from this property.

9. "Seconds of time-domain (T-D) integration" (sTDI).‡ Generally, the computing effort required by SIME per se is virtually negligible as compared to the tasks re-

---

*Note the difference between t_u and critical clearing time (CCT): CCT is the maximum time that a contingency may remain without the system losing its capability to recover a synchronous operation, whereas t_u is the time at which the system irrevocably loses synchronism (that will happen if the contingency's clearing time is larger than CCT). Numerically, t_u is quite larger than CCT.

†Strictly speaking, a system could be either stable or not. However, the adjective "very unstable" is adequate since the SIME provides a stability margin that can be used to rank unstable cases according to their severity, which is a measure of the excess of critical machines' kinetic energy and, in preventive control, is directly related to the amount of active power change required to stabilize the system. Cases with small unstable margins require changing small amounts of power to be stabilized, and are thus good candidates to apply preventive control (security redispatch, see Section B.3.2); very unstable cases (which are referred to in the cited paragraph) do not have a standard margin, need a large amount of power to be stabilized, and are better controlled using OLEC (see Section B.3.3). Severity of the contingency is thus very important information required to decide whether preventive or emergency control is going to be used in preventive transient stability assessment and control.

‡This parameter is used when the temporal information provided to SIME comes from time-domain simulations.

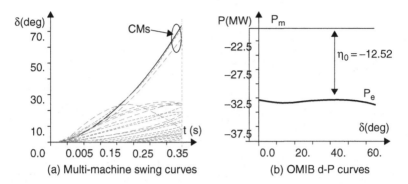

<u>Figure B-5.</u> Example of a severely stressed case: contingency 1b applied to the EPRI 88-machine system. Clearing time t_e = 95 ms; PC = 5600 MW, t_u = 395 ms. Adapted from [34].

quired by the temporal information. When the information used by SIME comes from T-D computations, sTDI is a handy "measure" of the overall computing effort of SIME-based techniques; it represents the time required by the T-D program to run the simulations and, thus, renders comparisons of computing performance (almost) independent of the computer used and of the system size.

10. It should be emphasized that the computations of SIME per se are performed online, during either the time-domain transient stability computations or the real-time datasets acquisition, as appropriate.

11. An objection made sometimes is: "What if the generators split into more than two groups"? The answer is quite straightforward: SIME relies on the fact that the loss of synchronism becomes irrevocable at t_u, that is, as soon as the ma-

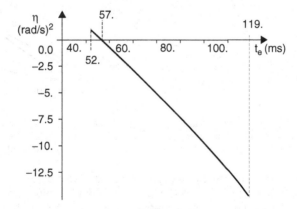

<u>Figure B-6.</u> Typical variation of the standard stability margin with the fault-clearing time t_e. Drawn on the EPRI 627 machine system. Adapted from [28].

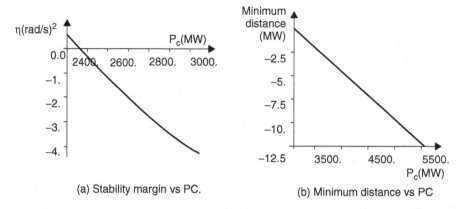

(a) Stability margin vs PC.   (b) Minimum distance vs PC

Figure B-7. Typical variation of the standard stability margin and of its substitute with PC. Drawn on the EPRI 88-machine system, contingency # 1b. Adapted from [34].

chines split irrevocably into two groups (see Section B.1). This does not imply that the machines of each group will not split further after t_u. An example is shown in Figure B-8, where an unstable simulation on the New England 39-machine IEEE test system apparently causes the machines to split into three coherent groups [Figure B-8(a)]. SIME objectively detects the system instability at $t_u = 0.6$ s where it also identifies the groups of critical and noncritical machines [Figure B-8(b)]. It must be noted that the group of noncritical machines

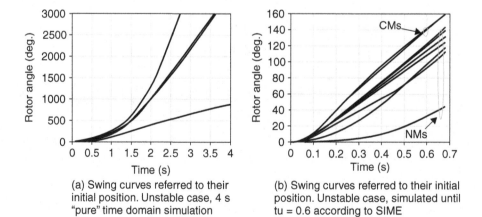

(a) Swing curves referred to their initial position. Unstable case, 4 s "pure" time domain simulation

(b) Swing curves referred to their initial position. Unstable case, simulated until tu = 0.6 according to SIME

Figure B-8. Comparison of the swing curves on an unstable case, with $t_e = 150$ ms. Machine angles are referred to their initial position. The simulations are drawn on the New England test system using detailed modeling (fourth-order synchronous machine model with IEEE Type 1 excitation control system. Speed governors are not included. Adapted from [36].

splits further into two groups, but this happens well after the time to instability (0.6 s). It is also worth mentioning that the critical group does not necessarily correspond to the largest angular separation (or "angular distance" between adjacent machine angles); for example, in the case of Figure B-8(b) it corresponds to the fifth "largest distance."

12. SIME may be seen as a generalization of the extended equal-area criterion (EEAC). Indeed, like EEAC, it relies on the OMIB transformation of the multimachine system temporal information. But, unlike EEAC, which uses "time-invariant" OMIBs by freezing their parameters for long time periods and using a classical simplified model for the multimachine system, SIME uses "time-varying" or "generalized" OMIBs, obtained by continuously refreshing their parameters and, hence, the parameters of the equal-area criterion, without requiring any modeling-simplifying assumption [27].

B.3 PREVENTIVE SIME

Preventive SIME, in general, encompasses a number of techniques that aim at enabling the power system to withstand any anticipated plausible contingency, should it occur. Basically, it drives a time-domain transient stability program and combines time-domain information about the multimachine system dynamics with the OMIB equal-area criterion to identify the critical machines and the stability margins. This paramount information paves the way to techniques capable of addressing various transient-stability issues such as analysis, including contingency filtering, ranking, and assessment; near-optimal preventive control; open-loop emergency control; and transient-oscillations damping assessment and control.

B.3.1 Transient Stability Analysis

In conventional transient stability studies, analysis aims at evaluating contingency severity in terms of stability limits. The stability limits of current concern are critical clearing times (CCTs) and power limits (PLs). As opposed to the CCTs, which are uniquely defined,* the PLs are expressed rather loosely in terms of the maximum power that the system may sustain without losing synchronism. Hence, the assessment depends upon a large number of factors, such as generation patterns of the system power plants, as further pointed out in remark 3, Section B.3.1.1.

CCTs and PLs are "measures" of the system robustness. The choice between them is a matter of operating practices, which in turn are linked to system specifics and physical limitations. Generally, CCTs are more popular in Europe, whereas PLs are preferred in the United States. In all cases, CCTs offer additional benefits: they are instrumental for contingency screening techniques and provide a handy reference for comparing the performance of various methods.

---

*CCT = maximum time that a contingency may remain without the system losing its capability to recover synchronous operation.

B.3.1.1 Critical Clearing Time Calculation

By definition, for the given operating conditions and contingency scenario, the stability limit (CCT or PL) corresponds to the zero-stability margin value. The limit search consists of performing successive transient-stability simulations of unstable scenarios with decreasing severity, and computing the corresponding margins until reaching the zero-margin value. A straightforward iterative procedure relying on item 8 in Section B.2.2 consists of linearly extrapolating the successive stability margins. Generally, the procedure requires a small number of transient stability simulations, about two to four. Note that whenever multiswing phenomena are considered, it is necessary to run a simulation that remains stable over the entire integration period. This is the most, and only, time-consuming T-D simulation required by SIME.

The lower right-hand part of Figure B-9 illustrates the search of the CCT in a stability case run on the fully modeled 94-machine Hydro-Québec power system* and shows that the first linear extrapolation between the two negative margins ($\eta_1 = -6.987$ and $\eta_2 = -3.976$) provides an initial, approximate CCT value.[†] As can be seen, the successive simulations are performed with decreasing clearing times, and the margins corresponding to two successive simulations are extrapolated or interpolated linearly. The simulations stop when the margin cancellation is obtained with sufficient accuracy. A more accurate value is further obtained by running a stable simulation with a clearing time smaller than the approximate CCT = 167 ms, and then by interpolating between the resulting margin ($\eta_3 = +1.67$) with the previous less negative margin ($\eta_2 = -3.976$). The CCT calculated with SIME is 175.5 ms, whereas the ST-600 program run alone gives a CCT (ST-600) = 176 ms. Thus, in this case, good accuracy required three stability simulations, of which one was stable, whereas two unstable simulations were enough for a good approximation of first-swing CCT.[‡] It is interesting to assess the computing effort for searching approximate and accurate CCTs, in terms of sTDI (see item 9 in Section B.2.2). As shown in Figure B-9:

- Approximate CCT computation: 478 + 522 ms = 1024 ms = 1.024 sTDI
- Accurate CCT computation: 1.024 + 15 = 16.024 sTDI

Again, it must be noted that the stable simulation took the lion's share. It is, therefore, interesting to avoid it via various "loopholes," like those suggested by the FILTRA scheme described in Section B.3.1.3. The reader might wish to compare these computing times with those required by a purely time-domain program to accomplish the corresponding tasks. In addition, please note that:

- The efficiency of a CCT search depends upon the proper choice of initial contingency-clearing time. Indeed, a too small fault-clearing time could yield a stable simulation that SIME cannot treat at this stage. Conversely, too large an initial

---

*In these simulations, SIME was coupled with the ST-600 transient-stability program of Hydro-Québec.
[†]In Figure B-9, the approximate CCT is 167 ms and the accurate one is 156 ms.
[‡]Recall, however, that, in general, unstable margins extrapolation does not guarantee absence of multiswing instabilities (whose CCT is smaller than the first-swing CCT, but whose time to instability is larger).

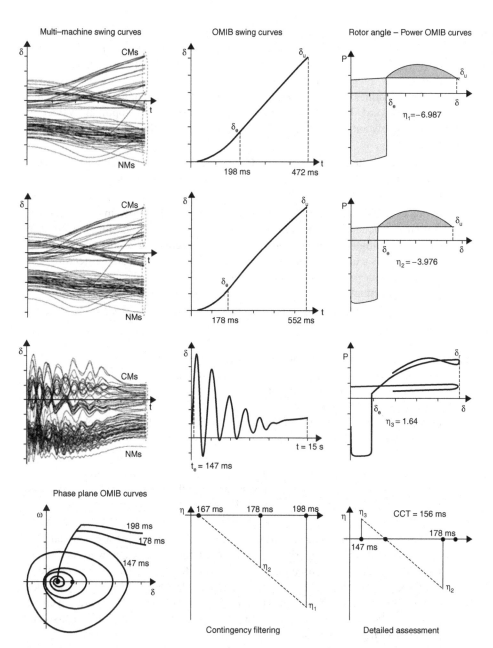

Figure B-9. Accurate and approximate SIME computations of critical clearing times. Also shown are multimachine swing curves and SIME's three basic representations of transient stability phenomena: OMIB swing curves, P-δ curves, and phase-plane curves. The simulations were performed on the Hydro-Québec 94-machine system. Adapted from [12].

fault-clearing time would yield an unduly unstable simulation, requiring additional calculations to recover the "interesting" range of fault-clearing times. The difficulty is circumvented thanks to pragmatic criteria that automatically stop the during-fault T-D simulation and switch to the postfault one [28]. Note that the numerical values of these criteria are determined offline, but their tuning is quite straightforward. Note also that the range of variation of "optimal" values is rather narrow and independent of the system size (whether having 3 or 600 machines).

- It may happen that successive simulations yield different groups of CMs. This makes impossible the comparison of their corresponding margins. But this question is easily addressed via appropriate algorithmic adjustments [28].

- In practice, calculating the CCT of a given contingency is much more straightforward than calculating the power limit. Indeed, the problem of calculating the CCT consists of simulating successive scenarios with decreasing contingency-clearing times and has a unique solution,* whereas a power limit calculation requires exploring various patterns of critical and noncritical machines' generation, and may lead to a potentially (very) large number of solutions (see Section B.3.2).

- The above iterative procedure is valid for computing first-swing as well as multi-swing CCTs.

B.3.1.2 OMIB Representations of System Dynamics

A useful SIME by-product is provided by the three OMIB-based representations of the multimachine system dynamics illustrated in Figure B-9 and defined as follows:

- *OMIB swing curve.* Provides a synthetic description of the multimachine system swing curves that compresses into a single-swing curve [compression rate: n (the number of system machines) to 1].†

- *OMIB power-angle representation.* Another type of rich interpretation or exploration of the multimachine system behavior. It is the basis of SIME-based stabilization procedures and emergency SIME

- *OMIB phase-plane representation.* The plot on the phase plane (δ, ω) of OMIB trajectories corresponding to various operating conditions‡ of a given stability case. It provides a particularly compact way of describing the power system dynamics, by reducing the problem dimensionality from N (the total number of dynamic state variables) to 2 (compression rate: $N/2$). Observe how nicely are described the dynamic phenomena (which, as shown in Figure B-12, are governed by over 1000 state variables), and how clearly can be described the back-swing and multiswing phenomena.

---

*Apart from the special, and rare, case of multiple solutions due to the nonlinear character of transient stability.

†The development of a Prony-based approach to online transient oscillations damping assessment and control described in Section B.3.4 takes advantage of this representation.

‡For example, various contingency clearing times, as in Figure B-9, or various prefault generation patterns, as shown in Figure B-12.

Of course, the above OMIB-based representations are complemented with the traditional multimachine swing curves, provided by the SIME driven time-domain program and stopped at t_u or t_r, as appropriate.

B.3.1.3 Contingency Filtering, Ranking, and Assessment (FILTRA)

Although effective, the SIME-based CCT computation procedure of Section B.3.1.1 may become prohibitive for very large numbers of contingencies, as may be the case in practice. Indeed, in the context of preventive control, it is important to assess all "plausible" contingencies, so as to avoid missing a dangerous one whose occurrence would threaten the stability of the system. On the other hand, for a properly designed power system, most of the considered contingencies are "uninteresting," that is, harmless.

This suggests that it is important to develop contingency filtering techniques able to screen out the "uninteresting" contingencies, while ranking the remaining "potentially interesting" ones. The filtering-ranking phase should be followed by an assessment phase to scrutinize the truly severe contingencies. SIME provides effective filtering techniques, thanks to its rich information and the property of quasilinear variation of margins with stability conditions. This allows getting a first-guess CCT via two negative margin extrapolations (see Section B.3.1.1). Using this property together with customized adjustments to power systems specifics yields a large variety of FILTRA techniques developed in [12,28].

Figure B-10 illustrates the general organization of a two-block FILTRA scheme that classifies a contingency in one of the following four classes:

1. *Harmful* (*H*) if the contingency drives the system to instability, that is, if its CCT is smaller than the clearing time (the time response of system protections).
2. *Potentially Harmful* (*PH*) if it is "almost" harmful, that is, milder than harmful, but likely to become harmful under slightly modified operating conditions.
3. *First-Swing Unstable* (*FSU*) or *Stable* (*FSS*) if, under the given clearing scenario, it drives the system to first-swing instability or stability.
4. *Harmless* (*HS*), if it is FSU but neither Harmful nor Potentially Harmful.

B.3.2 Near-Optimal Transient-Stability Control

This section addresses how transient-stability control seeks countermeasures that enable a system to withstand otherwise "harmful" contingencies by modifying its prefault operating conditions. Simply stated, such control countermeasures, or actions, "stabilize harmful contingencies."

Section B.3.2.1 describes several SIME-based, purely preventive control techniques that rely on rescheduling the generated powers by shifting generation from CMs to NMs.* This goal may be achieved in many ways, thus offering the choice among numerous stabilization patterns. When there are many harmful contingencies, one may

---

*In the case of up-swing instabilities. For back-swing instabilities, generation power is shifted from NMs to CMs.

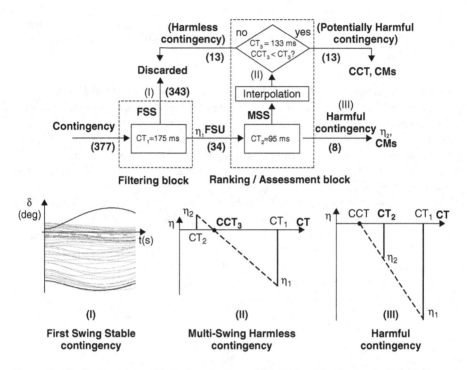

<u>Figure B-10.</u> General two-block organization of FILTRA and schematic description of the various classes of contingencies. Illustration based on the Hydro-Québec 94-machine power system. Adapted from [12].

wish to stabilize them simultaneously. Both individual and simultaneous contingency stabilizations are considered.

The control relying on a SIME-driven OPF algorithm provides near-optimal solutions with respect to selected objectives and is further described in Section B.3.2.2. An open-loop emergency control designed offline and applied online if the anticipated contingency actually occurs is also proposed, as shown in Section B.3.3. Finally, Section B.3.4 describes SIME-based techniques for transient oscillation damping assessment and control.

B.3.2.1 Preventive Contingency Stabilization

B.3.2.1.1 SINGLE-CONTINGENCY STABILIZATION. Physical reasoning suggests that in order to stabilize a multimachine system subject to a harmful contingency, the strategy would be to "pull" the more advanced machines "closer" to the less advanced ones. However, this "qualitative" reasoning does not provide clear quantitative answers to the important questions: (a) "which" machines to "pull," and (b) "by how much?"

SIME addresses these issues in a systematic way, through the following reasoning and ensuing iterative procedure:

- To stabilize an unstable case, cancel its (negative) margin.
- According to EAC, to cancel the margin decrease the accelerating area and/or increase the decelerating area in the OMIB $\delta - P$ plane; for example, by reducing the OMIB mechanical power by ΔP_{OMIB}, as shown in Figure B-1.
- To get a first-guess ΔP_{OMIB}, use the compensation scheme proposed in [28], which expresses it in terms of the margin to cancel, $\Delta P_{\text{OMIB}} = f(\eta)$
- To "transform" ΔP_{OMIB} to multimachine system parameters, use the following expression (which neglects the transmission losses):

$$\Delta P_{\text{OMIB}} = \Delta P_C = \sum_{i \in CMs} \Delta P_{C_i} = -\Delta P_N = -\sum_{j \in NMs} \Delta P_{N_j} \quad (5)$$

where ΔP_{C_i} denotes the change in mechanical (active) power of the ith CM, ΔP_C the total generation change in CMs, and ΔP_N the total generation change in NMs.

- Using the above first guess* $\Delta P_C = \Delta P_{C_0}$ together with the initial, prefault CMs generation, P_{C_0}, initialize the iterative process sketched in Figure B-11 to get the actual generation decrease in CMs.[†]
- Distribute ΔP_C among the system's CMs and the corresponding increase among NMs according to a selected generation pattern, as discussed in the following section.

B.3.2.1.2 GENERATION RESCHEDULING PATTERNS. Equation (5) suggests that there are many degrees of freedom for distributing ΔP_N among NMs; and whenever there are many CMs, there are many degrees of freedom and corresponding patterns for distributing ΔP_C. From the large number of possible patterns for distributing ΔP_C, we mention the following four:

1. Pattern #1: ΔP_C distributed on CMs proportionally to the product $d_i \times B_i$; here d_i denotes the "distance" of the angle of the ith CM with respect to the angle of the most advanced NM, and M_i its inertia coefficient
2. Pattern #2: ΔP_C distributed among some of the CMs
3. Pattern #3: ΔP_C distributed on all CMs according to their inertia coefficients
4. Pattern #4: ΔP_C equally distributed on all CMs

---

*A sound first guess ΔP_{C_0} is often suggested by the power system conditions.

[†]The iterative procedure illustrated in Fig. B-11 is summarized in the following steps. For a given negative margin η_0:

- decrease = P_{C_0} by ΔP_{C_0} to get $P_{C_1} = P_{C_0} - \Delta P_{C_0}$.
- decrease the active power of CMs by ΔP_{C_0} and increase by the same amount the active power of the NMs.
- perform successively a power flow to compute the new operating conditions and a stability run to compute the corresponding stability margin, η_1.
- perform a linear extrapolation or interpolation, as appropriate, to get a first-guess power limit $P_{C\text{Lim}}$.
- compute the new power decrease ΔP_{C_1} using $P_{C\text{Lim}}$ and repeat steps (ii) and (iii) using the new ΔP_{C_1} value, and if $\eta_2 = 0$, stop; otherwise, adjust as appropriate and continue the process.

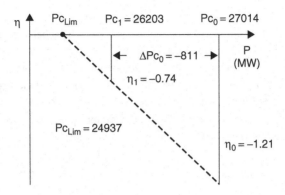

<u>Figure B-11.</u> First steps of the stabilization procedure applied to contingency # 30 simulated on the EPRI 88-machine system. (ΔP_C is equally distributed on all CMs.) Adapted from [34].

From a practical point of view, patterns #1, #3, and #4 are difficult to implement, at least when the number of CMs is large, but pattern #2 appears to be acceptable.

Concerning the distribution of generation among NMs, the patterns may be dictated by various constraints, related to market or electric considerations; for example, one may seek maximum profit or maximum transfer capability on a given corridor. However, some of the resulting solutions may conflict with stability and, hence, be unacceptable. For example, think of the case in which the total power increase is reported on the most advanced NM which may become critical and cause loss of synchronism.

In the absence of particular constraints or objectives, the total generated power could be distributed proportionally to the inertia coefficients of the NMs. Another, and more effective, solution consists of using an OPF program, as described in Section B.3.2.2.

B.3.2.1.3 ILLUSTRATIONS AND OBSERVATIONS. We illustrate the stabilization procedure described above on the EPRI 88-machine system (434 buses, 2357 lines). The base-case prefault operating condition has a total generation of 350,749 MW. The contingency considered is a three-phase short circuit applied at a 500 kV bus and cleared by 95 ms after its occurrence by tripping one line. Since the CCT is 66 ms, this leads to loss of synchronism, with 36 CMs going apart from the remaining 52 NMs.

Figures B-2 and B-3 in Section B.2 illustrate interesting aspects of this case. In particular, they suggest that this is a moderate instability, since its stabilization requires a small decrease in the initial generation of CMs (from 26,162 MW to 25,377 MW, that is, about 1.44% of the initial value).* Actually, this is consistent with the general observation that stabilization often requires just a small percent reduction of the active power generated by CMs, except for some extreme operating conditions and unusually severe contingencies. Such an extreme case is displayed in Figure B-12.

---

*Incidentally, only two iterations were sufficient for the iterative procedure.

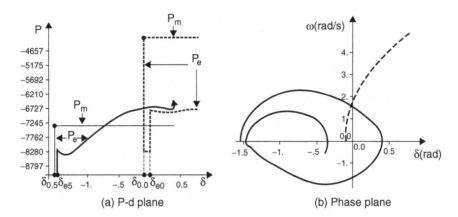

Figure B-12. Preventive control of a very severe contingency, applied on the EPRI 88-machine system. Illustration of the prefault initial conditions and of the last (fifth) iteration of the iterative procedure. Dotted lines: total plant prefault power of the unstable case Pc_0 = 5600 MW. Solid lines: total plant prefault power of the stabilized case Pc_5 = 2423 MW. Total seconds of time-domain integration required to stabilize the case: sTDI = 8.385s, including a 5 s stable simulation. Adapted from [34].

Figure B-12 depicts a simulation performed with the same EPRI 88-machine system, except that, here, the three-phase short circuit is cleared by opening two lines, but still 95 ms after its occurrence. In this case, there are seven CMs, initially generating 5600 MW. Stabilizing this case would require a CM generation decrease of 5600 − 2423 = 3177 MW, that is, more than 50% of the initial CM generation; most probably, this countermeasure would be deemed too expensive to apply preventively. An interesting alternative solution is presented in Section B.3.3.

B.3.2.1.4 MULTICONTINGENCY SIMULTANEOUS STABILIZATION. The single-contingency stabilization procedure described above may easily be extended to the simultaneous stabilization of a set (or all) of harmful contingencies. The algorithm consists of focusing on the most harmful contingencies, as described in the following:

1. For a given harmful contingency: compute the decrease in generation of the group of CMs, following the procedure of Section B.3.2.1.1, if this group is composed of a single CM, impose this decrease on that machine; if it is composed of many CMs, distribute this decrease according to above-mentioned pattern #2.
2. Proceed similarly with all harmful contingencies.
3. For each CM, choose the maximum generation decrease among those computed in steps 1 and 2.
4. Compute the total generation decrease obtained for all CMs, and compensate by an equal generation increase in NMs.

5. Run a power-flow program, followed by SIME to assess the new margin of all contingencies, and decide whether to iterate or not.

The above description suggests that stabilizing all the contingencies simultaneously is as straightforward as stabilizing them individually, and (almost) equally expensive. Extensive simulations with a large number of power systems and stability cases corroborate this observation.

The second observation is that, generally, the number of sound patterns is very large, which provides many possibilities for the proper choice of CMs.

The third observation is that there is a significantly large number of degrees of freedom for the choice of NMs. Different solutions may, therefore, be appraised in order to achieve preassigned objectives. The following section advocates the use of an OPF algorithm with various objective functions, as, for example, maximizing transfer power on a given transmission corridor (cutset) connecting two areas.

B.3.2.2   OPF for Near-Optimal Transient-Stability Control

TRANSIENT-STABILITY-CONSTRAINED (TSC) OPF TECHNIQUES. Several attempts have been made to imbed transient-stability constraints within OPF. They yielded two main approaches, depending upon how they handle these constraints. In the following, we refer to them as "global" and "sequential," and outline their principle.

- *Global approach.* For each time step of the T-D simulation to stabilize, the power system transient-stability model is converted into an algebraic set of equations. The set of nonlinear algebraic equations resulting from the whole T-D simulation is then included in the OPF as a stability constraint, forming a (generally huge) single nonlinear programming problem [18,20,6]. Hence, the name given to this approach, as opposed to the following one.

- *Sequential approach.* The transient-stability constraints are directly converted into the conventional constraints of a standard OPF program, for example, transient-stability constraints are expressed as new limits of active generation power. This approach does not affect the size of the power system model or the complexity of the OPF solution method and can be used in conjunction with any conventional OPF program [33,37,13,22,3,34].

Reference [34] discusses extensively the pros and cons of the two approaches, then selects and develops the SIME-based sequential approach described below.

SIME-BASED SEQUENTIAL APPROACH. In principle, the SIME-based TSC techniques may comply with either of the above approaches.* So far, however, the sequential method has prevailed. Note that if in theory this approach cannot guarantee optimality,

---

*In particular, these techniques could replace advantageously some of the stability criteria used within the global approach.

in practice it provides interesting near-optimal solutions, thanks to the transparency and multifaceted information conveyed by these TSC techniques. Besides, the sequential approach may easily comply with market requirements for the choice of CMs and NMs on which to redispatch generation.

Furthermore, the method can be used in conjunction with any standard OPF algorithm, avoiding embedding dynamic variables. Finally, the size and the complexity of the TSC–OPF problem are comparable with those of the standard OPF and the SIME-based TSC problems.

Figure B-13 illustrates the principle of the sequential approach within the integrated online TSA&C software.

INTEGRATED TSC–OPF SOFTWARE. The major computational steps of the Integrated TSC–OPF software are as follows:

- The state estimator provides the data to OPF, which sets up an operating state that achieves a predefined objective while meeting all the static security constraints.
- Transient stability of this optimized operating state is assessed with respect to an initial list of contingencies by the following functions: the contingency filtering ranking and assessment (FILTRA) function, block 1; and the transient-stability assessment and control function, composed of blocks 2 to 4.
- FILTRA discards the "uninteresting" contingencies (i.e., most of the contingencies of a generally very large initial contingency list) and sends the harmful ones to the transient-stability assessment and control block, along with the corresponding margins and CMs.
- Using this information, the transient-stability control function (block 2) determines the corresponding control actions necessary to stabilize the set of harmful contingencies simultaneously, following the procedure of Section B.3.2.1.4.
- The generation rescheduling on NMs is then performed using a standard OPF program (block 3), which can also meet additional technical or economical objective functions.
- The stability of the new operating state is assessed in block 4, for each one of the initially harmful contingencies. The state is declared to be stable if it is stable for all contingencies, and the process stops. If not, the overall cycle is repeated until all harmful contingencies have been stabilized. Generally, 2 to 3 iterations are sufficient.

In addition, it must be noted that:

- In the above procedure, SIME is in charge of generation shifting from each CM, whereas OPF decides how to redispatch the total generation shift among NMs. One could as well think of OPF taking care of the CM's generation shift within rules predetermined by SIME.

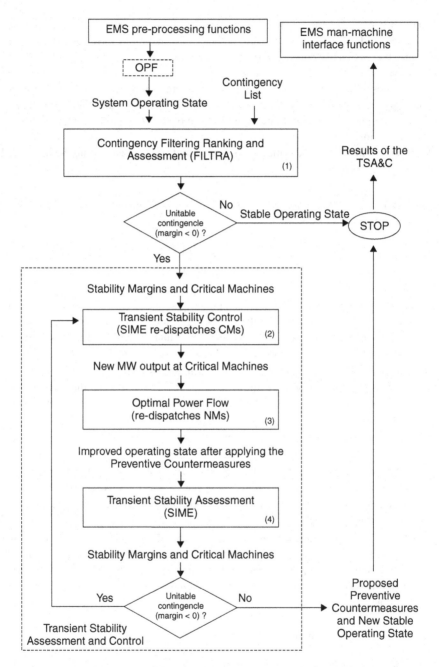

Figure B-13. Integrated TSA–OPF software. Adapted from [36].

- The approach is very flexible, and may match different market rules. In particular, it may help the system operator solve congestion management problems in the balancing market, an hour or half an hour ahead of real-time operation.

Extensive simulations carried out on various power systems, such as the Mexican interconnected system, corroborate the efficiency of this TSC–OPF software [36,34].

B.3.3 Open-Loop Emergency Control (OLEC)

B.3.3.1 OLEC Scope and Principle

In order to stabilize particularly severe contingencies, preventive control may require expensive actions that the system operator may be reluctant to trigger for containing postulated contingencies which, most probably, would not occur. The transparency prevailing in the deregulated electric industry renders such control even more burdensome.

Limiting to the minimum the preventive control actions fosters a more extensive use of emergency control. Here, the countermeasures are automatically triggered *after* a contingency has actually occurred and, possibly, been cleared by appropriate protection schemes. Two SIME-based emergency control techniques have been elaborated during the last years: a purely emergency control, referred to as Emergency SIME, as described in Section B.4; and a mixed preventive–emergency control, referred to as "open-loop emergency control" (OLEC), which is addressed in the following.

In order to control extremely severe contingencies that may require excessive preventive countermeasures, OLEC advocates *mitigating* them with emergency countermeasures, triggered only if the postulated contingency actually occurs. This is realized by performing *partial preventive generation rescheduling,* under the assumption that this partial control will be complemented with *real-time emergency generation tripping,* should the contingency occur. The principle of the resulting procedure is summarized as follows [36,35]:

1. For an anticipated harmful contingency, run the preventive SIME to determine the corresponding (negative) margin and CMs.
2. If many CMs are equipped with a generation tripping scheme, determine the maximum and minimum number of units available for tripping.
3. Simulate generation tripping, assumed to be triggered after a delay of, say, 150 ms from the contingency inception*; under these new conditions, run again the preventive SIME to compute the new margin and related generation shifting from the remaining CMs to NMs. Repeat the procedure with all possible number of CMs to be tripped.
4. Run repeatedly the preventive control function to determine the best compromise between preventive and emergency countermeasures.

*Of course, the earlier the tripping action, the better the efficiency of the approach.

It must be noted that:

- Whenever necessary, that is, if generation tripping is too large or if generation rescheduling is too difficult to implement,* compute in the preventive mode the amount of load that should be automatically shed in the emergency state [39].
- OLEC is an accurate and flexible technique, thanks to the knowledge of CMs and stability margins. Another important advantage is the ability to assess and simulate the effect of realistic emergency control actions using detailed representation of the power system.

B.3.3.2 Illustration

We consider again the case of seven CMs, initially generating 5600 MW and 81 NMs, and the severe contingency consisting of the three-phase short circuit applied at a 500 kV bus and cleared 95 ms after its occurrence by opening two lines. To stabilize the system in the purely preventive mode, it was found that it is necessary to shift from CMs to NMs 3177 MW, thus bringing the prefault generation of the seven CMs down to 2423 MW. The OLEC procedure was next performed under the following conditions [35]:

- Two CMs generating 1590 MW will be tripped 150 ms after the contingency occurrence, *if* it occurs.
- The preventive control is, therefore, applied to the remaining five CMs, initially generating 4010 MW.

Figure B-14(a) depicts the dynamics of the system before tripping the two CMs, whereas Figure B-14(b) displays the dynamics at the end of the preventive stabilization procedure performed with the remaining five CMs involving seven successive simulations. The legend of Figure B-14(b) indicates that OLEC requires the total prefault generation of the seven CMs to be 4005 MW, which corresponds to a total decrease on the seven CMs of 1595 MW. Obviously, the 1595 MW decrease is significantly smaller than the 3177 MW reduction imposed by the purely preventive mode.

Figure B-15 shows another interesting illustration of the rescheduling process of the five remaining machines; observe the gradual improvement of the power system stability as the prefault generation of the five CMs gradually decreases from 4010 MW to 2415 MW.

Reference [35] describes in detail how various salient parameters, such as triggering time delay, number of generators to trip, and so on, affect the OLEC.

*Generation tripping always causes a power imbalance between generation and load. If the amount of power to be shed or the number of machines to be tripped is small, generating reserves are able to compensate for this imbalance. In the case of a larger imbalance due to the tripping of a more important number of machines, the system could become unstable, as a consequence of this imbalance, by frequency instability. In these cases, generation tripping must be combined with load-shedding schemes, in order to avoid frequency instability. On the other hand, If generation rescheduling is difficult to perform, the excess of kinetic energy causing transient instability could be eliminated by shedding loads in selected areas in order to stabilize the system.

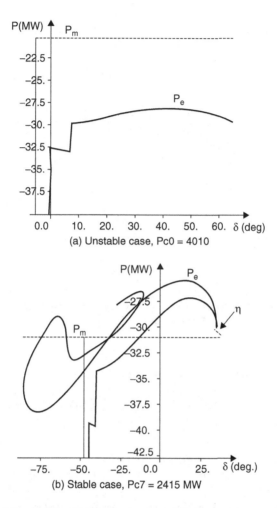

Figure B-14. Stabilization by OLEC: illustration of the two extreme simulations (iterations # 0 and 7). Total plant prefault power P_{c0} = 5600 MW (a) Unstable case. (b) Stabilized case, for a total plant prefault power of 4003 MW. Total seconds of time-domain
integration required to stabilize the case: sTDI = 10.377 s, including a 5 s stable simulation. Adapted from [36].

B.3.4 Transient Oscillations Damping Assessment and Control

The online analysis and control of power system transient-oscillations damping is another interesting application of SIME. The issue is of great concern, especially in re-
structured electricity markets in which extended interconnections and MW transfers
that push the grid to its limits have increased the complexity of system operations.

Damping, in general, is often analyzed by Prony techniques. Single- and multichan-
nel Prony algorithms explore the system damping, considering one or several genera-

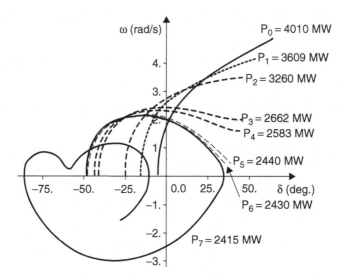

<u>Figure B-15.</u> OMIB phase plane representation of the seven simulations performed by OLEC, under the assumption that two generators would automatically be tripped if the contingency actually occurs. Adapted from [35].

tor angle curves at a time. The accuracy of Prony calculations strongly depends on the proper choice of the generator curves [4,17], but this choice may become problematic for systems composed of a large number of machines. SIME overcomes easily this difficulty, hence, the idea to combine it with Prony methods.

B.3.4.1 Oscillations Damping Assessment

Obviously, Prony analysis can benefit greatly from the key SIME outcomes, namely:

- Compression of the multimachine swing curves into an OMIB
- Identification of the CMs along with their respective relevance

As usual, the "relevance" of a CM is assessed in terms of its angle (absolute or relative to a reference, for example, to the angle of the most advanced noncritical machine), evaluated at t_u, the time to instability. Note that, by construction, SIME determines the OMIB and CMs on unstable simulations, whereas, generally, Prony analysis assesses damping characteristics in a stable simulation. However, observation 2 in Section B.2.2 allows reconciling this apparent incompatibility by considering the CMs of a "borderline unstable case" to be the "relevant machines" (RMs) for a "borderline stable case." Knowledge of the RMs along with their relevance allows Prony analysis to assess the impact (sensitivity) of the (or of some selected) RM(s) on the overall power system damping.

The above considerations lead to the following procedure:

1. For a given contingency, run SIME to compute its critical clearing time (CCT) or power limit (PL) as appropriate. This computation requires two or three unstable simulations and one stable simulation.
2. For the least unstable simulation above, identify the CMs and corresponding OMIB; further, determine the relevance of each one of the CMs.
3. On the stable simulation used in step 1, perform Prony analysis to assess damping of the OMIB swing curve.

B.3.4.2 Oscillations Damping Control

Studies of oscillation damping following large disturbances are generally restricted to analysis aspects, probably because of the difficulty in identifying CMs, along with their degree of influence. The SIME-based control techniques described so far can readily be extended to stabilize unstable oscillation damping, or to improve stable ones; the control of transient undamped oscillations simply consists of "overstabilizing" the contingencies. The following iterative procedure derives readily:

1. Combine SIME and Prony programs to assess the transient oscillation damping of the case of concern.
2. If damping is found to be insufficient (damping ratio very small or even negative), use the information about the CMs provided in the last unstable simulation in order to redispatch generation, generally by shifting a very small amount of power from critical to noncritical machines.
3. After redispatching, run a power flow (or better, an OPF) in order to find the new operating state of the system, and perform a new simulation with SIME.
4. If the system is stable, use the CMs identified in the last unstable simulation to compute the OMIB swing curve and the new damping of the system. If this meets the damping performance criteria stop; otherwise go to step (2).

B.3.4.3 Integrated Damping Assessment and Control Function

Figure B-16 displays the general organization of an integrated scheme. It makes Prony analysis a systematic technique with negligible computing requirements as compared to time-domain ones.

B.3.4.4 Illustrations

Various aspects of transient oscillations damping assessment and control functions are described below, mainly by means of figures. Numerical results and more detailed account of these simulations may be found in [32].

PLANT MODE OSCILLATIONS DAMPING ASSESSMENT AND CONTROL. A plant mode transient-oscillation case was analyzed on the EPRI 627-machine test system after applying a three-phase short circuit. The case is analyzed and controlled by SIME-based

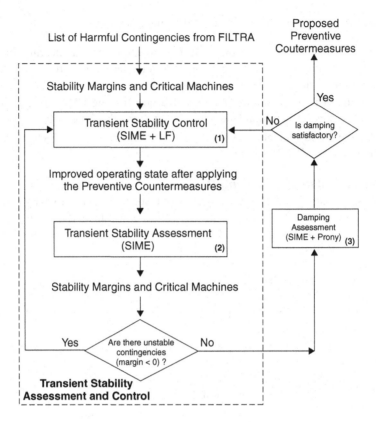

Figure B-16. Integrated SIME-based transient stability and damping assessment and control online function. Adapted from [36].

TSA&C and Prony software. The SIME software was coupled with the EPRI ETMSP time-domain program [9]. Prony analysis was performed using the spectrum function of the EPRI Output Analysis Program (OAP).

Figure B-17 displays simulation results produced by SIME. Observe that the CMs identified in the unstable simulation, labeled units 2075 and 2074, are easily found to be relevant machines in the stable simulation, too.* Observe also that, in this case, SIME does not really help Prony for the purpose of analysis since the CMs are easily identifiable from the multimachine swing curves, but it becomes very efficient for the control of oscillations damping.

In accordance with the damping assessment procedure, the Prony analysis was executed on the OMIB swing curve of the stable case [Figure B-17(d)]. In addition, as a matter of testing the validity of the simulation, Prony analysis was also performed on the two relevant generators, units 2075 and 2074.

---

*According to the legends of these figures, stabilizing the system requires a small decrease in the CMs' generation (from 2204 MW to 2128 MW).

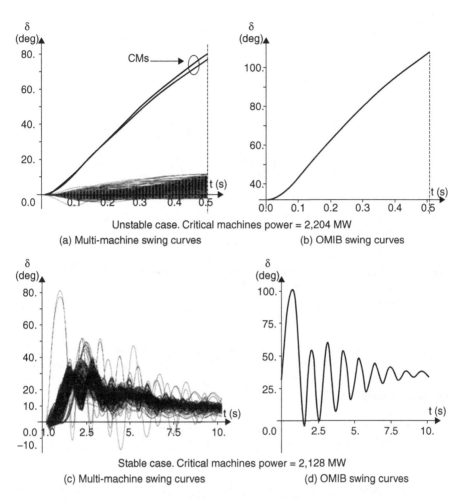

Unstable case. Critical machines power = 2,204 MW

(a) Multi-machine swing curves   (b) OMIB swing curves

Stable case. Critical machines power = 2,128 MW

(c) Multi-machine swing curves   (d) OMIB swing curves

Figure B-17. Unstable and stable cases simulated on the EPRI 627-machine test system. (Contingency clearing time of both cases: 67 ms.) Adapted from [36,32].

Figure B-18 displays the OMIB and relevant machines' swing curves. Observe that the frequency of the three curves is almost the same, whereas the value of the damping ratio of the OMIB is in between that of machines 2075 and 2074.

The actual damping ratio was found to be 8.23%. SIME easily calculated the damping ratio increase in terms of the CMs' generation decrease, as shown in Figure B-19.

Figure B-20 displays another interesting description of the damping variation with CMs' generation on the OMIB phase plane.

INTERAREA OSCILLATION DAMPING ASSESSMENT AND CONTROL. The simulations described here concern the interarea oscillatory instability mode and use the EPRI 88-

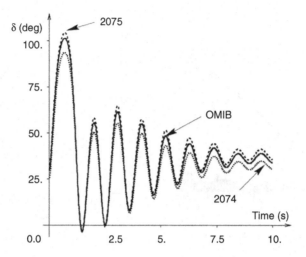

Figure B-18. Comparison of the swing curves of the OMIB and the relevant machines (2075 and 2074) of the stable case simulated on the EPRI 627-machine test system. Adapted from [32].

machine system described earlier. As a reminder, 36 CMs in this case are separated from the remaining 52 NMs. Figure B-21 shows the swing curves of the unstable case and of the stabilized case explored on a maximum integration period of 10 s that were used for Prony analysis.

The graph shown in Figure B-22 was drawn for an "optimal window" (ranging from 4.18 to 9.78 s) and illustrates the accuracy of Prony damping assessment: the

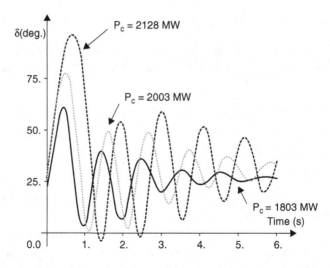

Figure B-19. OMIB swing curves for selected simulations. Adapted from [36,32].

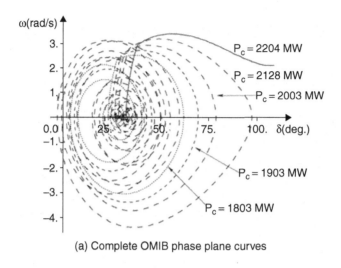

(a) Complete OMIB phase plane curves

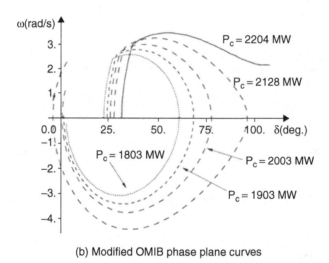

(b) Modified OMIB phase plane curves

Figure B-20. OMIB phase plane curves. Adapted from [36].

OMIB swing curve fits (almost) perfectly with the curve computed by Prony analysis in the "optimal window." It also suggests the imminence of instability. Actually, instability arises at 11.4 s, as shown by the curves illustrated in Figure B-23, which are from a simulation conducted under the same conditions as in Figures 21(c) and 21(d) but for a larger integration period, which allows shedding more light into these seeming peculiarities [32]. Figure B-23 displays the OMIB swing curves of two extreme cases, one "fully unstable" (having the largest negative damping) and the other one "fully stabilized," drawn for a maximum integration period of 15 s.

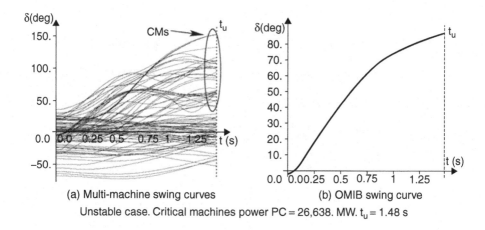

(a) Multi-machine swing curves   (b) OMIB swing curve

Unstable case. Critical machines power PC = 26,638. MW. t_u = 1.48 s

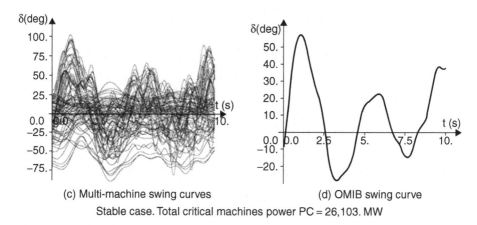

(c) Multi-machine swing curves   (d) OMIB swing curve

Stable case. Total critical machines power PC = 26,103. MW

Figure B-21. Unstable and stable cases simulated on the 88-machine system. Adapted from [32].

B.4 EMERGENCY SIME

B.4.1 Emergency SIME in Brief

In contrast to preventive SIME, which aims at reinforcing the power system's stability on the assumption that contingencies might occur, Emergency SIME aims at designing and triggering control actions after a contingency has actually occurred but still early enough to avoid loss of synchronism that would otherwise by caused by this contingency.

More precisely, emergency SIME (E-SIME) is a real-time, closed-loop emergency control technique, which aims at containing, in real time, the loss of synchronism of a system subject to dangerous contingencies. Given that the time to instability of severe

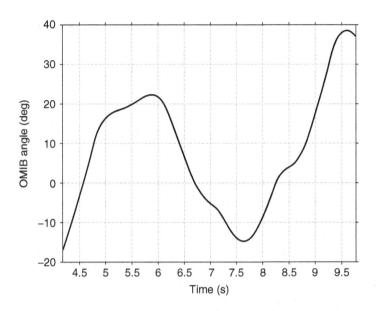

<u>Figure B-22.</u> Comparing the original OMIB swing curve with the one estimated by Prony analysis, for the "stable" case simulated on the 88-machine system [Figs 3.19(d) and 3.21]. Adapted from [32].

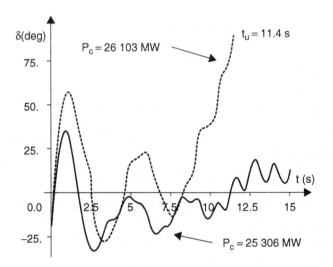

<u>Figure B-23.</u> Stabilizing an interarea oscillation by decreasing the power of the CMs from 26,103 MW to 25,306 MW. Adapted from [32].

contingencies may be very short (fractions of a second), the use of real-time measurements is instrumental for this ambitious objective.

Below, we briefly identify the key objectives, predictions, and steps of the corresponding procedure. The following paragraphs describe the way to accomplish them. A detailed account may be found in [42,10,16].

B.4.1.1 Objectives
Upon a disturbance inception and its clearance, E-SIME pursues the following objectives:

- It predicts whether the system is stable or driven to instability.
- If driven to instability, it predicts how unstable the system is going to be.
- It predicts where and how much corrective action to take (preassigned type of corrective action).
- It continues assessing whether the executed corrective action has been sufficient or whether to proceed further

B.4.1.2 Key Predictions
The key predictions entailed in E-SIME are as follows:

- First step: predict (in)stability.
- Second step: predict the severity of instability.
- Third step: predict and trigger appropriate control actions.

This three-step procedure is repeated in a closed loop until the complete system is stabilized.

B.4.1.3 Main Tasks
The above predictions are accomplished by performing the following tasks, on the basis of real-time measurements received at regular time intervals, Δt:

- Predicting the evolution of the individual machines' rotor angles to identify candidate critical machines,
- Computing the corresponding OMIB parameters δ, ω, γ, and P_a at $(t_i - 2\Delta t, t_i - \Delta t, t_i)$,
- Computing the OMIB curve P_a–δ and its solution, δ_u (the unstable angle),
- Computing the corresponding margin and time to instability, t_u, and
- Computing the control action necessary to cancel the margin.

Note that the computing effort required by the above tasks is negligible.

B.4.1.4 Technological Means
In order to perform the above tasks, E-SIME relies on the following assumptions and technological performance:

- Phasor measurement units (PMUs) placed at the (main) power plant stations
- The network of PMUs must be supported by a high-speed communication infrastructure [5] to transmit information about the status of the (main) power plants
- Suggested rate of receiving measurements: $\Delta t = 1$ cycle
- Communication delays not higher than 20–70 ms are practically acceptable [5]
- The measurements start being used immediately after the disturbance clearance

B.4.2   Predictive Transient Stability Assessment

The predictions identified in Section 4.1.2 are illustrated in Figure B-24.

B.4.2.1   Predicting the OMIB Structure

After the disturbance has been cleared:

- Perform a Taylor series expansion of the individual machines' rotor angles to predict their evolution, say, 100 ms ahead.
- Rank the predicted machine rotor angles, identify the largest angular distance between two successive angular deviations and declare the corresponding machines to be the "candidate critical machines," the remaining ones being the "candidate noncritical machines."
- Build accordingly the "candidate OMIB."

B.4.2.2   Predicting the P_a–δ Curve

This consists of computing the parameters of the "candidate OMIB" and, in particular, its accelerating power, P_a, and rotor angle, δ, using (at least) three successive measurement sets acquired at $t_i - 2\Delta t_i$, $t_i - \Delta t_i$, t_i.

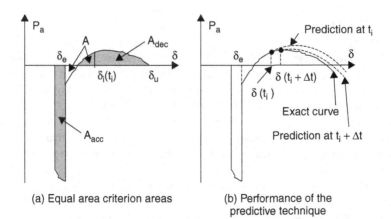

(a) Equal area criterion areas   (b) Performance of the predictive technique

Figure B-24. Prediction techniques using the OMIB P_a–δ curves. Adapted from [10].

The equation

$$P_a(\delta) = a\delta^2 + b\delta + c \tag{5}$$

is written for the above three different times and then solved for a, b, and c.*

B.4.2.3 Predicting Instability
This step begins by searching for a solution of

$$P_a(\delta_u) = a\delta_u^2 + b\delta_u + c \tag{6}$$

to determine whether the OMIB reaches the unstable conditions. If it does not, repeat the calculations in Sections B.4.2.1 through B.4.2.3 by using a new measurement set. Otherwise, the "candidate OMIB" has been found, for which the method computes successively:

- The unstable angle δ_u, which is the solution of Equation (6)
- The unstable margin

- $$\eta = -\int_{\delta_i}^{\delta_u} P_a d\delta - \tfrac{1}{2} M\omega_i^2 \tag{7}$$

- The time to instability

- $$t_u = t_i + \int_{\delta_i}^{\delta_u} \frac{d\delta}{\sqrt{(2/M)\int_{\delta_i}^{\delta} -P_a d\delta + \omega_i^2}} \tag{8}$$

where δ_i stands for $\delta(t_i)$ and ω_i for $\omega(t_i)$. Note that the practical importance of parameter t_u is even greater in the context of emergency control than for preventive control techniques.

B.4.2.4 Validity Test
An important procedural step consists of verifying the validity of the above predictions. E-SIME provides a handy test, thanks to the following observation: Under given contingency and operating conditions, the value of the (negative) margin should be constant, whatever the time step. Of course, this applies also to the time to instability t_u.

Hence, the computations of predictive transient-stability assessment should be pursued with successive Δt_i's until reaching a (almost) constant value of the margin and the corresponding time to instability, unless the latter is too short. Actually, there may be a trade-off between the above-mentioned validity test and the time to instability. Indeed, the shorter the time, the earlier the corrective action should be triggered, even before complete convergence to a constant margin value, if necessary.

*Subsequently, refine the estimated curve by using newly acquired sets of measurements and by processing a (weighted) least-squares technique, which has been shown to be particularly robust.

Let us also note that the use of Taylor series expansion in Section B.4.2.1 makes sense thanks to the OMIB transformation. In contrast, predicting the parameters, such as, the accelerating power, of all the system generators would have led to totally unreliable results.

B.4.3 Emergency Control

Once instability has been detected and its size calculated, the task of control is twofold:

1. Assess where and how much corrective action to take (preassigned type of corrective action).
2. Continue assessing whether the executed corrective action has been sufficient or whether to proceed further.

The principle of emergency control is illustrated in Figure B-25. It suggests that to stabilize the system one should increase the decelerating area [denoted as A_{dec} in Figure B-24(a) and A_{dec2} in Figure B-25] so as to cancel the negative margin.

Note that there is always an additional time delay before triggering the corrective action. This delay results from the summation of three components: the time needed to receive the real time measurements, the time to transmit the order to the power plant, and the time to apply the corrective action. Note also that after sending the order to activate the action, E-SIME continues monitoring and controlling the system in closed-loop fashion, until achieving power system stabilization. Note again that the prediction of the time to instability may influence the control decision (size of control or time to trigger).

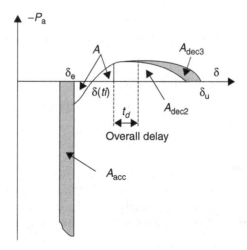

Figure B-25. Principle of emergency control. Adapted from [10].

The technique used when the emergency control is generation tripping comprises two tasks:

1. *Identification of the machine(s) to be tripped.* The candidate machines are among the critical machines that have been previously identified from the predicted (and ranked) swing curves of the individual machines. The machine(s) to be tripped are chosen to be the most advanced one(s).
2. *Computation of the stability margin of the controlled system.* The margin obtained after the above generation tripping must be calculated in order to assess whether the system thus controlled is stabilized or further control action is still needed.

Observe that because of communication and triggering-order delays, the measurements used to predict the status of the *controlled* system are, during a certain period, taken *on the uncontrolled system.* References [10, 11, 28] develop a technique to circumvent this difficulty.

It is important to remark that E-SIME provides only the total amount of generation to be tripped. Therefore, it may happen that the control problem cannot be solved by tripping generation at only one power plant, in which case a coordinated control action should be envisioned. There are two possible approaches.

The first is to trip at once all the estimated amount of generation to be tripped, as soon as the first unstable margin appears, by tripping the most advanced machines that produce the estimated generation to be tripped (this is the case when the estimated time to instability is tight with respect to the delays in control action application).

The second strategy consists of tripping the machines one by one, starting with the most advanced machine, then monitor the system state and trip the second most advanced machine if the computed stability margin is still negative and the time to instability is large enough. The approach continues until the margin becomes positive, but if after tripping the first machine the time to instability is found to be tight with respect to the delays, it trips all the remaining generation (estimated to be tripped) at once.

B.4.4   E-SIME Framework

The E-SIME tasks are schematically structured in the closed-loop control framework illustrated in Figure B-26.

B.4.5   Illustration

The E-SIME using generation tripping control is illustrated with a real-life power system model taken directly from the South-Southeast Brazilian System [10]. Time-domain simulations run with the ST600 transient-stability program [38] are used to simulate the temporal information.

The South-Southeast Brazilian system comprises 63 machines, 1180 buses and 1962 lines (more details about this system may be found in [10]). The system is

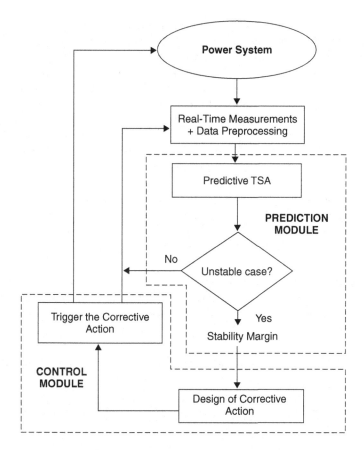

Figure B-26. General E-SIME framework. Adapted from [28].

equipped with automatic control devices, whose objective is to avoid overloads larger than 50% on line(s) and/or transformer(s) resulting from the outage of other transmission equipment following an important disturbance.

The generation-tripping scheme is applied to the Itaipu transmission system (8 hydraulic machines of 700 MW, at the 60 Hz side of Itaipu). The measurements are artificially created by time-domain simulations at the rate of 20 ms. The contingency is a three-phase short circuit. The corrective control action consists of disconnecting n machines (n should not exceed five so as to ensure a minimum of three machines in service) at the 60 Hz side of the Itaipu hydro station. The control order of this scheme is triggered 70 ms after the fault clearance (20 ms for the order to reach Itaipu and 50 ms for activating the control action). The measurements are injected at the rate of 20 ms.

Two cases are simulated. In the first case, the contingency (three-phase short circuit) is applied at the Itaipu side, whereas in the second case it is applied at the Foz do Iguacu side. In both cases, the short circuit is cleared by opening the faulted line after 80 ms.

In the first case, the critical clearing time is found to be 72 ms (i.e., slightly smaller than the assumed clearing time of 80 ms). The first set of measurements is supposed to be acquired at 95 ms, followed by sets acquired every 20 ms: 115, 135, and 155 ms. The stability prediction at about 155 ms identifies a set of 62 critical machines, but the following time steps (175, 195, etc.) identify the critical machines to be composed of the eight machines at Itaipu. Table B-1 groups the results obtained from $t_i = 155$ ms up to 435 ms and conveys the following information:

- At $t_i = 235$ ms, the margin, η, is shown to have reached a (almost) constant, negative value. This makes reliable the prediction of system's imminent instability ($t_u = 591$ ms) if no control action is taken. It is, therefore, decided to shed generation at Itaipu. More precisely, two machines to be shed are identified. The control (decided at 235 ms) is supposed to be triggered 150 ms later, at 385 ms.
- Hence, from $t_i = 255$ ms on, the predicted value of the margin obtained after tripping two machines is given in Table B-1. (For illustration, we also report additional information on the uncontrolled system characteristics from 255 ms on. Observe that the values of the time to instability and of the margins—positive as well as negative—remain almost constant.)

Table B-1. Transient stability assessment and control for the South-Southeast Brazilian system [10]

t_i (ms)	δ_u (rad)	t_u (ms)	η/M (rad/sec)^2	η/M after tripping
155	—	—	>0	—
175	2.08	743	−0.54	—
195	2.07	727	−0.58	—
215	1.96	596	−1.21	—
235	1.96	591	−1.23	—
Corrective decision is taken (2 units shed)				
255	1.95	585	−1.28	1.82
275	1.93	576	−1.34	1.79
295	1.93	575	−1.34	1.73
315	1.93	571	−1.35	1.74
335	1.93	572	−1.41	1.78
355	1.93	572	−1.41	1.72
375	1.94	579	−1.35	1.78
Corrective action is applied				
395	1.93	577	−1.38	1.76
415	1.94	578	−1.40	1.78
435	1.94	578	−1.40	1.79

- Obviously, E-SIME was able to stabilize the (otherwise unstable) case by tripping two out of the eight machines at Itaipu. The control action is actually activated at 385 ms after the contingency inception, 305 ms after its clearance. Note that this time is smaller than the time to instability, which is about 580 ms.

The second case corresponds to an extremely severe contingency whose critical clearing time is about 42 ms and its time to instability about 280 ms. Obviously, E-SIME alone could not stabilize such a violent instability, since its action could not be triggered fast enough. Actually, the existing automatic tripping procedure for this case consists of tripping five machines 70 ms after the line opening. For this particular case, E-SIME is used in order to find the minimum number of machines to shed at $t = 80 + 70 = 150$ ms. A set of simulations has been conducted with different clearing times and different numbers of machines to be shed [10].

Figure B-27 describes how E-SIME may be used to adjust the existing automatic control scheme. This illustration corresponds to a clearing time of 80 ms and four (instead of five) generators to trip 150 ms after the disturbance. Observe that E-SIME shows that tripping one additional generator later on may stabilize the system (provided that the time delay decreases to 125 ms).

Further explorations revealed that if the contingency can be cleared as early as 54 ms after its inception, the system may be stabilized by tripping four (instead of five) generators. The economy of one generator with respect to the actual automatic tripping procedure is made possible thanks to the E-SIME and its guarantee to monitor the system in a closed-loop fashion. This shows that E-SIME is not able to control extremely

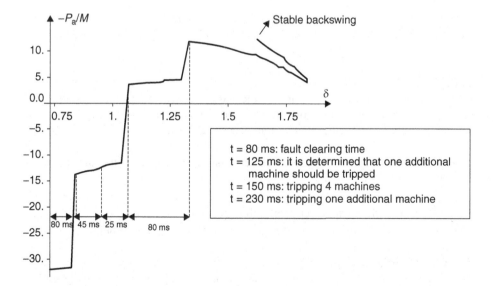

Figure B-27. P_a–δ curve with successive control actions. Adapted from [10].

severe contingencies (whose times to instability are smaller than 450 ms, corresponding to critical clearing times below, say, 70 ms), and leads to the idea of combining it with OLEC (see [35]).

B.4.6 Discussion

B.4.6.1 Salient Features

Summarizing the main features identified so far, we note that E-SIME is:

- A closed-loop control technique,
- Contingency independent; the contingency characteristics (location and nature) are implicitly included in the real-time measurements,
- Independent of power system modeling,
- Independent of parameter uncertainties, and
- Sensitive to the accuracy of the real-time measurements.

Concerning the *time to control,* it is expected that with today's technology, the mean time between fault clearance and execution of the first control action is about 450 ms. This results from the following approximate assessment of time delays needed:

- Assessing instability: $\sim$ 10 to 15 measurement sets $\Rightarrow$ $\sim$ 250 to 350 ms,
- Computing margin and corresponding control action $\Rightarrow$ $\sim$ 0 ms, and
- Sending and triggering the control action $\Rightarrow$ $\sim$ 50 to 100 ms.

This suggests that E-SIME alone could not control extremely severe contingencies. A recently proposed technique, combining E-SIME with OLEC [30] provides an alternative solution to very stringent cases.

B.4.6.2 Still Open Questions

E-SIME has been simulated in various dedicated places where the current control action is generation tripping. The consistently good results obtained have also uncovered a need for further improvements of various aspects such as:

- Centralized control at the control center versus decentralized at the main controllable power plants, and
- Global versus local processing (collecting measurements from all power plants versus from the local power plants "of concern").
- How should one design control actions other than generation tripping?

B.4.6.3 Concluding Remarks

The very principle of E-SIME is fully convincing. It has been extensively tested by considering data from real power plants equipped with generation tripping system protection schemes*: Itaipu in Brazil [10], Churchill Falls in Hydro-Québec [42], and WSCC (now WECC) [11]. A European project (EXAMINE) [7] was devoted to test its practical application in the southern region of Italy. The quality of the results obtained leaves no doubt about the effectiveness of the method, whose specifics, gathered in various places, are summarized below.

Advantages

- By using real-time measurements (neglecting noise and inherent errors) the stability assessment of E-SIME is more accurate than that of any time-domain program, since measurements include the complete power system dynamics response.
- The emergency control action is determined in real time, according to the severity of the instability problem in a closed-loop mode; this means that, after triggering a control action, E-SIME continues assessing system dynamics. It can, therefore, apply additional control if the first one was not sufficient to stabilize the power system.
- E-SIME, and emergency controls in general, are the preferred form of stability controls in countries or areas in which wholesale electricity markets have been implemented, since the electric power trading arrangements set by bilateral contracts and auction markets are modified only in cases when these controls are really needed; for instance, after the occurrence of an actual severe contingency.

Challenges

E-SIME assessment and control should be performed *early enough* to avoid power system instability. Since this method is applied to control transient instabilities caused by very severe contingencies, the available time for E-SIME is usually very short. This time is still shorter in practice since there are measurement and communication delays, and measurements are not received in a continuous way, but at a discrete sampling rate. The system transient instabilities and hardware characteristics imply that the method could not be applied to extremely severe contingencies, which would drive the system to instability in less than 450 ms.

Since the early 1990s, the development and application of new equipment (PMUs), resulting from important advances in communications and electronics, provided a new set of fast synchronized measurements. The impressive and continuous hardware development has not been followed by the development of software (and theoretical methods) required to take full advantage of the measurements so as to implement real-time controls for all kinds of stability problems. Nowadays, there is a true race to develop comprehensive practical methods able to use this huge amount of new, reliable

---

*Generation tripping is used in all kinds of units, but specially in hydro-generator units, like most of the above-cited plants. This is because the risk of damage to the unit from a sudden trip is not as great compared to the other types of plants.

information. E-SIME, developed in the mid-1990s, is one of the very few methods that could reach practical application in the near future. In an apparent contradiction, this method is able to control the fastest form of power system instability, but this is only possible because of the use of SIME representation and the information of power system dynamics.

An additional technological challenge is that classical PMUs do not deliver information directly usable by E-SIME, not only because they are connected to the high-voltage buses instead of the machines' buses, but also because they measure electrical variables (voltages and currents that can experience discontinuity under switching in the network), whereas the quantities needed are machines' rotor angles, speeds, accelerations (mechanical variables), and electrical powers. In order to make PMU information directly usable, the following approaches have been recently proposed:

- Compute rotor angles and speeds from phasor measurements, relying on the classical generator model, and relate phasors to reactances (step-up transformer and generator) [29].
- Estimate generator variables from PMU data using artificial neural networks [1].
- Meter the rotor angle and speed, together with phasors, using a new technological solution [19].

The use of some new specific techniques could increase the time available to perform E-SIME assessment and control and to improve this method practical applicability, for example:

- Instead of using the signal indicating that the line has been disconnected to start E-SIME analysis, use directly the signal that the transmission-line relay sends to the breakers in order to trip the line.
- The communication time delay is continuously decreasing. Nowadays, delays shorter than 50 ms could be achieved.

Final remark

The first developments of E-SIME date back to the late 1990s [42]. At that time, E-SIME was considered to be too expensive, if not too demanding with respect to the cost and performance of the technology available; for example, PMUs were too expensive to install massively, their performance was still poor, and fiber optic communications was still in its infancy.

Today, however, the available technology appears to meet E-SIME's requirements. Thus, a decade after its genesis, E-SIME has a tool of paramount practical importance.

B.5 POSTFACE

This appendix has described a general transient stability methodology, combining temporal information about power system dynamics with direct techniques, and yielding two main transient-stability assessment and control approaches: preventive and emergency. The essential features of these approaches are summarized below.

B.5.1 Preventive SIME

Preventive SIME (P-SIME) gets temporal information from T-D simulations provided by a transient-stability program. The constraint imposed by its coupling with a T-D program, which might be viewed as a handicap, is by and large compensated by the resulting outcomes. They are summarized in the following three sections.

B.5.1.1 Advantages Specific to T-D Programs

Coupling SIME with a T-D program makes it as accurate as the T-D program, and as flexible in terms of modeling. It is able to analyze in detail:

- Power system components, including generators, static and dynamic loads, static VAr compensators, FACTS devices, and their controls,
- Any type of contingency, for example, three-phase, single-phase, unbalanced faults, or a combination of the tripping of several system components, and
- Any type of instability, such as first-swing or multiswing, upswing or backswing, plant or interarea mode.

B.5.1.2 Advantages Specific to the Equal-Area Criterion (EAC)

The EAC provides:

- Stability margins;
- Clear representation of system dynamics, enhancing the analysis and design of emergency discrete controls commonly applied in the electric industry; and
- Straightforward stability assessment.

B.5.1.3 Advantages Resulting from the Hybridization of T-D Simulations with EAC

The hybridization of T-D simulations with EAC provides for the:

- Identification of the critical machines, information of paramount importance in very many applications;
- Assessment of the "time to instability," which gives a clear indication of a system's degree of instability;
- Additional representations of system dynamics, providing clear views of complex transient stability phenomena; and
- Objective detection of instability/stability conditions and ensuing criteria for early termination of T-D simulations. These criteria, furnished by closed-form expressions, are valid for all power systems, irrespective of their size, type, and specifics, in contrast to T-D methods, which use pragmatic criteria that are not necessarily totally faithful.

The above information is based on an impressive number of SIME-based security-function components, addressing a much larger variety of transient-stability aspects, in particular, near-optimal preventive control, transient-oscillations damping assessment and control and, finally, a special scheme of open-loop emergency control (OLEC), which mitigates preventive control actions to render them more acceptable in practice. These various techniques can be combined to form comprehensive security assessment and control, able to comply with the specifics of the various study contexts: planning, operational planning, and online assessment and preventive control.

Stated otherwise, P-SIME enlarges substantially the domain of application of a T-D method by adding control techniques, without sacrificing either the accuracy or the flexibility of the T-D method, and, in addition, speeding it up. On the other hand, thanks to the tight interplay of the equal-area criterion with the T-D method, SIME gets rid of all kinds of simplifying assumptions and inaccuracies linked to multimachine direct approaches.

It is essential to emphasize that, unlike pure direct methods, SIME is *not* an approximate method. Rather, it combines the advantages of T-D and direct methods, while avoiding their difficulties and enlarging substantially their possibilities. The developed systematic and efficient control techniques are certainly achievements of paramount importance.

Of course, many other interesting applications may be thought of, as for example the exploration of the influence of FACTS devices on system stability [15]. More generally, the one-machine system, which in the old days was used in sensitivity studies to assess the approximate influence of various system controls (PSS, etc.), may now advantageously be replaced by P-SIME.

P-SIME has been applied to a large number of power systems, having various sizes, specifics, and modeling, such as Hydro-Québec, Electricité de France, Belgian Power System, WSCC, Mexican Interconnected System, Brazilian South-Southeast and Brazilian Interconnected Systems, Greek Interconnected System, and Italian Interconnected System. In these simulations, P-SIME was coupled with various T-D programs (Matlab, ETMSP, Eurostag, ST600, Simpow, and SICRE). More recently, it has been implemented within the OMASES European project [23].

Without exception, P-SIME has shown to have very good characteristics, in terms of speed, accuracy, and effectiveness. To summarize, P-SIME is today fully mature and ready to be used in a large number of transient stability assessment and control applications.

B.5.2 Emergency SIME

This newly developed approach has a unique feature: it succeeds in assessing and controlling *in real time* transient instabilities, which are known to develop extraordinarily fast. Certainly, the method is still perfectible and depends on IT performances. The rapid progress in IT leaves no doubt about method's future improvements and success for real-world implementations.

B.6   REFERENCES

[1] Del Angel A., Guerst P., Ernst D., Glavic M., and Wehenkel L., "Estimation of Rotor An-
     gles of Synchronous Machines Using Artificial Neural Networks and Local PMU-based
     Quantities," *Neurocomputing,* Vol. 70, pp. 2668–2678, 2007.

[2] Aylett P. D., "The Energy-Integral Criterion of Transient Stability Limits of Power Sys-
     tems," *Proceedings of IEE,* Vol. 105C, pp. 527–536, 1958.

[3] Bettiol A., Ruiz-Vega D., Ernst D., Wehenkel L. and Pavella M., "Transient Stability-
     Constrained Optimal Power Flow," *Proceedings of the IEEE Budapest PowerTech,* Bu-
     dapest, Hungary, 1999.

[4] CIGRE Task Force 38.01.07, "Analysis and Control of Power System Oscillations,"
     Technical brochure, 1996.

[5] CIGRE Task Force 38.02.13, "Advanced Angle Stability Controls," Technical brochure,
     1999.

[6] Chen L., Ono A., Tada Y., Okamoto H., and Tanabe R., "Optimal Power Flow Con-
     strained by Transient Stability," *Proceedings of POWERCON 2000,* University of West-
     ern Australia, Perth, Australia, 2000.

[7] Diu A., and Wehenkel L., "EXaMINE—Experimentation of a Monitoring and Con-
     trol System for Managing Vulnerabilities of the European Infrastructure for Electric
     Power Exchange," *Proceedings of IEEE PES Summer Meeting 2002,* Chicago, June
     2002.

[8] El-Abiad A. H. and Nagappan K., "Transient Stability Region of Multimachine Power
     Systems," *IEEE Transactions on PAS,* Vol. 85, pp. 169–179, 1996.

[9] Electric Power Research Institute (EPRI) (1994), Extended Transient Midterm Stability
     Program Version 3.1, User's manual. Final EPRI report No. EPRI TR-102004, Projects
     1208-11, 12, 13, 1994.

[10] Ernst D., Bettiol A., Zhang Y., Wehenkel L., and Pavella M., "Real-Time Transient Sta-
     bility Emergency Control of the South-Southeast Brazilian System," in *Proceedings of
     the V SEPOPE,* Salvador, Brazil, Paper IP044, 1998.

[11] Ernst D. and Pavella M., "Closed-Loop Transient Stability Emergency Control," *Pro-
     ceedings of IEEE/PES Winter Meeting,* January, Singapore, 2000.

[12] Ernst D., Ruiz-Vega D., Pavella M., Hirsch P. and Sobajic D., "A Unified Approach to
     Transient Stability Contingency Filtering, Ranking and Assessment". *IEEE Transactions
     on Power Systems,* Vol. 16, No. 3, pp. 435–444, 2001.

[13] Fouad A. A., and Jianzhong, T., "Stability Constrained Optimal Rescheduling of Genera-
     tion," *IEEE Transactions on Power Systems,* Vol. 8, No. 1, pp. 105–112, 1993.

[14] Fortescue C. L., "Transmission Stability. Analytical Discussion of Some Factors Entering
     into the Problem," *AIEE Transactions,* Vol. 44, pp. 984–994, 1925.

[15] Ghandhari M., Andersson G., Pavella M., and Ernst D., "A Control Strategy for Control-
     lable Series Capacitor in Electric Power Systems," *Automatica,* Vol. 37, No. 10, pp.
     1575–1583, 2001.

[16] Glavic M., Ernst D., Ruiz-Vega D., Wehenkel L., and Pavella M., "E-SIME—A Method
     for Transient Stability Closed-Loop Emergency Control: Achievements and Prospects,"
     in *Proceedings of Bulk Power System Dynamics and Control—VII,* August 19–24,
     Charleston, South Carolina, 2007.

[17]   Gibbard M. J., Martins N., Sanchez-Gasca J. J., Uchida N., Vittal V., and Wang L., "Recent Applications of Linear Analysis Techniques," *IEEE Transactions on Power Systems*, Vol. 16, No. 1, pp. 154–162, 2001.

[18]   Gan D., Thomas R. J., and Zimmerman R. D., "A Transient Stability Constrained Optimal Power Flow," *Proceedings of Bulk Power System Dynamics and Control IV—Restructuring*, IREP 1998, pp. 83–89, 1998.

[19]   Jin, Y-Q., Qin C., Wu F., Han J-D., Xu Q., Yan D-J., and Ju P., "Wide Area Measurement System Based on Rotor Angle and Phasor Measurement with Applications," in *Proceedings of Bulk Power System Dynamics and Control-VII*, Charleston, South Carolina, 2007.

[20]   La Scala M., Trovato M., and Antonelli C., "On-Line Preventive Control: An Algorithm for Transient Security Dispatch," *IEEE Transactions on Power Systems*, Vol. 13, No. 2, pp. 601–610, 1998.

[21]   Magnusson P. C., "Transient Energy Method of Calculating Stability," *AIEE Transactions*, Vol. 66, pp. 747–755, 1947.

[22]   Momoh J. A., Wang Y. and Chattopadhyay D., "Stability Constrained Optimal Rescheduling Using Compromise Programming," *Proceedings of the 28th North American Power Symposium*, MIT, Boston, 1996.

[23]   Massucco S., Wehenkel L., Bihain A., Cirio D., Fiorina M., Lopez R., Lucarella D., Ruiz-Vega D., Vournas C., and Van Cutsem, T., "OMASES: A Dynamic Security Assessment Tool for the New Market Environment," In *Proceedings of Bologna Power Tech Conference*, 2003.

[24]   Park R. H., and Banker E. H., "System Stability as a Design Problem," AIEE Transactions, Vol. 48, pp. 170–193, 1929.

[25]   Pai M. A., *Power System Stability. Analysis by the Direct Method of Lyapunov.* North-Holland Publishing Company, Amsterdam, 1981.

[26]   Pai M. A., *Energy Function Analysis for Power System Stability.* Kluwer Academic Publishers, Boston, 1989.

[27]   Pavella M., "Generalized One-Machine Equivalents in Transient Stability Studies," PES Letters, in *IEEE Power Engineering Review*, Vol. 18, No.1, pp. 50–52, 1998.

[28]   Pavella M., Ernst D., and Ruiz-Vega D., *Transient Stability of Power Systems: A Unified Approach to Assessment and Control.* Kluwer Academic Publishers, 2000.

[29]   Rovnyak S., Liu C-W., Lu J., Ma W., and Thorp J., "Predicting Future Behavior of Transient Events Rapidly Enough to Evaluate remedial Control Options in Real Time," *IEEE Transactions on Power Systems*, Vol. 10, No. 3, pp. 1195–1203, 1995.

[30]   Ruiz-Vega D., Glavic M., and Ernst D., "Transient Stability Emergency Control Combining Open-Loop and Closed-Loop Techniques," In *Proceedings of IEEE PES General Meeting*, June, Toronto, Canada, 2003.

[31]   Ruiz-Vega D., Messina A. R., and Pavella M., "A Novel Approach to the Assessment of Power System Damping." Presented at the 14th Power Systems Computation Conference PSCC2002, Sevilla, Spain, 2002.

[32]   Ruiz-Vega D., Mesina A., and Pavella M., "Online Assessment and Control of Transient Oscillation Damping," *IEEE Transactions on Power Systems*, Vol. 19, No. 2, pp. 1038–1047, 2004.

[33]   Ribbens-Pavella M., Murthy P. G., Howard J. L., and Carpentier J. L., "On-Line Transient Stability Assessment and Contingency Analysis," presented at International Conference on Large High Voltage, CIGRE, 1982.

[34] Ruiz-Vega D., and Pavella M., "A Comprehensive Approach to Transient Stability Control. Part I: Near Optimal Preventive Control," *IEEE Transactions on Power Systems,* Vol. 18, No. 4, pp. 1446–1453, 2003.

[35] Ruiz-Vega D., and Pavella M., "A Comprehensive Approach to Transient Stability Control. Part II: Open Loop Emergency Control," *IEEE Transactions on Power Systems,* Vol. 18, No. 4, pp. 1454–1460, 2003.

[36] Ruiz-Vega D., "Dynamic Security Assessment and Control: Transient and Small Signal Stability." Dépôt légal D/2002/0480/19, ISSN 0075-9333. Collection des Publications de la Faculté des Sciences Appliquées No. 213. Institut de Mécanique et Génie Civil, Université de Liège, Liège, Belgium, 2002.

[37] Sterling J., Pai M. A., and Sauer P. W., "A Methodology of Secure and Optimal Operation of a Power System for Dynamic Contingencies," *Journal of Electric Machines and Power Systems,* Vol. 19, No. 5, 1991.

[38] Vallete A., Lafrance F., Lefebvre S., and Radakovitz R., "ST600 Programme de Stabilité: Manuel d'utilisation Version 701," Hydro-Quebec, Vice-présidence technologie et IREQ, 1987.

[39] Winter W. H., and Cowbourne D. R., "The Bruce Load and Generation Rejection Scheme," *Proceedings of the CIGRE-IFAC Symposium on Control Applications for Power System Security,* Paper 207-03, Florence, Italy, 1983.

[40] Xue Y., Van Cutsem Th., and Ribbens-Pavella M., "A Simple Direct Method for Fast Transient Stability Assessment of Large Power Systems," *IEEE Transactions on Power Systems,* Vol. 3, No. 1, pp. 400–412, 1988.

[41] Xue Y., and Pavella M., "Extended Equal Area Criterion: An Analytical Ultra-Fast Method for Transient Stability Assessment and Preventive Control oOf Power Systems," *International Journal of Electric Power and Energy Systems,* Vol. 11, pp. 131–148, 1989.

[42] Zhang Y., Wehenkel L., and Pavella M., "A Method for Real-Time Transient Stability Emergency Control," *Proceedings of CPSPP'97, IFAC/CIGRE Symposium on Control of Power Systems and Power Plants,* Beijing, China, pp. 673–678, 1997.

NOTATION

All abbreviations, acronyms and symbols are fully defined at the place they are first introduced. As a convenience to the reader, we have collected below some of the more frequently used ones in several places.

Abbreviations and Acronyms

CCT	critical clearing time; also denoted t_c
CM	critical machine
EAC	equal-area criterion
E-SIME	emergency SIME
NM	noncritical machine
OLEC	open-loop emergency control
OMIB	one-machine infinite bus
OPF	optimal power flow

PMU	phasor measurement unit
SIME	single-machine equivalent
sTDI	seconds of time-domain integration
TSA	transient stability assessment
TSA&C	transient stability assessment and control
T-D	time domain

Basic symbols

P_m	OMIB mechanical power
P_e	OMIB electrical power
P_a	OMIB accelerating power
M	OMIB inertia coefficient
δ	OMIB rotor angle
δ_u	OMIB unstable angle
ω	OMIB rotor speed
t_u	time to (reach) instability (conditions)
η	margin defined by SIME
P_C	active power generation of the group of critical machines
P_N	active power generation of the group of noncritical machines

APPENDIX **C**

DETECTION AND EVALUATION OF STABILITY-CONSTRAINED TRANSMISSION PATHS

Marius Pomarleanu and Savu C. Savulescu

C.1 INTRODUCTION

In the context of electricity market operations, a primary concern is the ability to transfer power across vast interconnected networks while meeting a broad range of operating reliability constraints. A common scenario consists of compensating load increases and/or generation outages in a system area by increasing the generation elsewhere. In order to ensure that the grid does not get too close to its stability limits, it is, therefore, important to evaluate the maximum transfer capability across the "links" that interconnect the areas involved in such transactions.

A "link" identifies a group of transmission lines that form a topological cutset, that is, their removal splits the network in two areas, one on each side of the link. The maximum power that can be transferred across a link is limited by thermal and stability constraints. The stability limit of a link can be quantified by the further loading of the link, that is, the additional amount of power that can be sent from one side of the link to the other side, without causing instability. This indicator, which is referred to as the *stability reserve* of the link, can be expressed either in MW or in percentage from the maximum link loading. In a sense, the concept of "stability constrained link" is similar to the concept of "congestion path," with the difference that the former is concerned with stability, rather than thermal, violations.

Real-Time Stability Assessment in Modern Power System Control Centers. Edited by Savu C. Savulescu
Copyright © 2009 The Institute of Electrical and Electronics Engineers, Inc.

"Stability-constrained links" may appear in any multiarea power system in which large MW blocks are transferred between weakly interconnected areas. This is often the case in longitudinal transmission networks that span distinct system areas with significant load-generation imbalances, although stability-constrained links may appear in other topological configurations as well. The analysis of recent blackouts due to instability revealed that most of them followed a similar pattern:

- Large MW blocks get transferred from areas with inexpensively priced energy toward areas in which the load demand has increased due to an actual increase in load, or perhaps because one or several local generating units are scheduled for maintenance, or simply because the local generation is too expensive.
- As a result, certain links get loaded closer and closer to their stability limits and their stability reserves get smaller and smaller.
- At this moment, a generation or transmission outage takes place. Typically, such incidents evolve into cascading outages.
- Since the link was already operating with a small stability reserve, the physical phenomena leading to blackout are triggered and the widespread disturbance becomes unavoidable.

The detection of critical links is an intrinsically difficult proposition. To begin with, the search of all the possible links entails a graph-topological procedure that may identify hundreds of thousands of links even for moderately sized networks. Once the full set of links has been determined, a stability criterion should be used to compute the maximum transfer capability of each link. Then, the ability should exist for the links to be ranked in the order of their stability margins. In addition, it should also be possible to evaluate user-defined links. Recognizing the fact that current network analysis technologies neither provide tools for quickly identifying the security-constrained links nor have the ability to recommend what generating units should be acted upon in order to avoid the blackout, an algorithm has been developed that solves this complex problem thoroughly and expeditiously.

C.2 APPROACH

C.2.1 Background

The evaluation of the operating reliability of transmission networks in normal or emergency operating conditions is a complex undertaking, to say the least. One of the major difficulties stems from the wide variety of aspects that must be evaluated, thus making it difficult, if not impossible, to reduce the number of simulations to just a few standard scenarios. As a minimum, the analyst should take into account the:

- System conditions prior to a disturbance, actual or postulated,
- Location, duration, and size of the disturbance,

- System's trajectory from the initial state to the critical state, that is, the scenario envisioned for stressing or worsening the system conditions.

Traditional operating reliability studies encompass load-flow and contingency analysis and stability assessment. An important goal of stability assessment is to determine whether the system can withstand a list of major, yet credible, contingencies. This is the realm of *transient stability analysis.* An equally important objective is to evaluate the risk of approaching instability in small steps, for example, via small load changes accompanied by slow changes in the bus voltages that may trigger a voltage collapse, or by gradual load changes that may eventually cause one or several generators to get out of synchronism. Traditionally, this topic has been addressed by *steady-state stability analysis.* At the present time, the industry refers to it as *voltage stability analysis,* but this terminology is neither uniformly understood nor universally accepted. In order to avoid any confusion, throughout this appendix we will use the terminology voltage and steady-state stability analysis.

There is no way to handle all the aspects of stability at once. Each one requires detailed models and adequate tools tailored to the physical phenomena being evaluated. The problem becomes even more complex when the target is a vast interconnected system. The large amount of data, the significant computing times, and the technical skills required to interpret the results render the analysis difficult.

The latter point is not academic. Even if computational speed is somehow achieved and the stability calculations are performed "online", that is, off-line with real-time data, or, maybe, truly in real time, the results would subsequently be used by system dispatchers and security engineers who may have neither the time nor the skills required to perform lengthy assessments. This opens the door for methods which, even if based on simple procedures and estimations entailing modeling assumptions and algorithm simplifications, produce results that are easy to interpret and understand.

C.2.2 Steady-State Stability at a Glance

Steady-state stability is the stability of the system under conditions of gradual or relatively slow changes in load [5]. Accordingly, the steady-state stability limit (SSSL) of a power system is "a steady-state operating condition for which the power system is steady-state stable but for which an arbitrarily small change in any of the operating quantities in an unfavorable direction causes the power system to loose stability" [14].

This definition does not provide the support needed to classify the events that take place in a power system network that may impact the stability and identify the causes of instability [12]. Other aspects must also be considered, such as:

- Physical nature of instability,
- Most important system parameter in which the instability can be observed,
- Severity of the disturbance considered in the simulation
- Most adequate method for computing and predicting stability, and

- Equipments, processes, and scenarios that must be taken into account in order to achieve stability.

The phenomena pertaining to the realm of steady-state stability of vast interconnected systems are extremely complex. Accordingly, specialized tools have been tailored to address natural stability versus stability maintained by fast voltage controllers, local versus global stability, and the stability of power transfers across links between system areas versus voltage stability.

The conventional method of the small oscillations for estimating the steady-state stability [19,20,26] consists of examining the eigenvalues of the characteristic equation associated with the system of differential equations that describe the free transient processes after a small disturbance takes place in an automatically controlled power system. The necessary and sufficient condition for steady-state stability is that all the real parts of the eigenvalues be negative [26]. The analysis encompasses the following steps:

- Describe the transient processes in the form of a system of nonlinear differential equations.
- Linearize the equations around the solution point by expanding them into a Taylor series and retaining only the linear (first-order) terms.
- Calculate the main (characteristic) minor determinants and develop the characteristic equation.
- Determine the sign of the real roots and the sign of the real part of the complex roots of the characteristic equation.

The approach is laborious and is replaced by determining relationships between the roots and the coefficients of the characteristic equation. Venikov calls these relations steady-state stability criteria [26] and classifies them into *algebraic* (Routh–Hurwitz), *frequency-domain* (Nyquist), and *practical*. A necessary condition for steady-state stability is derived from the Hurwitz criterion by evaluating the sign of the last term of the characteristic equation, which is the Jacobian determinant D. A change of sign from positive to negative (all Hurwitz determinants are positive) with further loading of the system indicates aperiodic instability. The instability in the form of self-oscillations, however, remains unrevealed by this method [26].

If the generators are radially connected to a nodal point, and if, based on practical considerations, it may be assumed that some operating variables are constant, the condition $D = 0$ leads to "practical criteria" that are valid within certain limits, for example, the *synchronizing power criterion* $dP/d\delta > 0$, which assumes constant frequency and constant voltage at the nodal point, and the *reactive power voltage and steady-state stability criterion* $d\Delta Q/dV < 0$ which assumes that the frequency is constant and the power balance is maintained at the load node [26]. The practical stability criteria have enabled the development of techniques and procedures that make it possible to quickly evaluate the degree of operating stability of a power system.

Since steady-state instability occurs when a change of the state variables causes the Jacobian matrix associated with the dynamic state equation to become singular, one might think that instability happens in only one way regardless of how the system conditions were stressed, or "worsened," in order to reach the limit. Published references, however, indicated that the network characteristics that generate the *most probable system-stressing patterns* should be taken into account when selecting the procedures and indicators used for steady-state stability analysis.

In this context, for vast interconnected systems it is essential to assess stability when large blocks of power are transferred across the network. This, in turn, requires evaluating the maximum transfer capability across the link, that is, between the areas that get involved in the transactions, when a reduction in generation in one area is compensated by increasing the generation elsewhere.

Potentially, there are many such links in the network, some with adequate margins to further increase the MW transfer without risk of instability, but some others in which a further loading of the link might cause steady-state instability. The early identification of such stability-constrained links is imperative for the operating reliability of the transmission system.

C.2.3 Stability Reserve of the Link

The stability reserve, that is, the steady-state stability margin, of the link can be expressed in terms of the *further link loading,* that is, the additional amount of MW that can be transferred between the areas separated by the link before reaching steady-state instability. The stability reserve of the link is an *indicator,* and is computed by applying a steady-state stability criterion in conjunction with a system-stressing procedure that entails increasing the MW transfer between the areas on the two sides of the link until steady-state instability is reached. In order to identify the stability-constrained links, a systematic search of all the possible links must first be performed, followed by the evaluation of each link and, finally, by ranking them in the order of their stability reserves. Given a solved load-flow case or a state estimate, the approach consists of:

- Detecting all the links within the transmission system,
- Computing the steady-state stability reserve for each link,
- Ranking the links in the order of their steady-state stability reserves, and
- Recommending a control strategy, that is, raising or lowering the MW output of certain machines, which can help increase the stability margin of the link.

There are situations in which certain links are known beforehand. Therefore, it should also be possible to perform the same array of calculations on user-defined links. The computations should be fast enough to enable the fast evaluation of the power transfers between areas known a priori to have stability limitations, as often as needed, for each transaction, offline and in real time. Once such information has been obtained, it can be further processed by a multiarea stability computational engine to determine how far from instability are the areas separated by the link.

C.2.4 Algorithm

The first step of the algorithm consists of parsing the original power system transmission network and identifying all the available links. Figure C-1 depicts schematically a link that separates the power system network into two areas. In order to determine how much additional power can be transferred between Area 1 and Area 2 before reaching the steady-state stability limit of the link, a network transformation must be applied to convert the original system to a simple network consisting of two equivalent generators interconnected through an equivalent link.

This is achieved by building a special type of REI equivalent with two REI generators and two REI loads. The generators situated in Area 1 are aggregated in the equivalent REI generator REI G1, whereas the generators situated in Area 2 are aggregated in the equivalent REI generator REI G2. Similarly, the loads in each area are aggregated in the REI L1 and REI L2, respectively. This is shown in Figure C-2.

As far as the modeling and aggregation of generators in a REI generator bus, the following comments should be made:

- For practical reasons, only the machines generating in excess of a user-specified MW threshold are aggregated. The other generators are represented through a constant impedance.
- The internal transient reactances x'_d of the generators are included in the model. Since the electromotive forces (emfs) behind these reactances are constant dur-

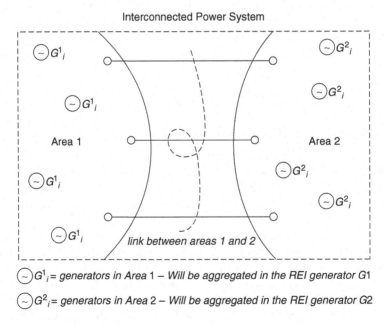

$\textcircled{\sim}G^1{}_i$ = generators in Area 1 – *Will be aggregated in the REI generator G1*

$\textcircled{\sim}G^2{}_i$ = generators in Area 2 – *Will be aggregated in the REI generator G2*

Figure C-1. Original system separated into two areas by a link.

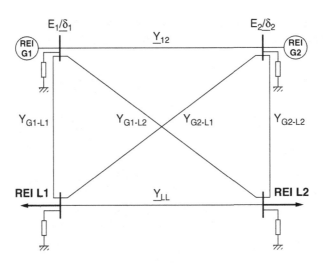

<u>Figure C-2.</u> Original system replaced with REI equivalents in each one of the areas separated by the link.

ing the time span considered for steady-state stability analysis, the bus voltage moduli at the REI generator buses are constant. The angles of these voltages, however, are allowed to vary.

The rules for aggregating the loads are as follows:

- The reactive parts of the loads, including those that may appear on generator buses, are replaced with bus-to-ground reactive shunts.
- The real parts of the loads are grouped in REI equivalent load centers by using Paul Dimo's standard zero-power energy balance and REI equivalencing procedure.

After the REI reduction has been performed, the equivalent network shown in Figure C-2 can be further simplified by noting that:

- The equivalent admittances between the REI generator in one area and the REI load in the other area, designated as $Y_{G1\text{-}L2}$, $Y_{G2\text{-}L1}$, and Y_{11} in Figure C-2, are negligible when compared with Y_{12}. This corresponds to the physical reality that there is practically no direct power transfer between the generators in one area and the loads in the other area; rather, all the power transfers take place across the equivalent link, that is, across the admittance Y_{12} in Figure C-2.
- The equivalent admittance Y_{LL} between the REI equivalent load buses is negligible, which means tat there is no power transfer between the loads in Area 1 and the loads in Area 2.

The final model consisting of two generators connected by a link is depicted in Figure C-3, where:

$$P_{12} = \frac{E_1^2}{Z_{12}} \sin \alpha_{12} + \frac{E_1 E_2}{Z_{12}} \sin(\delta_{12} - \alpha_{12}) \tag{1}$$

$$\underline{E}_1 = E_1 \arg \delta_1 \qquad \text{and} \qquad \underline{E}_2 = E_2 \arg \delta_2 \tag{2}$$

are the emf of generators G1 and G2:

$$\delta_{12} = \delta_1 - \delta_3 \tag{3}$$

$$\alpha_{12} = 90° - \psi_{12} \tag{4}$$

and

$$\underline{Z}_{12} = Z_{12} \arg \psi_{12} \tag{5}$$

is the reciprocal (transfer) impedance of the equivalent link, that is, the transmission system between equivalent generators G1 and G2.

From the maximum power transfer theorem [20], the maximum power that can be transferred from Area 1 toward Area 2 is calculated with:

$$P_{12max} = \frac{E_1^2}{Z_{12}} \sin \alpha_{12} + \frac{E_1 E_2}{Z_{12}} \tag{6}$$

at the critical angle given by

$$\delta_{12crt} = 90° + \alpha_{12} \tag{7}$$

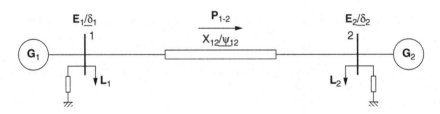

$E_1\underline{/\delta_1}$ = complex internal voltage of REI generator G1

$E_2\underline{/\delta_2}$ = complex internal voltage of REI generator G2

$X_{12}\underline{/\psi}_{12}$ = transfer impedance between 1 and 2

L_1 = equivalent load in Area 1 (including losses)

L_2 = equivalent load in Area 1 (including losses)

<u>Figure C-3.</u> Power transferred between Area 1 (equivalent generator G1) and Area 2 (equivalent generator G2) across the equivalent link.

According to Venikov, the stability reserve of the link in Figure C-1 is given by

$$R = \frac{P_{12max} - P_{12}}{P_{12max}} \tag{8}$$

The alternate expression of the steady-state stability indicator R given in (9),

$$R = 1 - \frac{P_{12}}{P_{12max}} \tag{9}$$

provides the percentage amount of additional amount of power that can be carried across the link up to the limit of steady-state stability. By substituting P_{12} and P_{12max} from (1) and (6) we get

$$R = \frac{1 - \sin(\delta_{12} - \alpha_{12})}{1 + \dfrac{E_1}{E_2} \sin \alpha_{12}} \tag{10}$$

Let us note that formulae (1) through (10) were originally developed for the very simple case of two generators connected by a line (Venikov) but here we applied them to an equivalent network consisting of two equivalent generators connected through an equivalent link. It may be useful to show how the values of $\underline{E}_1 = E_1 \arg \delta_1$ and $\underline{E}_2 = E_2 \arg \delta_2$ are derived in this case.

The power generated by the machines in the Area 1 that are aggregated in G1 is given by

$$\underline{S}_1 = \underline{E}_1 \times \underline{I}_1^* = \sum_{g=1}^{m} \underline{S}_{g1} = \underline{E}_1 \sum_{g=1}^{m} \underline{I}_1^* = \underline{E}_1 \left(\sum_{g=1}^{m} \frac{\underline{S}_{g1}}{\underline{U}_{g1}} \right) \tag{11}$$

where:

$\underline{S}_1$ = complex power generated by the REI generator G1
$\underline{E}_1$ = complex emf of REI generator G1
$\underline{I}_1$ = complex current of REI generator G1
$\underline{S}_{g1}$ = complex power generated by the machine g that is aggregated in the REI generator G1
$\underline{I}_{g1}$ = complex current of generator g
$\underline{U}_{g1}$ = complex voltage at the terminals of generator g

Based on these notations we obtain

$$\underline{E}_1 = \frac{\displaystyle\sum_{g=1}^{m} \underline{S}_{g1}}{\displaystyle\sum_{g=1}^{m} \frac{\underline{S}_{g1}}{\underline{U}_{g1}}} = E_1 \arg \delta_1 \tag{12}$$

The generators $m + 1 \dots n$ in Area 2 are aggregated in the REI equivalent generator G2:

$$
\underline{E}_2 = \frac{\displaystyle\sum_{g=m+1}^{n} \underline{S}_{g2}}{\displaystyle\sum_{g=m+1}^{n} \frac{\underline{S}_{g2}}{\underline{U}_{g2}}} = E_2 \arg \delta_2 \tag{13}
$$

The transfer impedance $\underline{Z}_{12}$ of the equivalent link results from the REI network reduction process.

The accuracy of this technique has been verified by comparing its computational results with those obtained by using a conventional steady-state stability program [2]. In all the cases that were evaluated, the difference between the stability reserve of the link computed with the method described herein and the additional amount of power that could be transferred between the two areas determined with the conventional stability program was in the range of 10% of the stability reserve, that is, if the conventional stability program computed a stability reserve of 30%, the value determined with this algorithm was in the range of 27% through 33%. An early implementation of this method has been used for quite sometime, and the results obtained to date confirm its solidity and suitability for detecting stability constrained links in a large transmission system.

C.2.5 Topological Search of all the Links in a Network

The identification of all the links in the transmission system is performed by using a graph topological search technique. Starting from each node of the graph (bus in the transmission network) new edges (lines and/or trafos) are added until a link, that is, a group of edges that separate the network into two separate areas, is identified. The algorithm proceeds as follows:

1. Build the list of all the nodes and mark the nodes where there are generators.
2. Initialize the first node by selecting the first generator node in the list.
3. Assign this node to Area 1 and build the list of edges that connect it with the external area.
4. Define the first link.
5. Remove this node from the list of generators which, at this stage, designates the generators in Area 2.
6. Continue the connectivity search in Area 1 and retain the next generator node.
7. Create a new list containing the edges between nodes in Area 1 and those in the external area.
8. Define a new link that separates the 2 generator node (already assigned to Area 1) from the rest of the system by starting from list of edges selected earlier.

9. Update the list of generators in the external system.

10. Add a new node in Area 1 and repeat the previous steps until 90% of the total number of nodes have been parsed, thus assuring an acceptable level of redundancy.

11. Repeat steps 2 through 10 of the algorithm by initializing the search at the second generator node, then the third, and so on, until all the links have been identified.

Throughout this process, the numbers of generators assigned to Area 1 and Area 2 change continuously but their sum remains constant. New links that encompass previously identified links are disregarded.

Let us note that the aggregation of generators is based on topological considerations rather than type (thermal, hydro, etc.) or ownership. The percentage of 90% of generator nodes identified in an area is used as a criterion for having completely covered the area.

C.2.6 Detection of Stability-Constrained Links

For each one of the links identified during the topological search described above, a two-REI generator equivalent is built in accordance with the algorithm described in Section C.2.4. E_1, E_2, δ_1, and δ_2 are computed from Equations (12) and (13) and δ_{12} is obtained from Equation (3). The value of the transfer impedance between the equivalent generators G1 and G2 is Z_{12} arg ψ_{12} and the angle α_{12} is given by (5).

We can now determine directly from Equation (10), for each link, the value of its stability reserve:

$$R = \frac{1 - \sin(\delta_{12} - \alpha_{12})}{1 + \dfrac{E_1}{E_2} \sin \alpha_{12}}$$

The last step of the procedure consists of ranking all the links in the decreasing order of their stability reserves. For practical purposes, only the most dangerous links (50–100), if any, are retained in the final list of stability constrained links.

C.2.7 Remedial Action

Once the stability reserve of the link has been computed, the procedure determines how the generators in both areas should be acted upon in order to unload the link, that is, to reduce the transfer across it. Let us say that the power is transferred from Area 1 toward Area 2. Increasing the stability reserve of the link, that is, reducing the amount of MW that crosses it, can be achieved by increasing the generation in Area 2 and simultaneously lowering the generation in Area 1. The generators in each area are identified and the required control action increase/decrease is displayed in the output results.

C.2.8   User-Defined Links

The procedure also provides the ability to evaluate user-defined links. In this case, rather then searching through all the topological combinations, whose number grows rapidly when the system size increases, the program starts directly from the links defined by the user and evaluates them by following the procedure described in Section C.2.4.

C.3   CONCLUSIONS

The methodology described in this Appendix solves a contemporary, and extremely important, problem: the stability of power transfers across vast interconnected systems. It determines the stability reserves of all the links in the network and ranks them in the order of their distance to instability, thus allowing the program user to identify those links that are potentially dangerous.

The solution technique is fast and robust and is suitable both for offline and in real time. The procedure was designed to work within the same data environment used by real-time stability tools and has been implemented both as a stand-alone application and seamlessly integrated with a fast steady-state stability-assessment program.

The speed, functionality, and versatility of this technique recommend it for system and market operations, and for a broad range of operational and planning studies.

C.4   REFERENCES

[1]   Anderson, P. M., and Fouad A. A., *Power System Control and Stability,* The Iowa University Press, Ames, Iowa, 1990.

[2]   Arie E., Pomârleanu M., Bejuscu L., and Botgros M., "Determinarea Rezervei de Stabilitate Statică a Sistemelor Electroenergetice Complexe," *Studii şi Cercetări de Energetică şi Electrotehnică,* Vol. 23, No. 3, 1973.

[3]   Barbier, C., and Barret, J. P., "An Analysis of Phenomena of Voltage Collapse on a Transmission System," *Revue Générale de l'Electricité,* Vol. 89, No. 7, pp. 3–21, 1980.

[4]   Borremans, P., Calvaer A., de Reuk J. P., Goossens J., et al., *Voltage Stability—Fundamental Concepts and Comparison of Practical Criteria*—Report 38, 11, CIGRE, 1984.

[5]   Crary, S. B., *Power System Stability,* General Electric Series, Schenectady, New York, Copyright 1945, Third Printing October 1955.

[6]   Dimo, P., "Etude de la Stabilité Statique et du Réglage de Tension," *Revue Générale de l'Electricité,* Paris, Vol. 70, No. 11, pp. 552–556, 1961.

[7]   Dimo, P., *Nodal Analysis of Power Systems,* Abacus Press, Kent, England, 1975.

[8]   Dimo P., Pomârleanu M., and Bejuscu L., "Indicator Sintetic al Stabilităţii Statice a Ansamblului Sistemului Electro-energetic," *Generalizarea Experienţei de Exploatare,* vol. 134.

[9]   DyLiacco, T. E., Savulescu, S. C., and Ramarao, K. V., "An On-line Topological Equivalent for a Power System," *IEEE Transactions on PAS,* Vol. PAS-97, No. 9, pp. 1550–1563, 1978.

[10] Erwin, S. R., Oatts, M. L., and Savulescu, S. C., "Predicting Steady-State Instability," *IEEE Computer Applications in Power,* Vol. 4, No. 7, pp. 15–22, 1994.

[11] Kundur, P., and Morison, G. K., "Classes of Stability in Today's Power Systems," *IEEE Transactions on Power Systems,* Vol. 8 T-PWRS, No. 3, pp. 1159–1171, 1993.

[12] Kundur P., *Power System Stability and Control,* McGraw-Hill, New York, 1994.

[13] Kundur, P., *Introduction to Techniques for Power System Stability Search,* A special publication of the Power System Dynamic Performance Committee of the IEEE PES, TP-138-0, pp. 1–3, 1999.

[14] Navarro-Perez, R., and Prada, R. B., "Voltage Collapse or Steady-State Stability Limit," in *Proceedings of International Seminar on Bulk Power System Voltage Phenomena II,* edited by L. H. Fink, pp. 75–84, 1993.

[15] Oatts, M. L., Erwin, S. R., and Hart, J. L., "Application of the REI Equivalent for Operations Planning Analysis of Interchange Schedules," *IEEE Transactions on PAS,* Vol. PAS-109, No. 5, pp. 547–555, 1990.

[16] Pomarleanu, M., "The Selection and Ranking of Power System Dangerous Cuts," presented at Symposium on Bulk Power System Dynamics and Control IV—Restructuring, Santorini, August 24–28, 1998.

[17] Pomarleanu, M., and Bejuscu, L., "Selectarea Operativa a Secţiunilor Unui Sistem Electroenergetic cu Rezervele de Stabilitate Statica Cele Mai Mici," *Energetica,* Vol. 47, No. 2, 1999.

[18] Tinney, W. F., and Powell, W. I., "The REI Approach to Power Network Equivalents," presented at PICA'77 Conference, May, Toronto, Canada, 1977.

[19] Sauer P. W., and Pai, M. A., "Power System Steady-State Stability and the Load-Flow Jacobian," *IEEE Transactions in Power Systems,* Vol. 5 T-PWRS, No. 4, pp. 1374–1381, 1989.

[20] Sauer, P. W., and Pai, M. A., "Relationships between Power System Dynamic Equilibrium, Load-Flow, and Operating Point Stability," in *Real Time Stability in Power Systems,* Ed. S. C. Savulescu, Springer Verlag, Norwell, MA, 2006.

[21] Savulescu, S. C., "Equivalents for Security Analysis of Power Systems," *IEEE Transactions on PAS,* Vol. Pas-100, No. 5, 1981, pp. 2672–2682, 1981.

[22] Savulescu S. C., *Grafuri şi Reţele Electrice,* Editura Tehnică, Bucureşti, 1994.

[23] Savulescu, S. C. (Ed.), *Real Time Stability in Power Systems,* Springer Verlag, Norwell, MA, 2006.

[24] Savulescu, S. C., "Fast Assessment of the Distance to Instability. Theory and Implementation," in *Real Time Stability in Power Systems,* Ed. S. C. Savulescu, Springer Verlag, Norwell, MA, 2006.

[25] Seshu S., and Reed M. B., *Linear Graphs and Electrical Networks,* Addison Wesley, Reading, Massachusetts, 1961.

[26] Venikov, V. A., *Transient Processes in Electrical Power Systems,* English Translation, MIR Publishers, Moscow, 1977.

[27] Vournas, C. D., Sauer, P. W., and Pai, M. A., Relationships between Voltage and Angle Stability of Power Systems, *Electrical Power and Energy Systems,* Vol. 18, No. 8, pp. 493–500, 1996.

[28] Wu, F. F., and Narasimhamurti, N., Necessary Conditions for REI Reduction to be Exact, presented at IEEE PES Winter Meeting 1979, Paper A 79 065-4, 1979.

[29] IEEE, PES Task Force on Terms and Definitions, "Proposed Terms and Definitions for Power System Stability," *IEEE Transactions on PAS,* Vol. PAS-101, No. 7, 1982.

[30] http://www.quickstab.com.

[31] IEEE, *Guide for Synchronous Generator Modeling Practices in Stability Analysis,* PES Publication P1110/D11, sponsored by the Joint Working Group on determination and Application of Synchronous Machine Models for Stability Studies, May 1990.

[32] CIGRE, *Indices Predicting Voltage Collapse Including Dynamic Phenomena,* prepared by CIGRE, WG 38.02., TF 11, 1994.

INDEX

Printed in the United States
By Bookmasters